Hazards and Disasters Series

Hydro-Meteorological Hazards, Risks, and Disasters

Series Editor

John F. Shroder
Emeritus Professor of Geography and Geology
Department of Geography and Geology
University of Nebraska at Omaha
Omaha, NE 68182

Volume Editors

Paolo Paron PhD
Department of Water Science and Engineering
River Basin Development Chair
UNESCO-IHE, Institute for Water Education Delft, Netherlands

Giuliano Di Baldassarre PhD
Department of Earth Sciences Program for Air,
Water and Landscape Sciences
Uppsala University, Uppsala, Sweden

AMSTERDAM • BOSTON • HEIDELBERG • LONDON • NEW YORK • OXFORD
PARIS • SAN DIEGO • SAN FRANCISCO • SINGAPORE • SYDNEY • TOKYO

Elsevier
Radarweg 29, PO Box 211, 1000 AE Amsterdam, Netherlands
The Boulevard, Langford Lane, Kidlington, Oxford OX5 1GB, UK
225 Wyman Street, Waltham, MA 02451, USA

Library of Congress Cataloging-in-Publication Data
Application submitted

British Library Cataloguing in Publication Data
A catalogue record for this book is available from the British Library

ISBN: 978-0-12-394846-5

For information on all Elsevier publications visit our web site at http://store.elsevier.com

This book has been manufactured using Print on Demand technology. Each copy is produced to order and is limited to black ink. The online version of this book will show color figures where appropriate.

Contents

Contributors

Konstantinos M. Andreadis, Jet Propulsion Laboratory, California Institute of Technology, Pasadena, CA, USA

Stefano Balbi, Dipartimento di Economia and Venice Centre for Climate Studies, Università Ca' Foscari di Venezia, Venezia, Italy; Basque Centre for Climate Change (BC3), Bilbao, Spain

S.F. Balica, s.f.balica@gmail.com

Paul D. Bates, School of Geographical Sciences, University of Bristol, Bristol, UK

Gerardo Benito, Museo Nacional de Ciencias Naturales, Spanish Research Council (CSIC), Madrid, Spain

Claudio Biscaro, Dipartimento di Economia and Venice Centre for Climate Studies, Università Ca' Foscari di Venezia, Venezia, Italy; Institut für Organisation und Globale Managementstudien, Johannes Kepler Universität, Linz, Austria

Thom Bogaard, UNESCO-IHE Institute for Water Education, Delft, The Netherlands; Delft University of Technology, Civil Engineering and Geosciences Water Management, Water Resources, Delft, the Netherlands

S. Bohms, SGT Inc., Contractor to USGS EROS Center, Sioux Falls, SD, USA (work performed under G10PC00044)

M. Budde, U.S.Geological Survey (USGS) Earth Resources Observation and Science (EROS) Center, Sioux Falls, SD, USA

Amy Dabrowa, School of Geographical Sciences, University of Bristol, Bristol, UK

Giuliano Di Baldassarre, UNESCO-IHE Institute for Water Education, Delft, The Netherlands; Department of Earth Sciences, Uppsala University, Uppsala, Sweden

Andrés Díez-Herrero, Instituto Geológico y Minero de España, (IGME), Geological Survey of Spain, Madrid, Spain

Q. Dinh, Dipartimento di Elettronica, Informazione e Bioingegneria, Politecnico di Milano, Italy

Animesh K. Gain, Dipartimento di Economia and Venice Centre for Climate Studies, Università Ca' Foscari di Venezia, Venezia, Italy

Carlo Giupponi, Dipartimento di Economia and Venice Centre for Climate Studies, Università Ca' Foscari di Venezia, Venezia, Italy

Shreedhar Maskey, UNESCO-IHE Institute for Water Education, Delft, The Netherlands

Vahid Mojtahed, Dipartimento di Economia and Venice Centre for Climate Studies, Università Ca' Foscari di Venezia, Venezia, Italy

Micah Mukolwe, UNESCO-IHE Institute for Water Education, Delft, The Netherlands; Delft University of Technology, Civil Engineering and Geosciences Water Management, Water Resources, Delft, the Netherlands

Jeffrey C. Neal, School of Geographical Sciences, University of Bristol, Bristol, UK

I. Popescu, UNESCO-IHE, Institute for Water Education, Delft, The Netherlands

Magdalena Rogger, Institute of Hydraulic Engineering and Water Resources Management, Vienna University of Technology, Vienna, Austria

J. Rowland, U.S.Geological Survey (USGS) Earth Resources Observation and Science (EROS) Center, Sioux Falls, SD, USA

Guy J.-P. Schumann, Jet Propulsion Laboratory, California Institute of Technology, Pasadena, CA, USA

G.B. Senay, U.S.Geological Survey (USGS) Earth Resources Observation and Science (EROS) Center, Sioux Falls, SD, USA

Dimitri P. Solomatine, UNESCO-IHE, Institute for Water Education, Delft, the Netherlands; Water Resources Section, Delft University of Technology, Delft, the Netherlands

Patricia Trambauer, UNESCO-IHE Institute for Water Education, Delft, The Netherlands

N.M. Velpuri, ASRC InuTeq LLC, Contractor to USGS EROS Center, Sioux Falls, SD, USA (work performed under G13PC00028)

J.P. Verdin, U.S.Geological Survey (USGS) Earth Resources Observation and Science (EROS) Center, Sioux Falls, SD, USA

Alberto Viglione, Institute of Hydraulic Engineering and Water Resources Management, Vienna University of Technology, Vienna, Austria

Kun Yan, UNESCO-IHE, Institute for Water Education, Delft, the Netherlands

C. Young, ERT Inc., Contractor to USGS EROS Center, Sioux Falls, SD, USA (work performed under G10PC00044)

Editorial Foreword

General hazards, risks, and disasters: Hazards are processes that produce danger to human life and infrastructure. Risks are the potential or possibilities that something bad will happen because of the hazards. Disasters are that quite unpleasant result of the hazard occurrence that caused destruction of lives and infrastructure. Hazards, risks, and disasters have been coming under increasing strong scientific scrutiny in recent decades as a result of a combination of numerous unfortunate factors, many of which are quite out of control as a result of human actions. At the top of the list of exacerbating factors to any hazard, of course, is the tragic exponential population growth that is clearly not possible to maintain indefinitely on a finite Earth. As our planet is covered ever more with humans, any natural or human-caused (unnatural) hazardous process is increasingly likely to adversely impact life and construction systems. The volumes on hazards, risks, and disasters that we present here are thus an attempt to increase understandings about how to best deal with these problems, even while we all recognize the inherent difficulties of even slowing down the rates of such processes as other compounding situations spiral on out of control, such as exploding population growth and rampant environmental degradation.

Some natural hazardous processes such as volcanoes and earthquakes that emanate from deep within the Earth's interior are in no way affected by human actions, but a number of others are closely related to factors affected or controlled by humanity, even if however unwitting. Chief among these, of course, are climate-controlling factors, and no small measure of these can be exacerbated by the now obvious ongoing climate change at hand (Hay, 2013). Pervasive range fires and forest fires caused by human-enhanced or -induced droughts and fuel loadings, megaflooding into sprawling urban complexes on floodplains and coastal cities, biological threats from locust plagues, and other ecological disasters gone awry; all of these and many others are but a small part of the potentials for catastrophic risk that loom at many different scales, from the local to planet girdling.

In fact, the denial of possible planetwide catastrophic risk (Rees, 2013) as exaggerated jeremiads in media landscapes saturated with sensational science stories and end-of-the-world, Hollywood productions is perhaps quite understandable, even if simplistically shortsighted. The "end-of-days" tropes promoted by the shaggy-minded prophets of doom have been with us for centuries, mainly because of Biblical verse written in the early Iron Age during remarkably pacific times of only limited environmental change. Nowadays, however, the Armageddon enthusiasts appear to want the worst to validate their death desires and prove their holy books. Unfortunately we are all entering times when just a few individuals could actually trigger societal breakdown by error or terror, if Mother Nature does not do it for us first. Thus we enter contemporaneous times of considerable peril that present needs for close attention.

These volumes we address here about hazards, risks, and disasters are not exhaustive dissertations about all the dangerous possibilities faced by the ever-burgeoning human populations, but they do address the more common natural perils that people face, even while we leave aside (for now) the thinking about higher-level existential threats from such things as bio- or cybertechnologies, artificial intelligence gone awry, ecological collapse, or runaway climate catastrophes.

In contemplating existential risk (Rossbacher, 2013), we have lately come to realize that the new existentialist philosophy is no longer the old sense of disorientation or confusion at the apparently meaninglessness or hopelessly absurd worlds of the past, but instead an increasing realization that serious changes by humans appear to be afoot that even threaten all life on the planet (Kolbert, 2014; Newitz, 2013). In the geological times of the Late Cretaceous, an asteroid collision with Earth wiped out the dinosaurs and much other life; at the present time, in contrast, humanity itself appears to be the asteroid.

Misanthropic viewpoints aside, however, an increased understanding of all levels and types of the more common natural hazards would seem a useful endeavor to enhance knowledge accessibility, even while we attempt to figure out how to extract ourselves and other life from the perils produced by the strong climate change so obviously underway. Our intent in these volumes is to show the latest good thinking about the more common endogenetic and exogenetic processes and their roles as threats to everyday human existence. In this fashion, the chapter authors and volume editors have undertaken to show you overviews and more focused assessments of many of the chief obvious threats at hand that have been repeatedly shown on screen and print media in recent years. As this century develops, we may come to wish that these examples of hazards, risks, and disasters are not somehow eclipsed by truly existential threats of a more pervasive nature. The future always hangs in the balance of opposing forces; the ever-lurking, but mindless threats from an implacable nature, or the heedless bureaucracies countered only sometimes in small ways by the clumsy and often feeble attempts by individual humans to improve our little lots in life. Only through improved education and understanding will any of us have a chance against such strong odds; perhaps these volumes will add some small measure of assistance in this regard.

Hydrometeorological risks, hazards, and disasters: The chapters presented in this volume represent some new thinking, as well as summaries of a lot of the best that have come before in dealing with the ever-recurring storms and floods, as well droughts and the commonly associated heat waves. Editors Di Baldassarre and Paron have brought together 11 good chapters from a variety of authors that give us a succinct view of these hazards, risks, and disasters that add further to our understandings. Although many kinds of storms were not included in this volume, mainly because of publication time constraints, still the material that has been presented will give a good idea of some of the most common recurrent problems that we humans must learn to put up with better from atmospheric sources of problems.

John (Jack) Shroder
Editor-in-Chief
July 14, 2014

REFERENCES

Hay, W.W., 2013. Experimenting on a Small Planet: A Scholarly Entertainment. Springer-Verlag, Berlin, 983 p.

Kolbert, E., 2014. The Sixth Extinction: An Unnatural History. Henry Holt & Company, NY, 319 p.

Newitz, A., 2013. Scatter, Adapt, and Remember. Doubleday, NY, 305 p.

Rees, M., 2013. Denial of catastrophic risks. Science 339 (6124), 1123.

Rossbacher, L.A., October 2013. Contemplating existential risk. Earth, Geologic Column 58 (10), 64.

Foreword

It is a sobering fact that hydrometeorological extremes account for the overwhelming majority of natural hazards in the world today, and this realization provides a persuasive rationale for their detailed study. For example, in Chapter 10 of this book Maskey and Trambauer reproduce data from the EM-DAT International Disaster database for 1900–2013 that shows that floods, droughts, and storms have been responsible for approximately 20 million deaths over this period and $1.75 trillion in economic losses. This former figure represents approximately 90% of all deaths due to natural hazards.

Looking to the future, an increase in global population to 9 billion and perhaps beyond will further increase hydrometeorological risks as population densities increase, demand for food (and hence irrigation) goes up, the urban population expands, and people migrate to coastal and riverine floodplains. Climate change may further exacerbate these trends, and although the latest Special Report on Extremes by the Intergovernmental Panel on Climate Change suggests that it is as yet difficult to attribute changes in the probability distribution of hydrometeorological extremes to anthropogenic factors, good physical reasons based on atmospheric thermodynamics exist to believe that artificial climate change will alter extreme event frequency into the future.

Against this background, the present volume edited by Paolo Paron and Giuliano Di Baldassarre is both important and timely. In this the editors have assembled a range of contributions that provide a comprehensive review of the current state of the science in the field of hydrometeorological hazards and risks. This breadth of coverage is rather unique, as more typically each hydrometeorological hazard is considered separately. By adopting a more comprehensive approach, Paron and Di Baldassarre more clearly establish the similarities and differences in the science between each hazard area in a way that allows best practice to be shared and novel ideas to cross-fertilize into other disciplines.

The book showcases chapters on floods and storms and droughts. Each hazard is afforded an in-depth coverage. In the chapters on floods and storms a number of contributions arise from research conducted within the KULTURisk programme; a €3.25-million project funded by the European Commission to advance ideas of risk resilience in disaster management. The book begins with a chapter by Viglione and Rogger that outlines the various types of floods and provides an analysis of a real hazard representing each type. This type of postdisaster forensic analysis is exceptionally important in setting the context for our understanding of flood events, and provides the essential introduction for the chapters that follow. The following three chapters examine the physical science of flooding. First, Schumann et al. review the measurement, mapping, and modelling of floods, particularly emphasizing the use of remotely sensed data to provide adequate spatial coverage. Next, Benito and Díez-Herrero look at the use of paleoflood data to address issues associated with a flood instrumental record that is almost always short compared to the recurrence interval of typical events of interest. Finally, Yan et al. consider the use of global and low-cost terrain data sets to support flood studies. The section on flooding ends with three chapters on flood risk assessment by Balica et al., Giupponi et al., and Mukolwe et al. and a chapter by Dabrowa et al., which showcases training materials produced by the KULTURisk programme to support student learning and professional development in the area of computer modeling of flood risk.

The book then continues with two chapters examining drought risks. In the first of these, Senay et al. describe the U.S. Aid-funded Famine Early Warning System Network (FEWS-NET) that aims to provide timely information and products to support decision-making processes in drought-related hazard assessment, monitoring, and management. In the final chapter of the volume, Maskey and Trambauer examine the assessment of various aspects of hydrology related droughts and show that process based hydrological modeling has much to offer for drought risk assessment.

Taken as a whole, this volume provides a broad and detailed coverage of the current state of hydrometeorological hazard and risk science. The pervasive and pernicious nature of such hazards urgently requires fundamental science efforts such as these, and even small improvements in risk management have the potential to benefit many millions of people and save hundreds of millions of dollars of losses. In a world where governments and the public increasingly demand "science with impact," the rationale for improved efforts to understand hydrometeorological hazards, as described in this volume, is compelling.

Paul Bates
University of Bristol, UK

Section 1

Floods and Storms

Chapter 1

Flood Processes and Hazards

Alberto Viglione and Magdalena Rogger
Institute of Hydraulic Engineering and Water Resources Management, Vienna University of Technology, Vienna, Austria

ABSTRACT

Floods are classified into different types depending on where the water comes from and on their generating processes. Several types of floods are described in this chapter, including river floods, flash floods, dam-break floods, ice-jam floods, glacial-lake floods, urban floods, coastal floods, and hurricane-related floods. Examples of each flood type are provided and their dominant processes are discussed. Hydrological flood processes such as runoff generation and routing depend on the type of landscape, soils, geology, vegetation, and channel characteristics. They are driven and modulated by climate through precipitation and temperature. Also evapotranspiration and snow processes play a critical role determining, for example, before-event soil saturation. These processes vary widely around the world and, even at the same location, they vary between events. The chapter reviews methods for estimating the probability and magnitude of floods as a measure of the flood hazard. It is argued that understanding the flood processes for each of the flood types is a prerequisite for estimating the flood hazard reliably. This is particularly important if one expects the landscape or climate characteristics to change in the future.

1.1 INTRODUCTION

People have settled close to water bodies (rivers, lakes, and the sea) since the beginning of time and this has been for understandable reasons. Living close to water bodies was economically advantageous. Water bodies have long been the easiest transport corridors and the most important communication routes. Flood plains along rivers and near lakes were also attractive because of the fertility of the land and the easy access to irrigation water. Accessibility to the sea meant accessibility to (at that time) unlimited food availability. For all these reasons, the link between people and water bodies has always been strong and is still today (Di Baldassarre et al., 2013). However, living close to water bodies also involves the risk of flooding. Floods are among the most

Hydro-Meteorological Hazards, Risks, and Disasters. http://dx.doi.org/10.1016/B978-0-12-394846-5.00001-1

devastating natural (and sometimes human-produced) threats on Earth (Ohl and Tapsell, 2000). Floods involve inundations, i.e., submerged land from overflowing rivers and lakes when water overtops or breaks levees, from the sea because of high tides, and/or develop in otherwise dry areas due to accumulation of heavy rainfall. The risk at which people are exposed depends on many factors: the magnitude of flood events, how frequently they occur, the susceptibility of the people and their properties to be adversely affected, and their preparedness in the emergency situations caused by floods. In more technical worlds, *flood risk* is the result of the interactions between the *flood hazard* (which combines the flood probability and magnitude) and the *vulnerability* of the people and their properties. In this chapter, we focus on the flood hazard, whereas vulnerability is covered in Chapter 1.5. Both hazard and vulnerability very much depend on the type of flood and the processes determining it. In Section 1.2, floods of different types are discussed: river floods, flash floods, dam-break floods, ice-jam floods, glacial-lake floods, urban floods, coastal floods, and hurricane-related floods. We illustrate their process mechanisms through real world examples. For instance, most flood types are driven and modulated by climate, through precipitation and temperature, and by the landscape, since runoff generation and routing depend on soils, geology, vegetation, channel characteristics, etc. Also evapotranspiration and snow processes play a critical role, e.g., by controlling before-event soil saturation. These processes vary widely around the world and, even at the same location, they vary between events. Flood processes determine the way floods develop, their magnitude, volume, and speed. In Section 1.3, we discuss how the reliability of flood hazard estimation may be increased by understanding the flood generating processes of the different flood types. Finally, in Section 1.4 we discuss how the flood hazard may change in the future and how we can deal with it.

1.2 FLOOD TYPES AND THEIR PROCESSES

1.2.1 River Floods

In June 2013 the Upper Danube Basin (i.e., the German–Austrian part of the basin) was struck by a major flood (Blöschl et al., 2013a). The city centre of Passau (at the confluence of the Danube, Inn, and Ilz) experienced flood levels that were similar to the highest recorded flood in 1501, which is considered the "millennium flood" in central Europe (see Figure 1.1). Extraordinary flood discharges were recorded along the Saalach and Tiroler Ache at the Austrian-Bavarian border. The flood discharge of the Danube at Vienna, downstream of Passau, exceeded those observed in the past two centuries, in particular it exceeded the big August 2002 flood, till then referred to as the "century" flood in Austria.

The atmospheric situation of the event was a typical one for floods in the Upper Danube basin. A large-scale stationary atmospheric regime led to the

FIGURE 1.1 Flood marks on the Passau city hall. The 2013 flood mark is clearly visible and is significantly higher than the 1501 flood. This is probably due to the effect of waves, since the 2013 and the 1501 floods were of similar magnitudes. *From Blöschl et al. (2013b).*

blocking of a number of synoptic systems including the Azores and the Siberian anticyclone in the second half of May 2013. The moisture brought from the north-western Atlantic caused rainfall in the Upper Danube Basin from May 18 to 27. The cyclonic system, with its rotation and spatial extent, collected additional moisture from the Mediterranean, producing what van Bebber (1891) termed "Vb"-system, which caused persistent, heavy precipitation over the northern fringe of the eastern Alps, lasting from May 30 to June 4, 2013. Figure 1.2 shows the spatial pattern of precipitation for a period of seven days (May 29 to June 4, 2013). As indicated in the figure, precipitation was highest along the northern ridge of the Alps in Austria (Tirol, Salzburg, and Upper Austria) and very significant precipitation also occurred further in the north. Precipitation interpolated between the rain gauges based on weather radar exceeded 300 mm during this time period. The event consisted of two main precipitation blocks separated by a few hours of no or lower intensity rain (Blöschl et al., 2013a).

Moreover, May 2013 was one of the three wettest months of May in the past 150 years in the Upper Danube Basin. Air temperatures in the first three weeks of May were somewhat lower than the long-term average in the Upper Danube Basin and significant snowfall occurred at the high-elevation stations in the Alps. At the beginning of the event, the soils were wet throughout the Upper Danube Basin, although there was a pronounced north–south gradient with higher soil moisture in the north, and lower soil moisture in the south. Because of the relatively high antecedent precipitation, and therefore soil moisture, the event runoff coefficients were quite large in the Alpine

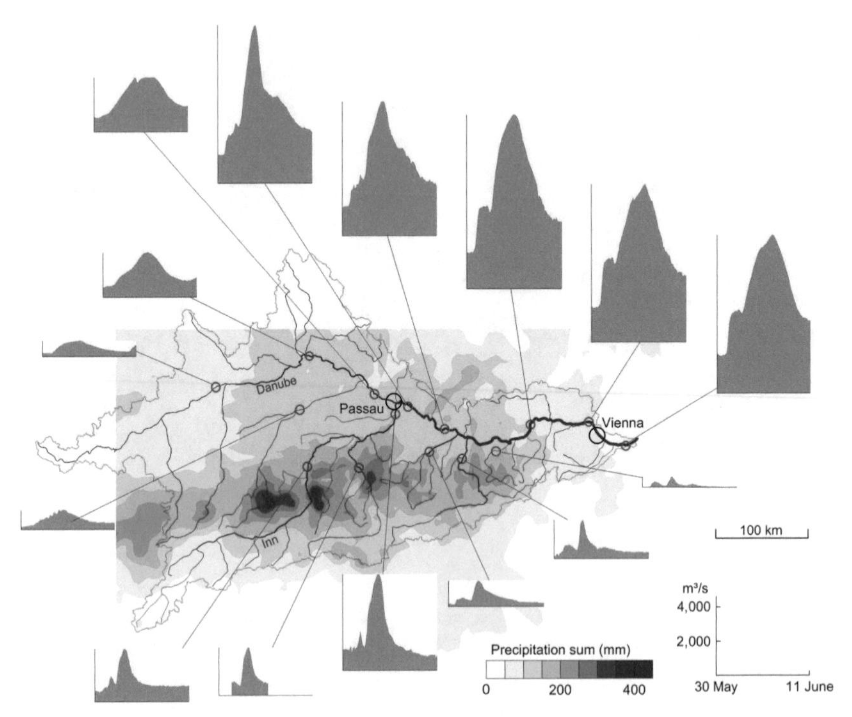

FIGURE 1.2 Total amount of the precipitation event and propagation of the June 2013 flood along the stream network of the Upper Danube Basin. Red circles indicate stream gauges. The scale shown on the bottom right relates to all hydrographs. *Redrawn from Blöschl et al. (2013a).*

catchments. However, when compared to runoff coefficients of other flood events in the same region, the runoff coefficients were not unusually high (Blöschl et al., 2013a). This is because part of the precipitation fell as snow and remained as snow cover until after the event in the highest parts of the catchment. In the Bavarian Danube catchment, instead, temperatures were above 0 °C in almost the entire catchment. However, because of the highly permeable soils and the large storage capacity in the catchment, only one-fourth of the precipitation contributed to the runoff in spite of the high antecedent soil moisture.

The spatiotemporal rainfall patterns of the 2013 flood, combined with differences in runoff response characteristics between the catchments (Gaál et al., 2012), produced complex patterns of runoff hydrographs within the Upper Danube Basin. Figure 1.2 gives an overview of the evolution of the flood within the basin. At the Bavarian Danube in the northwest of the basin, the flood response was delayed with relatively flat peaks. However, the total volume of the 2013 flood along the Bavarian Danube was exceptionally large because of the high rainfall and very high antecedent soil moisture. The Inn, coming from the Alps, exhibited a much faster response as is always the case with this type of regional floods (Blöschl et al., 2013a). The confluence of the Inn with the Bavarian Danube at Passau resulted in an amplification of the combined shape of the flood wave, significantly higher than in other big flood events in the area, because the flood wave of the Bavarian Danube arrived somewhat earlier than usual with smaller differences in the time lag between the Bavarian Danube and Inn waves. The inundation level in Passau was enormous (12.89 m), of the same order of magnitude as the 1501 flood event (BfG, 2013). After the confluence of the Bavarian Danube and the Inn at Passau, the 2013 flood wave traveled down the Austrian Danube, changing shape and shifting the timing, due to retention in the flood plains. Inflow from southern tributaries along the Austrian reach of the Danube, including the Traun, Enns, and Ybbs, gave rise to an early secondary peak, indicating that these tributaries peaked much earlier and hardly contributed to peak flows along the Danube.

The June 2013 event in the Upper Danube region allows pinpointing the dominant causal factors of a river flood: the atmospheric situation, the runoff generation, and the propagation of the flood wave along the main river and tributaries. For most river floods, the dominant processes are precipitation falling over an extended area for an extended period of time, runoff produced by saturation excess mechanism, and amplification of the flood wave due to synchronicity between tributary contributions. The combination of these three factors tend to produce high water levels of the river over an extended area, also downstream, where precipitation is not necessarily intense. The level of antecedent soil moisture is in many cases critically important, because it leads to saturation flow during the event, i.e., when the soils become saturated and the depression storage fills, all rainfall produces surface runoff on the hillslopes (Dunne, 1983). In catchments where a large amount of water is stored in the snowpack, when rain falls on an existing snow cover, moderate rainfall depths can cause enormous runoff depths in rivers because of the triggered snowmelt. Snowmelt occurs also during fair-weather periods often associated with a rapid increase in air temperature. These snowmelt floods usually occur over a period of one or two weeks in sequence, saturating the soils, continuously raising the flows, and finally causing a flood (Merz and Blöschl, 2003). Water from the hillslopes would run into channels and arrive at the river along different paths, potentially producing constructive resonance of the flood waves. For large rivers, all processes progress relatively slow along the river, and the high waters may last for days. However, when a levee breaks, a lot of water is released suddenly and the speed of the water at the breach can be compared with the speed of a flash flood or a dam-break flood, and the strength of the water may carry cars, trees, and even houses away.

1.2.2 Flash Floods

On October 25, 2011, heavy rainfall affected an area of ca. 1,000 km^2 between eastern Liguria and northern Tuscany (northwest Italy). In few hours, the storm rainfall caused thousands of shallow landslides, widespread erosive and depositional processes, and several local floods (see Amponsah, 2013). These led to 13 casualties, the evacuation of about 1,200 people, the interruption of both the A12 highway and the Genova–La Spezia railway, the closure of 43 percent of provincial roads, and the destruction of many bridges (Cevasco et al., 2013). Along the coast, the western sector of the Cinque Terre, "The Five Lands"—a UNESCO World Heritage Site, was affected by floods in Monterosso and Vernazza, causing 4 casualties and severe structural and economic damage (Figure 1.3).

The heavy precipitation was associated with an intense and quasi-stationary convective system that had developed near the coast. The low-level mesoscale flow patterns over the Ligurian Sea, along with orographic lift over the steep Apennines chain surrounding the coast, gave rise to pronounced convergence lines (Buzzi et al., 2013). The rainfall was typically convective with high temporal variability and maxima, as observed at specific stations, exceeded 450 mm in less than 12 h (Buzzi et al., 2013; Rebora et al., 2013). The maximum cumulative rainfall was recorded at the Brugnato rain gauge (Vara valley) with 539 mm/24 h and a peak of 153 mm/h (Cevasco et al., 2013).

Because of the intensity of rainfall, runoff production was due to both saturation excess and infiltration excess mechanisms on the hillslopes. This latter occurs when the rate of rainfall on the hillslopes exceeds the rate at which water can infiltrate into the ground (Horton, 1933; Dunne, 1983). Therefore surface runoff is produced even though the soil is unsaturated. Floods were associated also with landslides and debris flows. The strong

FIGURE 1.3 Flood waters rush into Vernazza's harbour (one of the "Cinque Terre" villages) on October 25, 2011. *Photo by Tom Wallace. From http://www.nbcnews.com/id/45307159/ns/travel-destination_travel/#.UuuFRDe9hQ0.*

spatial gradients of the precipitation had a major influence on flood response, with large differences in peak discharge between neighboring catchments. The specific peak flows were up to 20 $m^3/(s\ km^2)$ for catchments less than 30 km^2 of area. Floods were associated also with landslides, debris flow, and large woody debris. The magnitude of sediment transport processes, also quite variable among subbasins, seems to have been controlled both by peak water discharge and by local geomorphological conditions affecting sediment supply, i.e., occurrence of large landslides (Marchi et al., 2013). Peculiar land-use conditions characterize the Cinque Terre. The steep slopes have been almost completely terraced for vineyards and olive groves during the past millennium. Unfortunately, since the end of the 1800s, changing social and economic conditions have caused a progressive abandonment of cultivated terraces, with negative consequences for the maintenance of dry stone walls and therefore slope stability (Cevasco et al., 2013). The accumulation of water in the soils (by saturation excess) produced the large number of mud flows and landslides.

The technical authorities in charge of hydrometeorological forecast for the Liguria region predicted the scenarios with a lead-time of two days (Silvestro et al., 2012). However, the magnitude and the speed of the event were such that casualties could not be avoided.

For flash floods, speed is the keyword. Especially in areas with steep slopes, heavy rain can cause a riverbed that holds very little or no water at first, to suddenly brim with fast flowing water. The water level may rise very quickly. Along with the saturation excess mechanisms, the high intensity of rain may also produce surface runoff because of the infiltration excess mechanism. The amount of infiltration excess runoff depends on the rainfall intensity and the soil infiltration capacity. For instance infiltration excess is the dominant runoff process in arid areas where compacted soil prevents water infiltration, or in urban areas because of pavements (Section 1.2.6). The amount of water and the area flooded in a flash flood is relatively small compared to river floods. Major flash floods usually occur in small catchments. However, locally, the danger can be enormous because of the sudden onset and the high traveling speed of the water. In many cases flash floods cause mud flows and landslides, and the water flow can be powerful enough to transport large objects like rocks, trees, and cars.

1.2.3 Dam-Break Floods

The Banqiao Reservoir Dam is a large dam on the river Ru in the Henan province, China, one of the many dams of the Huai River system. The dam was first built in 1951 in order to control flooding as a response to the severe flooding in the Huai River Basin in 1949 and 1950. Its failure in 1975 caused more casualties than any other dam failure in history. It is considered one of the largest humanitarian disasters in the twentieth century. The dam failure killed about 26,000 people and 11 million people lost their homes. The number of deaths that occurred afterward due to illness and famine in the region is likely more than 200,000 people (Si and Quing, 1998). It took many years for the region to recover. The disaster illustrates the trade-off of dams designed for flood control. Although they allow eliminating or reducing the costs of small-to-moderate floods, this may come at the cost of a much larger flood causing catastrophic effects.

A succession of three successive heavy rainfall events occurred over the region, where the Banqiao Dam is located, on August 5, 6, and 7, 1975 (Si and Quing, 1998), following the collision of Super Typhoon Nina and a cold front. The typhoon was blocked for two days after landfall to the north of the tropical depression over the Henan province before its direction ultimately changed from northeastward to west (Wang, 2006; p. 170). As a result of this near stationary thunderstorm system, more than a year's rain fell within 24 h, a meter of water in three days, which weather forecasts failed to predict.

The Banqiao dam was an earthfill dam, i.e., an embankment of well compacted earth, with a clay-core wall (Zhang et al., 2009). The total capacity of the Banqiao dam was 492 million m^3, with 375 million m^3 reserved for flood storage. The height of the dam was at little over 116 m and the spillway was designed to pass floods expected every 1,000 years (Si and Quing, 1998). By August 8 the Banqiao Dam and the Shimantan Dam (a reservoir in the neighboring valley of the Hong River) had filled to capacity because the runoff exceeded the rate at which water could be disclosed through their spillways. Shortly after midnight (12:30 AM) the water in the Shimantan Dam reservoir on the Hong River rose 40 cm above the crest of the dam. Water spilled over the earthfill dam, eroding it and causing its collapse. The reservoir emptied its 120 million m^3 of water within 5 h. About half an hour later, shortly after

1:00 AM, the Banqiao Dam on the Ru River was crested. The spillways of the Banqiao dam were not able to handle the overflow of water, partially due to sedimentation blockage and that the sluice gate capacity was not enough to avoid overtopping. Some workers toiled amidst the thunderstorm trying to save the embankment. As the dam began to disintegrate one of them shouted "Chu Jiaozi" (The river dragon has come!), which is the sentence by which the catastrophic flood is remembered (Si and Quing, 1998). The crumbling of the dam created a surge of water 6 m high and 12 km wide. Behind this moving wall of water were 600 million cubic meters of more water. Altogether 62 dams broke. Downstream, the dikes and flood diversion projects could not resist such a deluge. They broke as well and the flood spread over more than a million hectares of farm land throughout several counties and municipalities. At the city of Huaibin, where the waters from the Hong and Ru Rivers come together, the floods produced by the Banqiao and Shimantan Dam failure joined. There was little or no time for warnings. The wall of water was traveling at about 50 km per hour (Si and Quing, 1998).

The causes for dam failures may be diverse. For earthfill dams such as the Banqiao Dam, failure mechanisms include overtopping, which was the case in 1975, and piping (Zhang et al., 2009). Overtopping is mainly due to insufficient spillway capacity and can cause large amounts of erosion on the downslope side of the dam, which may compromise the stability of the dam. For dam failures due to piping, the impact of inflow floods may not be extremely significant, although floods do increase the possibility of piping occurrence due to larger gradients of seepage flow. Cavities and cracks can develop within the dam due to differential settlement within the embankment, especially if the depth to bedrock is highly variable. The cavities and cracks can act as preferential conduits for water to flow through the dam and erode it from the inside out. Also other mechanisms may take place, such as slope stability and foundation failures. Often a combination of factors occurs and, in some cases, not of hydrologic nature. For instance, one of the causes of the Tirlyan reservoir failure in 1994 (Republic of Bashkortostan, Russia) was that one of the segment gates had been blocked years before because of concerns about sabotage, and could not be opened by the operators in time (Blöschl et al., 2013c). Another example is the Vajont Dam flood in 1963 (Northern Italy) which was produced by a massive landslide into the reservoir (Di Baldassarre et al., 2014). Strictly speaking this case was not a dam failure, since the dam structure did not collapse, and is still standing, but the huge wave produced by the landslide, resembling a tsunami wave, completely destroyed the villages in the valley downstream of the dam.

1.2.4 Ice-Jam Floods

The city of Montpelier, Vermont, USA was severely struck by an ice-jam flood on March 11, 1992. Unlike other flood types that are caused by severe rainfall events, this flood was generated by ice-induced backwater of the Winooski River. It was a typical ice-jam flood.

In the winter of 1992 temperatures were low and the Winooski river remained frozen with a solid ice cover till the beginning of March. In the second week of March, however, a storm system that developed over the mid-Atlantic Coast and moved in a north-easterly direction along a cold front caused an early spring thaw associated with above-freezing temperatures and rain (Denner and Brown, 1998).

The mild thaw weather, at the Winooski River caused a disruption of the ice cover and an increase in water levels due to snowmelt. Eventually, after a light rainfall event on March 10 ($\sim$20 mm), a 1.5-km-long ice jam formed along the bridges close to the city center, blocking the river channel and causing an ice-jam flood in the morning of March 11. Due to backwater effects downtown Montpelier was inundated up to a depth of 1.5 m (Denner and Brown, 1998). The flood hit the town with little warning causing damages of more than 4 million US$. The ice jam eventually broke in the afternoon of the same day destroying the Washington County Railroad Bridge by the large mass of moving ice and water (see Figure 1.4).

Ice-jam floods occur when the passage of ice along a reach is obstructed causing the incoming ice to accumulate, which results in a rise of the water levels upstream (Beltaos, 1995). Such ice-jam obstructions may occur along rivers at narrows, structures such as bridges, or at places where the slope changes. Similar obstructions may also occur at the outflow of lakes, especially glacial lakes. When ice jams break, a sudden increase in downstream water levels and velocities follows, similar to a dam-break flood (Beltaos, 1995), and the ice itself may collide with structures and cause

FIGURE 1.4 Washington County Railroad bridge destroyed by the pressure of ice and water during the March 11, 1992 ice-jam flood of the Winooski River. *Photo by Jackie Hurlburt. From: http://www.montpelier-vt.org/community/351/Flood-of-1992.html.*

damage (see Figure 1.4). Ice jams may therefore lead to flood problems in two ways: (1) due to backwater effects upstream of the ice jam as in the case of the Winooski River described above; and (2) due to the sudden surge wave after the failure of the ice jam. After an ice jam breaks loose, it may form again at a downstream location causing another flood event. Ice-jam floods usually occur in late winter or early spring during the ice breakup, but may also occur during freezing periods (Beltaos, 1995). They are aggravated by the fact that during the melt season additional water from snowmelt enters the stream and increases water levels. Ice-jam floods can cause water levels that far exceed even rare water floods at the local scale where they occur and are hard to forecast due to the rapidity with which an ice jam can build up (Beltaos, 1995). Ice jams are of considerable socio-economic concern and may have major impacts on riverside communities, aquatic life, infrastructure, navigation, and hydropower generation (Beltaos, 2007). They are typical for high-latitude countries with long winter periods.

1.2.5 Glacial-Lake Floods

On August 4, 1985, the Dig Tsho, a moraine-dammed glacial lake situated in Khumbu Himal, Nepal burst, causing massive destruction in the region downstream (Vuichard and Zimmermann, 1986, 1987). About 5 million m^3 of water were released destroying a newly built hydroelectric power plant, 14 bridges, 30 houses, many hectares of arable land and damaging the trail network. The peak discharge was estimated as 1,600 m^3/s (Vuichard and Zimmermann, 1987). The rupture of the moraine that dammed the glacial lake was caused by an ice avalanche in early August. A large ice mass became detached from the granitic wall after a long period of warm weather and splashed into the lake causing a sudden rise in the water level and producing a wave that overtopped the moraine dam, finally cutting a V-shaped breach (Vuichard and Zimmermann, 1986). The flood moved about 3 million m^3 debris within a distance of 40 km (Vuichard and Zimmermann, 1987).

Glacial-lake floods occur when the dams forming the glacial lakes fail and release the water producing a surge of water downstream (ICIMOD, 2011). For this reason, they have similar effects as the dam-break floods discussed in Section 1.2.3. The dams may form either due to the blockage of ice at the outflow of the glacier, similar to ice jams in rivers, or due to the presence of terminal moraine material. The dams may fail due to two principal mechanisms: (1) waves generated by rock, snow or ice avalanches, earthquakes, volcanic eruptions, or glacier calving; or (2) after weakening through seepage and internal erosion of the moraine that holds the impounded water (Kattelmann, 2003). Glacial-lake floods usually occur during the summer season when water levels are higher during ice and snowmelt and may be aggravated by rainfall events, especially in the regions that are affected by monsoon rains (Kattelmann, 2003). In contrast to snowmelt floods, which are one possible cause of river floods and even flash floods, the water levels of glacial lake floods tend to significantly decrease downstream because of the small source of water from the lake (Kattelmann, 2003). Glacial-lake floods are typical for mountainous regions with high elevations (>4,000 m) such as Nepal, Pakistan, Tibet, Alaska, Canada, or Iceland.

1.2.6 Urban Floods

The city of Hull (UK) suffered a major urban flood on June 25, 2007, which caused damage to over 8,600 homes and 1,300 businesses. The flood originated from a severe rainfall event with a depth of 100 mm that overwhelmed the city's drainage system (Coulthard and Frostick, 2010).

June 2007 was the wettest month recorded in the region since 1882 (Met office, 2007). On June 25, a deepening depression moved slowly across the United Kingdom bringing large amounts of rainfall to Lincolnshire, Yorkshire, and the Midlands, resulting in widespread flooding (Coulthard and Frostick, 2010) (Figure 1.5). The City of Hull measurement station recorded a record rainfall event of approximately 100 mm (Coulthard and Frostick, 2010).

The rainfall event led to an extensive flood in the city due to blocked roadside gullies and to issues with the conveyance of drained water in sewers and the performance of local pumping stations (Coulthard and Frostick, 2010). Due to its topographic position, with most of its area below sea level, the city is highly susceptible to pluvial floods. To convey storm water it relies on a pumped drainage system with no natural ways of drainage (Coulthard and

FIGURE 1.5 June 2007 flood in Burstwick, a village situated in the East Riding of Yorkshire about 13 km east of Hull. *Photo from http://www.ambiental.co.uk/riskcentral/flood-risk-in-hull/.*

Frostick, 2010). An extensive redesign of the drainage system was performed in 2001 that increased the flood storage and reduced the pumping capacity. Although some of the old pumping stations were reactivated in 2004, a subsequent modeling of the system after the 2007 flood indicated a slightly worse performance of the drainage system compared to the 2001 configuration (Coulthard and Frostick, 2010). The pluvial nature of the flooding led to slow rises in water levels across the city with flood waters rising up beneath houses through the underfloor cavities and foundations. Under these circumstances applied sandbags were of limited use (Coulthard and Frostick, 2010).

Urban flooding, or more specifically pluvial flooding, is a specific flood type caused by the lack of drainage in urban areas. In urban areas, where there is little soil that can store water, nearly all precipitation needs to be transported through the drainage system. Pluvial floods therefore occur when rainfall events exceed the capacity of storm water drains. This is usually associated with short storm events and high rainfall intensities, but may also occur for longer events with lower intensities, especially if the ground surface is impermeable, saturated, or frozen (Houston et al., 2011). They may develop unexpectedly at locations that are not obviously prone to flooding with minimal warning (Houston et al., 2011). Pluvial floods are usually associated with large economic damage (Jha et al., 2012), but a small number of casualties due to the nature of the flood. As described in the example above for the City of Hull, these kinds of floods are characterized by a slow rise in water table and low flow speeds with the water tables usually not reaching life-endangering heights. A health hazard may though occur for combined sewer system where a mixture of untreated sewage with storm water enters the streets and households (Houston et al., 2011). Depending on the location and local setting, urban or pluvial flooding can also occur in conjunction with river and coastal floods (Jha et al., 2012) or with groundwater floods, i.e., water rising up from the groundwater generally after long periods of sustained high rainfall.

1.2.7 Coastal Floods

The most catastrophic coastal flood in Europe in the last century is the 1953 flood, which hit the Netherlands, the east coast of England, Belgium, and Germany, causing over 2,100 casualties, of which more than 1,800 occurred in the Netherlands. It happened in the night or early morning, of February 1, 1953, when a storm surge was produced in the southern North Sea. In comparison with other major storms and floods in the Netherlands, the atmospheric depression leading to the 1953 storm field was not exceptionally deep, but the storm track was different and its propagation somewhat slower than usual. From the Atlantic, it moved east to the north of Scotland after which it curved sharply southward to propagate into the German Bight and then proceeded in a south-easterly direction over land (Gerritsen, 2005). This was a track much closer to the Netherlands than any of the preceding storm tracks on record. As a result, storm winds from the northwest were stronger and more sustained, leading to a higher and more sustained surge, i.e., offshore rise of sea water.

The maximum surge occurred at the time of spring-tide high water (Gerritsen, 2005). Because the time of the surge peak coincided with the time of spring-tide high water, the total water-level reached heights that, in many locations, exceeded those recorded ever before. The state of the dikes (i.e., heights, slopes, overall strength, and maintenance) in the Netherlands at the time of the flood was not very good. War-related construction in the period between 1940 and 1945 also affected the state of the dikes (Gerritsen, 2005). Military bunkers, and even whole complexes, were built into the dikes as part of the military defense system against an expected attack from the sea. Machine-gun units and manholes were dug, and piping was laid through the dikes (Slager, 1992). Since the dikes were in closed military areas, undermining by rabbits and moles occurred more than in other areas. After the war, many of these weak spots were insufficiently repaired by filling up the holes. In 1953, the weak spots mentioned above often proved to be the locations where the dike first gave way (Gerritsen, 2005). Most dikes started to collapse from the inside. The primary cause was wave overtopping, which led to penetration and subsequent saturation of the landward side with water. On steep slopes, this quickly led to sliding and collapse. During the night of February 1, and over the following days, some 150 dike breaches occurred in the sea dikes—the primary sea defenses, later followed by more breaches in the inner or sleeper dikes.

At the time, some 750,000 people lived in the affected areas of the Netherlands, of which about 136,500 ha were inundated (Gerritsen, 2005). As a result of the flood disaster, a total of 1,836 casualties occurred, tens of thousands of livestock perished, and approximately 100,000 people were evacuated. In the aftermath of the event, the Delta Works, a series of construction projects consisting of dams, sluices, locks, dikes, levees, and storm surge barriers, were specifically designed to prevent such a large-scale disaster happening again.

In general, a coastal flood occurs when the coast is flooded by the sea. The usual cause of such a surge is a severe storm, whose wind pushes the water up onto the land. Many coastal floods are produced by hurricanes (Section 1.2.8). The strong winds may create constructive interferences with the tides leading to spring tides. A flood starts when waves move inland on an undefended coast, or overtop or breach the coastal defense works like dunes and dikes, which was the case of the 1953 flood in the Netherlands. Very characteristic of a coastal flood is that the water level drops and rises with the tide. When a sea defense is breached, low tide is the time to repair the breach.

However, storms are not the only cause of coastal floods. Another cause are earthquakes resulting in tsunamis such as those in the Indian Ocean in 2004 or in Japan in 2011. Actually, the word "tsunami" means "harbor wave" in

Japanese. Even earthquakes of only moderate magnitude can produce waves of extraordinary height and power (Kanamori, 1972). When earthquakes occur beneath the sea, the water above the deformed area is displaced from its equilibrium position which is responsible for the subsequent wave formation (Fujii et al., 2011). Tsunamis' waves have a small amplitude and a very long wavelength offshore (many kilometers long and few centimeters high, which is why they generally pass unnoticed at sea). When the wave enters shallow water, it slows down and its amplitude (height) increases, a phenomenon known as "wave shoaling." Tsunamis cause damage by two mechanisms: (1) the smashing force of a wall of water traveling at high speed; and (2) the destructive power of the large volume of water running over the coast and often carrying a large amount of debris with it.

1.2.8 Hurricane-Related Floods

The city of New Orleans was struck by a major hurricane on August 29, 2005 that caused massive destruction due to extensive flooding. Hurricane Katrina was one of the deadliest and most damaging hurricanes in the history of the United States (Figure 1.6).

At the end of August 2005 Hurricane Katrina formed as a tropical storm over the Atlantic Ocean close to the Bahamas. It first hit Florida as a Category 1 hurricane on August 25, before entering the Gulf of Mexico. During its passage through the Gulf it developed into a Category 5 hurricane with wind speeds up to 280 km/h, but fortunately weakened to a Category 3 hurricane (200 km/h) before making its second landfall hitting Louisiana and New Orleans on August 29 (Jonkman et al., 2009).

The city of New Orleans was known to be vulnerable to hurricane floods due to its location in the Mississippi Delta in a polder below sea level entirely surrounded by levees (Jonkman et al., 2009). The flood protection system of the city was organized as a series of protected basins or bowls, each with its own perimeter levee system, which were dewatered by pumps (Nicholson et al., 2005). During its landfall on August 29, Hurricane Katrina caused a large onshore storm surge that resulted in massive destructions in the city. The storm surge produced numerous breaches and consequent flooding of 75 percent of the metropolitan areas of New Orleans. The levee and flood-wall failures mainly occurred due to overtopping and erosion caused by the storm surge (Nicholson et al., 2005). Three bowls of the city were flooded: the central part of the city (Orleans), New Orleans East, and St. Bernard.

Most severely affected was the area in St. Bernard where two major breaches in the flood walls resulted in a rapidly rising and fast moving flood with catastrophic consequences (Jonkman et al., 2009). In the other parts of the city, flooding occurred more gradually. In total an area of 260 km^2 was flooded with water levels rising up to 4 m. The dewatering of the city took over 40 days (Jonkman et al., 2009).

FIGURE 1.6 New Orleans, Louisiana in the aftermath of Hurricane Katrina, showing Interstate 10 at West End Boulevard, looking toward Lake Pontchartrain. The 17th Street Canal is just beyond the left edge of the image. The breach in the levee of that canal was responsible for much of the flooding of the city in the hours after the hurricane. *Photo by Kyle Niemi. From: http://en.wikipedia.org/wiki/File:KatrinaNewOrleansFlooded_edit2.jpg.*

Floods caused by hurricanes can be quite diverse in nature depending on the characteristic of the hurricane itself and the area that is struck by it. Hurricanes are tropical cyclones that develop over the Atlantic and Pacific Ocean (NOAA, 2001). At landfall they affect large areas with torrential rainfalls and storm surges. In the coastal areas, the storm surges are probably the most dangerous and destructive parts that can lead to coastal floods and, as shown in the case of New Orleans, destroy levees and flood walls leading to large floodings in cities located in these areas. The heavy rainfalls associated with hurricanes often cause inland floodings even hundreds of kilometers away from the coastal region. The danger of inland floodings depends on the storm speed. The slower the system moves, the more water can fall on the same location. This also depends on the orography and the antecedent soil moisture conditions (NOAA, 2001). Due to the large amounts of rainfall these inland floods may be much larger than ordinary river floods.

The extreme nature and variability of hurricanes makes it hard to foresee possible impacts and to extrapolate from one event to another complicating the design of appropriate flood protection measures.

1.3 FLOOD HAZARD PROBABILITIES

In the previous sections, different types of floods have been illustrated by presenting significant events. From both theoretical and practical perspectives, it is of interest to understand and estimate the probability of such events to happen. In this section, the focus is no longer on single events but on the entire set of floods that may threaten a location. In other words, we focus on the *flood hazard*, by which we indicate the combination of probability and magnitude of floods at a location. In engineering hydrology, flood hazard estimation is done by analyzing the magnitude of hydrologic/hydraulic variables related to flood events such as peak discharge, water level, or inundation extent, and how often these magnitudes occur in a given time period. Flood hazard estimation is traditionally performed based on extreme value statistics. Statistical models, i.e., distribution functions, are adjusted to ordered sequences of observed flood peaks and are used to extrapolate the flood magnitudes associated to very low exceedence probabilities (see, e.g., Gumbel, 1941). Many assumptions should apply for the method to give reliable results. For example, it is typically assumed that the flood data used are statistically independent and identically distributed, i.e., that each flood event is independent of those that came before it and that all floods have the same characteristics. The examples shown in Section 1.2 demonstrate that the second assumption may be grossly in error. Even floods of the same type, e.g., river floods, may be produced by different combinations of causal mechanisms, e.g., different atmospheric situations, different catchment initial conditions, different snow-related mechanisms such as accumulation and/or melting during the event, and so on. It has been rightly argued, in the literature, that statistical approaches for flood hazard estimation need to be complemented by the search for the causal mechanisms and dominant processes in the atmosphere, catchment, and river system (Klemeš, 1986; Merz and Blöschl, 2008a,b; Viglione et al., 2013; Merz et al., 2014). From a practical point of view, there are two ways of accounting for flood-producing mechanisms in estimating flood hazard at a location: i.e., (1) to classify observed flood events into classes and analyse them separately; or (2) to incorporate the knowledge on the flood-producing mechanisms in the statistical estimation.

Regarding the first approach, Hirschboeck (1987, 1988) and House and Hirschboeck (1997) performed a detailed analysis of causal mechanisms of floods in a number of catchments in Arizona and classified floods into three classes (tropical, convective, and frontal events). This classification allowed Hirschboeck (1987) and Alila and Mtiraoui (2002) to examine the flood statistics for each group of events and to derive more complex hydroclimatically defined probability distributions to characterize the flood frequency curves than those used before. Similarly, Merz and Blöschl (2003) developed a classification scheme based on process indicators, including the timing of the floods, storm duration, rainfall depth, snowmelt, catchment state, runoff response dynamics and the spatial coherence of floods. They found significantly different flood frequency statistics for different classes of river floods in Austria: long-rain floods, short-rain floods, flash floods, rain-on-snow floods, and snowmelt floods.

An alternative way of using process components in flood hazard estimation is by the derived flood frequency framework, proposed by Eagleson (1972) and later refined by Wood (1976), Fiorentino and Jacobellis (2001), Sivapalan et al. (2005), and Li et al. (2014) among others. In the case of floods in small- to middle-sized catchments, the derived flood frequency framework assumes that each independent flood peak in the complete data series is caused by an independent precipitation event. The relationship between precipitation (characterized by intensity and duration) and the resulting flood peak involves two kinds of transformations, i.e., runoff generation and runoff routing, and these are impacted by antecedent soil wetness. Since the event characteristics of precipitation (e.g., rainfall intensity and duration) are assumed random, and so is antecedent soil wetness, their probabilities can be directly combined (see, e.g., Viglione et al., 2009). Apel et al. (2006) also considered the probability of levee breaches, in a derived distribution approach, to determine their influence on the flood hazard of the river Rhine in Germany. Derived distribution analysis can then be used to obtain the cumulative distribution function of the population of flood peaks that occur in a catchment. A nice example is presented in Rogger et al. (2012, 2013), who used a derived distribution approach to explain how nonlinear catchment response related to a storage threshold may translate into a step change in the flood frequency curve for small Alpine catchments in Austria (Figure 1.7). Historically, such steep increases have often been treated as outliers, but an interpretation in terms of the flood generation processes may be more insightful and lead to a more accurate estimation of the flood hazard. There are also other cases in which different mechanisms dominating during different events may lead to step changes in the flood frequency curve. With increasing rainfall intensity or event rainfall depth that goes beyond a threshold, a switch from saturation excess to infiltration excess runoff may occur, or from subsurface storm flow to saturation (storage) excess, and both of these may be reflected in a sudden increase in

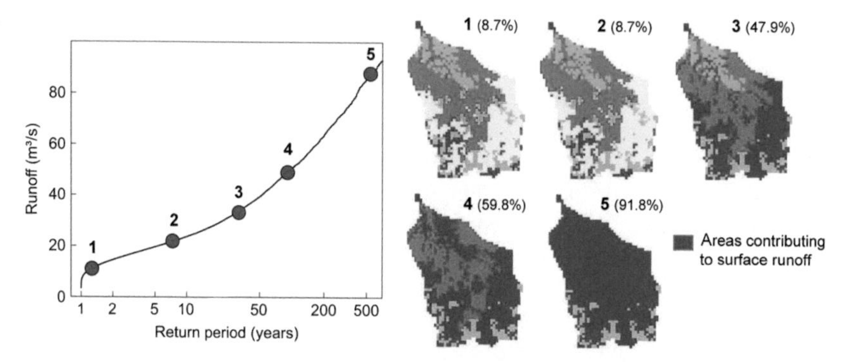

FIGURE 1.7 Runoff generation in the Weerbach catchment, Alpine Austria. (Right) Areas contributing to fast surface runoff for events of different magnitudes (1, low magnitude to 5, high magnitude). The percentage contributing area is given in parentheses. Blue indicates areas contributing to surface runoff; colors relate to different hydrological response units. (Left) The simulated flood frequency curve with these events indicated shows nonlinearity due to a change of processes. *From Rogger et al. (2012).*

runoff in the flood frequency curve (Sivapalan et al., 1990; Samuel and Sivapalan, 2008; Gioia et al., 2008).

Accounting for flood generating processes is not only useful for flood hazard estimation itself, but also for the evaluation of the associated uncertainties. Uncertainties in flood hazard estimation are due to many reasons (Merz and Thieken, 2005; Montanari, 2007; Montanari et al., 2009; Koutsoyiannis et al., 2007; Montanari, 2011). Hydrological processes have enormous spatiotemporal variability, which is difficult to capture. Since the characterization (of the probability) of rare events is of interest, long records of data are needed (Di Baldassarre and Uhlenbrook, 2012). Also, there may be uncertainties in the collected data (Montanari and Di Baldassarre, 2013). Researchers and practitioners have dealt with these uncertainties in different ways. Three types of approaches are commonly used. In the *traditional statistical approach*, uncertainty is estimated directly from the data (e.g., from the time series of annual peak discharges) and is a function of the record length (see e.g., Kuczera, 1999; Martins and Stedinger, 2000; Viglione et al., 2013). In the second approach, the causality of flood producing processes can be explicitly accounted for by propagating the inputs uncertainty to the outputs through models (e.g., Apel et al., 2008). The advantage of this method of *error-propagation* relies in the ability to attribute individual error sources based on causal relationships, but the difficulty remains that effects of unknown feedbacks cannot be identified. The third approach, the *comparative approach*, accounts for the total predictive uncertainty of estimation of flood hazards. In this approach, uncertainty is characterized by "blind-testing experiments." In places where a lot of information is available, prediction methods are applied based on only part of (or other) information and the result is then compared to the one obtained using all information available. This is the usual way to assess the performance of methods that predict flood runoff in ungauged locations (Bloschl et al., 2013d). Salinas et al. (2013), for example, assessed the performance of methods of prediction in ungauged locations reported in the literature for more than 3,000 catchments and interpreted their differences in terms of the underlying climate-landscape controls.

Since the three aforementioned approaches—i.e., traditional statistical uncertainty evaluation, error-propagation method, and comparative approach—take advantage of different types of information, using all of them, and eventually combining them, is the most appropriate way of dealing with uncertainty. This is in the spirit of flood frequency hydrology, a framework for understanding and estimating flood hazard by combining many pieces of information (Merz and Blöschl, 2008a,b; Viglione et al., 2013).

1.4 FLOODS IN A CHANGING WORLD

Recently there have been major floods around the world and it seems as if they had increased in number and magnitude. Take for example the extreme flood in the Upper Danube Basin discussed in Section 1.2.1. The June 2013 flood exceeded the largest flood runoff observed in the past two centuries in many parts of the Upper Danube Basin (Blöschl et al., 2013a) and came shortly after another record flood, the August 2002 event. Just at the time of writing this chapter, January 2014, a severe flash flood hit the French Riviera and Italian Liguria, not far from the Cinque Terre flood of 2011 discussed in Section 1.2.2. Webster et al. (2005) reported that, worldwide, the number of Category 4 and 5 hurricanes had increased 80 percent in the past 30 years (see also Elsner and Jagger, 2010).

A very important question, therefore, is whether floods have indeed increased in recent years. Analyses of flood data in the literature, however, are not conclusive (Hall et al., 2014). Historic proxy data such as flood marks and archival evidence generally suggest that, in Europe, recent floods have not exceeded the largest floods in the past five centuries (Brázdil and Kundzewics, 2006; Benito et al., 2003; Schmocker-Fackel and Naef, 2010) but, again, consistencies are not obvious. Also in the hurricane-science community, the large amount of uncertainty in the hurricane intensity data is recognized and the evidence of their increase is under debate (Holland, 2012). Part of the problem is the erratic nature of floods, as one big flood event does not indicate an increasing trend in flooding. Another part of the problem is the presence of natural decadal variations in the frequency and severity of floods (Montanari, 2003; Merz et al., 2014). Over a given observation period, a flood cycle of several decades may appear like a trend (Cohn and Lins, 2005). Figure 1.8 shows, in the top panel, the maximum annual floods of the Danube at Vienna for a period of 73 years. If one only examines this panel, the data suggest that there is an increasing trend of floods as five of the six largest floods that have

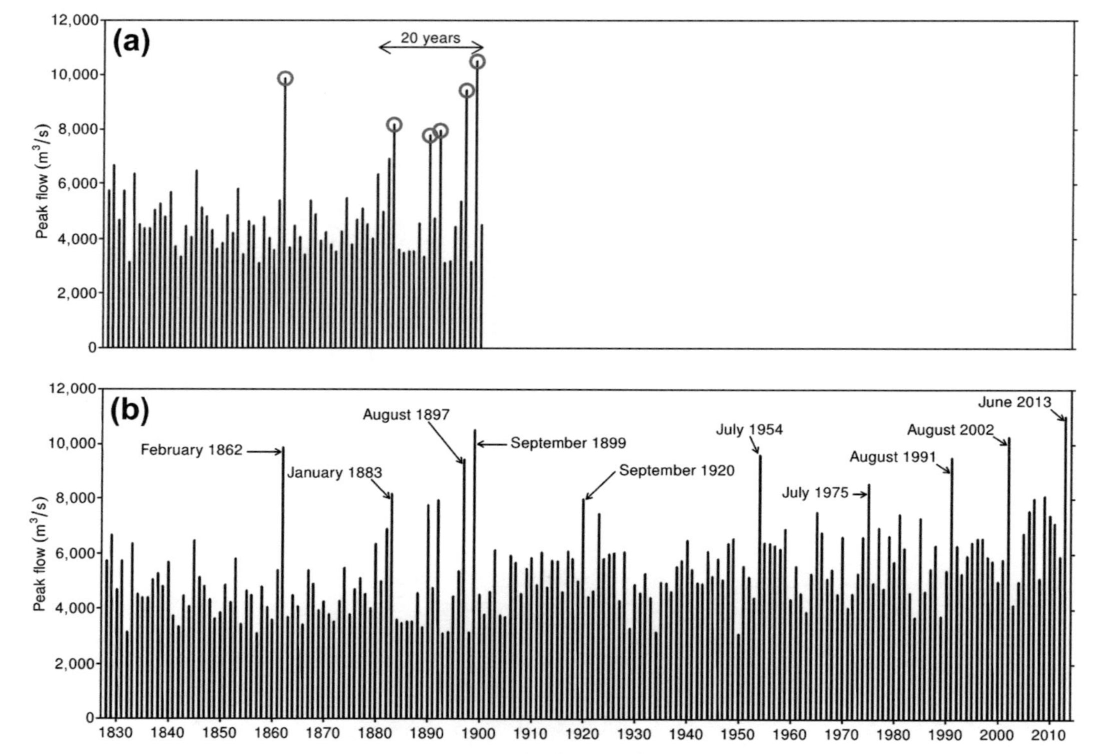

FIGURE 1.8 Annual maximum floods of the Danube at Vienna for 73 years (100,000 km^2 catchment area). (a) Five of the six largest floods have occurred in the last two decades. (b) Entire record 1828–2013. *Redrawn from Blöschl and Merz (2008).*

occurred at the end of the series (Blöschl and Montanari, 2010). However, the series shown in the top panel relates to the years 1828–1900. The lower panel shows the full series, indicating that the trend at the end of the late nineteenth century cannot be extrapolated to the future. The first half of the twentieth century did not have any large floods. What is needed are methods for detecting changes that account for the flood-generating mechanism. For instance, some of the main floods in the nineteenth century (e.g., February 1862 and January 1883) were ice-jam floods, whereas no major ice-jam flood has occurred in the last century due to river regulation.

Although the processes of individual flood events have been studied in great detail, identifying the causes of changes over time is much more difficult. For example, three main factors exist that potentially contribute to changes in river floods. These are related to climate, catchment processes, and the river network (Merz et al., 2012; Hall et al., 2014). Coastal floods may change due to changes in atmospheric phenomena, sea level rise and, maybe more importantly because of land subsidence due to groundwater pumping (Nicholls and Cazenave, 2010). The hazard that glacial lakes constitute to people and property in the valleys downstream has often been discussed in the light of global warming and the rapid retreat of glaciers (ICIMOD, 2011). The evacuated space between the retreating glacier and the lateral and end moraines may fill up with melt water, forming large ponds that may rapidly increase in size (Kattelmann, 2003).

Change in climate has attracted substantial recent discussion on potential increases in rainfall extremes (Blöschl and Montanari, 2010; Kundzewic, 2012). Changes in floods have been found to be correlated to climate characteristics such as precipitation and atmospheric circulation patterns (Petrow and Merz., 2009; Prudhomme et al., 2010; IPCC, 2013). Alexander et al. (2006) found slightly increasing trends of extreme precipitation contributions to annual precipitation at the global scale to support increasing flood trends. However, the issue is far from resolved and it is not clear how to model such changes reliably (Merz et al., 2011; Peel and Blöschl, 2011). It seems clear, however, that extreme floods do not arrive randomly but cluster in time into flood-poor and flood-rich periods consistent with the Hurst effect (Jain and Lall, 2001; Khaliq et al., 2009; Virmani and Weisberg, 2006; Szolgayová et al., 2014).

Changes in catchment processes include land-use changes. Numerous field studies (see Robinson et al., 2003) indicate significant flood increases as a result of deforestation. This is attributed to decreased interception storage in the canopy, decreased litter storage on the ground as well as changes in soil related to infiltration capacity and macropores. Land management and urbanization can also significantly affect flow paths and consequently floods (Wheater and Evans, 2009). Their effect is usually quite local as in urban floods, but less important for regional floods (Blöschl et al., 2007). Also, the importance of land-use tends to decrease with the magnitude of the events and increasing catchment size (Salazar et al., 2012).

Changes along the river network are more tangible. For a particular catchment, river regulations are usually very well documented and their effect can be calculated with good accuracy using hydrodynamic models. Levees have been built along many streams around the world in the past two centuries, thereby reducing potential retention volumes in the flood plains (e.g., Szolgay et al., 2008). Additionally, hydraulic structures such as reservoirs of run-on-river power plants, river regulation projects, and changes in stream morphology will contribute to changes in the flood propagation (Kresser, 1957). Also, as the flood waves change their celerities, the modified relative timings at confluences may alter the characteristics of the flood (Vorogushyn and Merz, 2013).

Analogously to flood hazard estimation at present (Section 1.3), understanding the interplay among processes and their change in time may be of benefit for flood hazard projection in the future. Two approaches of detecting/attributing change may be followed and eventually combined: data-based and modeling approaches. In the data-based approach, statistical methods are applied to observational flood data to detect whether significant changes have occurred and, subsequently, to data related to possible driving processes (climate, land-use, river works, pumping…) to attribute the changes. This is a useful approach if an extended and reliable database is available. For instance, the study of historical floods provides very valuable information for the period prior to systematic runoff observations, which typically started in the nineteenth or twentieth century. Even though documentary data (Brázdil et al., 2006, 2012) and flood marks (e.g., Macdonald, 2007) often suffer from spatial and temporal discontinuity, they allow inferring the frequency, seasonality, magnitude, causes, and impacts of events of a magnitude rarely witnessed within instrumental series (Barriendos et al., 2003). Also other proxy flood records, such as floodplain or lake sediments, offer the opportunity to extend the flood series further back in time (Werritty et al., 2006).

The modeling approach, in contrast, is the standard approach for assessing the sensitivity of floods to their drivers. Model-based scenario analyses involve hydrological model simulations using climatic variables as inputs and model parameters representing the land-use characteristics (Bronstert et al., 2002; Coulthard and Macklin, 2001). The simulations are then repeated with changed climate or land-use characteristics and the differences between the simulations are an indication of the sensitivity of floods to their drivers. Coupling the modeling approach to the data analysis, in the spirit of flood-frequency hydrology, may result into a more realistic detection of change in flood hazard and attribution to the causal mechanisms. Different methods may not yield exactly the same results because of differences in the method and the data used, but a comparative analysis for the same set of drivers and flood characteristics may allow one to learn from the differences between the methods to better understand their strengths and weaknesses (Merz and Blöschl, 2008a,b).

Even though it is recommendable to incorporate flood causal mechanisms in the estimation of flood hazards, the analysis of the flood causal factors can be extremely important per se, even if not used formally in statistical methods. Understanding the role of the combination of extreme factors that control the magnitude and characteristics of extreme floods may reveal what could possibly happen, even though the probability may be difficult to estimate. Consider, for example, the June 2013 flood in the Upper Danube River Basin described in Section 1.2.1. Among the factors that contributed to increasing the flood magnitude were relatively high antecedent soil moisture, little temporal shift between the flood peaks at the confluence of the Bavarian Danube and the Inn, and rainfall blocks close together resulting in a unique, large volume flood wave with relatively small peak attenuation. However, factors also existed that could have easily been more extreme. For instance, the precipitation event in June 2013 was considerably smaller than the precipitation event in September 1899 and in the Alpine area, a significant snowfall component existed that retained some of the water in the catchments. Rainfall as in 1899 with high antecedent soil moisture as in 2013 would likely produce a significantly larger flood (Blöschl et al., 2013a).

In the case of the urban flood described in Section 1.2.6, without the redesign of the drainage system of the city of Hull performed in 2001, the flood may have been less severe (Coulthard and Frostick, 2010). Also, if the coastal flood in the Netherlands (see Section 1.2.7) had happened in a different decade, and not in 1953, shortly after the Second World War, the number of weak spots in the dikes probably would have been lower (Gerritsen, 2005).

Disasters in general, and floods in particular, are due to combination of several unfortunate factors. This means that contingency matters (Blöschl et al., 2013c). In many cases these factors are hard to foresee. To anticipate surprises (and manage them), it is useful to explore possible events with just as much vigor as probable events (Kumar, 2011). One example in which possible mechanisms are taken into account is the flood risk management study of Wardekker et al. (2010), which explores imaginable surprises that they term "wildcards." The study proposes an uncertainty-robust adaptation strategy of strengthening the resilience (i.e., the ability to cope with adverse events) of the city of Rotterdam, using literature study, interviews, and a workshop. The "wildcards" or imaginable surprises for the area include thermohaline ocean circulation collapse, port freezing events, port malaria incidents, modified German water safety policy, enduring heat and drought, extreme storm, and failure of the storm surge barrier during an extreme storm. Their resilience approach is designed to make the system less prone to disturbances and capable of dealing with surprises. It may not be optimal in a cost-benefit sense but may result more robust than alternative approaches, even to those events that have not been foreseen. Such events are what Taleb (2007) called "black swan" events: unexpected events characterized by large impacts and for which failed predictions can only be explained in retrospective. Flood hazard estimation therefore is only one aspect of a more complete approach of flood-risk assessment. Reducing the vulnerability and increasing the resilience of the system by robust and flexible strategies may be of key importance in a changing world prone to black swan events (see Chapter 1.5).

REFERENCES

Alexander, L.V., Zhang, X., Peterson, T.C., Caesar, J., Gleason, B., Klein Tank, A., Haylock, M., Collins, D., Trewin, B., Rahimzadeh, F., Tagipour, A., Ambenje, P., Rupa Kumar, K., Revadekar, J., Griffiths, G., Vincent, L., Stephenson, D., Burn, J., Aguilar, E., Brunet, M., Taylor, M., New, M., Zhai, P., Rusticucci, M., Vazquez-Aguirre, J.L., 2006. Global observed changes in daily climate extremes of temperature and precipitation. J. Geophys. Res. 111, D05109. http://dx.doi.org/10.1029/2005JD006290.

Alila, Y., Mtiraoui, A., 2002. Implication of heterogeneous flood-frequency distributions on traditional stream-discharge prediction techniques. Hydrol. Processes 16, 1065–1084.

Amponsah, W., 2013. Learning from Extreme Hydrological Events: Analysis of the Flash Flood that Occurred on October 25–26, 2011 on the Magra River Basin (Italy) (Ms. thesis). University of Padova. July 2013.

Apel, H., Thieken, A.H., Merz, B., Blöschl, G., 2006. A probabilistic modelling system for assessing flood risks. Nat. Hazards 38 (1–2), 79–100.

Apel, H., Merz, B., Thieken, A.H., 2008. Quantification of uncertainties in flood risk assessments. Int. J. River Basin Manage. 6 (2), 149–162.

Barriendos, M., Coeur, D., Lang, M., Llasat, M.C., Naulet, R., Lemaitre, F., Barrera, A., 2003. Stationarity analysis of historical flood series in France and Spain (14th–20th centuries). Nat. Hazards Earth Syst. Sci. 3, 583–592. http://dx.doi.org/10.5194/nhess-3-583-2003.

Beltaos, S. (Ed.), 1995. River Ice Jams. Water Resources Publication, 13: 978-0-918334-87-9.

Beltaos, S., 2007. River ice breakup processes: recent advances and future directions. Can. J. Civ. Eng. 34 (6), 703–716.

Benito, G., Díez-Herrero, A., Fernández de Villalta, M., 2003. Magnitude and frequency of flooding in the Tagus Basin (Central Spain) over the last millennium. Clim. Change 58, 171–192.

BfG, 2013. Das Juni-Hochwasser des Jahres 2013 in Deutschland (The 2013 June flood in Germany), BfG Report no. 1793. Federal Institute of Hydrology, Koblenz, Germany.

Blöschl, G., Arn-Bardin, S., Bonell, M., Dorninger, M., Goodrich, D., Gutknecht, D., Matamoros, D., Merz, B., Shand, P., Szolgay, J., 2007. At what scales do climate variability and land cover change impact on flooding and low flows? Hydrol. Processes 21, 1241–1247. http://dx.doi.org/10.1002/hyp.6669.

Blöschl, G., Merz, R., 2008. Bestimmung von Bemessungshochwässern gegebener Jährlichkeit-Aspekte einer zeitgemäßen Strategie (Estimating design floods of a given return period-facets of a contemporary strategy). Wasserwirtschaft 98, 12–18.

Blöschl, G., Montanari, A., 2010. Climate change impacts – throwing the dice? Hydrol. Processes 24, 374–381. http://dx.doi.org/10.1002/hyp.7574.

Blöschl, G., Nester, T., Komma, J., Parajka, J., Perdigão, R.A.P., 2013a. The June 2013 flood in the Upper Danube Basin, and comparisons with the 2002, 1954 and 1899 floods. Hydrol. Earth Syst. Sci. 17, 5197–5212. http://dx.doi.org/10.5194/hess-17-5197-2013.

Blöschl, G., Nester, T., Komma, J., Parajka, J., Perdigão, R.A.P., 2013b. Das Juni-Hochwasser 2013-Analyse und Konsequenzen für das Hochwasserrisikomanagement. Österreichische Ingenieur- und Architekten-Zeitschrift 158. Jg., Heft 1–12.

Blöschl, G., Viglione, A., Montanari, A., 2013c. Emerging approaches to hydrological risk management in a changing world. In: Pielke Sr., R.A. (Ed.), Climate Vulnerability. Elsevier Inc., Academic Press, pp. 3–10. http://dx.doi.org/10.1016/B978-0-12-384703-4.00505-0 (Chapter 5).

Blöschl, G., Sivapalan, M., Wagener, T., Viglione, A., Savenije, H., 2013d. Runoff Prediction in Ungauged Basins – Synthesis across Processes, Places and Scales. Cambridge University Press, ISBN 9781107028180.

Brázdil, R., Kundzewicz, Z.W., Benito, G., 2006. Historical hydrology for studying flood risk in Europe. Hydrol. Sci. J. 51, 739–764. http://dx.doi.org/10.1623/hysj.51.5.739.

Brázdil, R., Kundzewicz, Z.W., Benito, G., Demarée, G., Macdonald, N., Roald, L.A., 2012. Historical floods in Europe in the past Millennium. In: Kundzewicz, Z.W. (Ed.), Changes in Flood Risk in Europe. IAHS Press, Wallingford, pp. 121–166.

Bronstert, A., Niehoff, D., Bürger, G., 2002. Effects of climate and land-use change on storm runoff generation: present knowledge and modelling capabilities. Hydrol. Processes 16, 509–529. http://dx.doi.org/10.1002/hyp.326.

Buzzi, A., Davolio, S., Malguzzi, P., Drofa, O., Mastrangelo, D., 2013. Heavy rainfall episodes over Liguria of autumn 2011: numerical forecasting experiments. Nat. Hazards Earth Syst. Sci. Discuss. 1, 7093–7135. http://dx.doi.org/10.5194/nhessd-1-7093-2013.

van Bebber, W.J., 1891. Die Zugstrassen der barometrischen Minima nach den Bahnenkarten der Deutschen Seewarte für den Zeitraum von 1870–1890. Meteorol. Z. 8, 361–366.

Cevasco, A., Brandolini, P., Scopesi, C., Rellini, I., 2013. Relationships between geo-hydrological processes induced by heavy rainfall and land-use: the case of 25 October 2011 in the Vernazza catchment (Cinque Terre, NW Italy). J. Maps 9 (2).

Cohn, T.A., Lins, H.F., 2005. Nature's style: naturally trendy. Geophys. Res. Lett. 32. http://dx.doi.org/10.1029/2005GL024476. Art. no. L23402.

Coulthard, T.J., Frostick, L.E., 2010. The Hull floods of 2007: implications for the governance and management of urban drainage systems. J. Flood Risk Manage. 3 (3), 223–231.

Coulthard, T.J., Macklin, M.G., 2001. How sensitive are river systems to climate and land-use changes? A model-based evaluation. J. Quat. Sci. 16 (4), 347–351.

Denner, J.C., Brown, R.O., 1998. March 11, 1992, ice-jam flood in Montpelier, Vermont. In: Perry, Charles, A., Combs, L.J. (Eds.), Summary of Floods in the United States, January 1992 through September 1993, US Geological Survey, water-supply paper 2499, pp. 1–286.

Di Baldassarre, G., Kemerink, J.S., Kooy, M., Brandimarte, L., 2014. Floods and societies: the spatial distribution of water-related disaster risk and its dynamics. WIREs Water vol 1, 133–139. http://dx.doi.org/10.1002/wat2.1015.

Di Baldassarre, G., Kooy, M., Kemerink, J., Brandimarte, L., 2013. Towards understanding the dynamic behaviour of floodplains as human-water systems. Hydrol. Earth Syst. Sci. 17, 3235–3244. http://dx.doi.org/10.5194/hess-17-3235-2013.

Di Baldassarre, G., Uhlenbrook, S., 2012. Is the current flood of data enough? A treatise on research needs for the improvement of flood modelling. Hydrol. Processes 26 (1), 153–158.

Dunne, T., 1983. Relation of field studies and modeling in the prediction of storm runoff. J. Hydrol. 65.1, 25–48.

Eagleson, P.S., 1972. Dynamics of flood frequency. Water Resour. Res. 8 (4), 878–898. http://dx.doi.org/10.1029/WR008i004p00878.

Elsner, J.B., Jagger, T.H., 2010. On the increasing intensity of the strongest Atlantic hurricanes. In: Hurricanes and Climate Change. Springer, Netherlands, pp. 175–190.

Li, J., Thyer, M., Lambert, M., Kuczera, G., Metcalfe, A., 2014. An efficient causative event-based approach for deriving the annual flood frequency distribution. J. Hydrol. 510, 412–423. http://dx.doi.org/10.1016/j.jhydrol.2013.12.035.

Fiorentino, M., Iacobellis, V., 2001. New insights about the climatic and geologic control on the probability distribution of floods. Water Resour. Res. 37 (3), 721. http://dx.doi.org/10.1029/2000WR900315.

Fujii, Y., Satake, K., Sakai, S.I., Shinohara, M., Kanazawa, T., 2011. Tsunami source of the 2011 off the Pacific coast of Tohoku Earthquake. Earth, Planets Space 63 (7), 815–820.

Gaál, L., Szolgay, J., Kohnová, S., Parajka, J., Merz, R., Viglione, A., Blöschl, G., 2012. Flood timescales: understanding the interplay of climate and catchment processes through comparative hydrology. Water Resour. Res. 48, W04511. http://dx.doi.org/10.1029/2011WR011509.

Gerritsen, H., 2005. What happened in 1953? the Big Flood in the Netherlands in retrospect. Philos. Trans. A: Math. Phys. Eng. Sci. 363 (1831), 1271–1291.

Gioia, A., Iacobellis, V., Manfreda, S., Fiorentino, M., 2008. Runoff thresholds in derived flood frequency distributions. Hydrol. Earth Syst. Sci. 12, 1295–1307. http://dx.doi.org/10.5194/hess-12-1295-2008.

Gumbel, E.J., 1941. The return period of flood flows. Ann. Math. Stat. 12 (2), 163–190.

Hall, J., Arheimer, B., Borga, M., Brázdil, R., Claps, P., Kiss, A., Kjeldsen, T.R., Kriaučiūnienė, J., Kundzewicz, Z.W., Lang, M., Llasat, M.C., Macdonald, N., McIntyre, N., Mediero, L., Merz, B., Merz, R., Molnar, P., Montanari, A., Neuhold, C., Parajka, J., Perdigão, R.A.P., Plavcová, L., Rogger, M., Salinas, J.L., Sauquet, E., Schär, C., Szolgay, J., Viglione, A., Blöschl, G., 2014. Understanding flood regime changes in Europe: a state of the art assessment. Hydrology and Earth System Sciences 18, 2735–2772. http://dx.doi.org/10.5194/hess-18-2735-2014.

Hirschboeck, K.K., 1987. Hydroclimatically-defined mixed distributions in partial duration flood series. In: Singh, V.P. (Ed.), Hydrologic Frequency Modeling: Proceedings of the International Symposium on Flood Frequency and Risk Analyses, May 14–17, 1986. D. Reidel, Norwell, MA, pp. 199–212. Louisiana State University, Baton Rouge.

Hirschboeck, K.K., 1988. Flood hydroclimatology. In: Baker, V.R., Kochel, R.C., Patton, P.C. (Eds.), Flood Geomorphology. John Wiley, Hoboken, NJ, pp. 27–49.

Holland, G.J., 2012. Hurricanes and rising global temperatures. Proc. Natl. Acad. Sci. 109 (48), 19513–19514.

Horton, R.E., 1933. The role of infiltration in the hydrologic cycle. Trans. Am. Geophys. Union 14, 446–460.

House, P.K., Hirschboeck, K.K., 1997. Hydroclimatological and paleohydrological context of extreme winter flooding in Arizona, 1993. In: Larson, R.A., Slosson, J.E. (Eds.), Storm-induced Geological Hazards: Case Histories from the 1992–1993 Winter Storm in Southern California and Arizona, Geological Society of America Reviews in Engineering Geology, vol. 11, pp. 1–24.

Houston, D., Werritty, A., Bassett, D., Geddes, A., Hoolachan, A., McMillan, M., 2011. Pluvial (Rain-Related) Flooding in Urban Areas: the Invisible Hazard. Joseph Rowntree Foundation, New York.

ICIMOD, 2011. Glacial Lakes and Glacial Lake Outburst Floods in Nepal. ICIMOD, Kathmandu.

IPCC, 2013. Summary for policymakers. In: Stocker, T.F., Qin, D., Plattner, G.-K., Tignor, M., Allen, S.K., Boschung, J., Nauels, A., Xia, Y., Bex, V., Midgley, P.M. (Eds.), Climate Change 2013: The Physical Science Basis. Contribution of Working Group I to the Fifth Assessment Report of the Intergovernmental Panel on Climate Change. Cambridge University Press, Cambridge, UK and New York, USA.

Jain, S., Lall, U., 2001. Floods in a changing climate: does the past represent the future? Water Resour. Res. 37, 3193–3205.

Jha, A.K., Bloch, R., Lamond, J., 2012. Cities and Flooding: a Guide to Integrated Urban Flood Risk Management for the 21st Century. World Bank Publications.

Jonkman, S.N., Maaskant, B., Boyd, E., Levitan, M.L., 2009. Loss of life caused by the flooding of New Orleans after Hurricane Katrina: analysis of the relationship between flood characteristics and mortality. Risk Anal. 29 (5), 676–698.

Kanamori, H., 1972. Mechanism of tsunami earthquakes. Phys. Earth Planet. Inter. 6 (5), 346–359.

Kattelmann, R., 2003. Glacial lake outburst floods in the Nepal Himalaya: a manageable hazard? Nat. Hazards 28 (1), 145–154.

Khaliq, M., Ouarda, T., Gachon, P., Sushama, L., St-Hilaire, A., 2009. Identification of hydrological trends in the presence of serial and cross correlations: a review of selected methods and their application to annual flow regimes of Canadian rivers. J. Hydrol. 368, 117–130.

Klemeš, V., 1986. Dilettantism in hydrology: transition or destiny? Water Resour. Res. 22 (9S), 177S–188S.

Koutsoyiannis, D., Efstratiadis, A., Georgakakos, K.P., 2007. Uncertainty assessment of future hydroclimatic predictions: a comparison of probabilistic and scenario-based approaches. J. Hydrometeorol. 8 (3).

Kresser, W., 1957. Die Hochwässer der Donau (The floods of the Danube). In: Schriftenreihe des österreichischen Wasserwirtschaftsverbandes, pp. 32–33 (Wien).

Kuczera, G., 1999. Comprehensive at-site flood frequency analysis using Monte Carlo Bayesian inference. Water Resour. Res. 35 (5), 1551–1557.

Kumar, P., 2011. Typology of hydrologic predictability. Water Resour. Res. 47, W00H05. http://dx.doi.org/10.1029/2010WR009769.

Kundzewicz, Z.W., 2012. Changes in Flood Risk in Europe. IAHS Press, Wallingford, UK, p. 544.

Macdonald, N., 2007. Epigraphic records: a valuable resource in re-assessing flood risk and long-term climate variability. Environ. Hist. 12, 136–140.

Marchi, L., Boni, G., Cavalli, M., Comiti, F., Crema, S., Lucía, A., Marra, F., Zoccatelli, D., 2013. The flash-flood of October 2011 in the Magra river (Italy): rainstorm characterization and flood response analysis. Geophys. Res. Abstr. 15. EGU2013-11125, EGU General Assembly 2013, Vienna, Austria.

Martins, E., Stedinger, J.R., 2000. Generalized maximum-likelihood generalized extreme-value quantile estimators for hydrologic data. Water Resour. Res. 36 (3), 737–744. http://dx.doi.org/10.1029/1999WR900330.

Merz, R., Blöschl, G., 2003. A process typology of regional floods. Water Resour. Res. 39 (12), 1–20. http://dx.doi.org/10.1029/2002WR001952.

Merz, B., Thieken, A.H., 2005. Separating natural and epistemic uncertainty in flood frequency analysis. J. Hydrol. 309 (1), 114–132.

Merz, B., Vorogushyn, S., Uhlemann, S., Delgado, J., Hundecha, Y., 2012. HESS opinions "More efforts and scientific rigour are needed to attribute trends in flood time series". Hydrol. Earth Syst. Sci. 16, 1379–1387. http://dx.doi.org/10.5194/hess-16-1379-2012.

Merz, B., Aerts, J., Arnbjerg-Nielsen, K., Baldi, M., Becker, A., Bichet, A., Blöschl, G., Bouwer, L.M., Brauer, A., Cioffi, F., Delgado, J.M., Gocht, M., Guzzetti, F., Harrigan, S., Hirschboeck, K., Kilsby, C., Kron, W., Kwon, H.-H., Lall, U., Merz, R., Nissen, K., Salvatti, P., Swierczynski, T., Ulbrich, U., Viglione, A., Ward, P.J., Weiler, M., Wilhelm, B., Nied, M., 2014. Floods and climate: emerging perspectives for flood risk assessment and management. Nat. Hazards Earth Syst. Sci. 14, 1921–1942. http://dx.doi.org/10.5194/nhess-14-1921-2014.

Merz, R., Blöschl, G., 2008a. Flood frequency hydrology: 1. temporal, spatial, and causal expansion of information. Water Resour. Res. 44, W08432.

Merz, R., Blöschl, G., 2008b. Flood frequency hydrology: 2. combining data evidence. Water Resour. Res. 44, W08433.

Merz, R., Parajka, J., Blöschl, G., 2011. Time stability of catchment model parameters – implications for climate impact analyses. Water Resour. Res. 47, W02531. http://dx.doi.org/10.1029/2010WR009505.

Met Office, 2007. UK Rainfall Statistics. Available from: http://www.metoffice.gov.uk/corporate/library/factsheets/factsheet09.pdf.

Montanari, A., 2003. Long-range dependence in hydrology. In: Doukhan, P., Oppenheim, G., Taqqu, M.S. (Eds.), Theory and Applications of Long-Range Dependence. Springer, New York, pp. 461–472.

Montanari, A., 2007. What do we mean by 'uncertainty'? the need for a consistent wording about uncertainty assessment in hydrology. Hydrol. Processes 21 (6), 841–845.

Montanari, A., Shoemaker, C.A., van de Giesen, N., 2009. Introduction to special section on uncertainty assessment in surface and subsurface hydrology: an overview of issues and challenges. Water Resour. Res. 45 (12).

Montanari, A., 2011. Uncertainty of hydrological predictions. Treatise Water Sci. 2, 459–478.

Montanari, A., Di Baldassarre, G., 2013. Data errors and hydrological modelling: the role of model structure to propagate observation uncertainty. Adv. Water Resour 51, 498–504. http://dx.doi.org/10.1016/j.advwatres.2012.09.007.

Nicholls, R.J., Cazenave, A., 2010. Sea-level rise and its impact on coastal zones. Science 328 (5985), 1517–1520.

Nicholson, P.G., et al., 2005. Preliminary Report on the Performance of the New Orleans Levee System in Hurricane Katrina on August 29, 2005. University of California at Berkeley.

NOAA National Oceanic and Atmospheric Administration, U.S. Department of Commerce, National Weather Service, 2001. Tropical Cyclones & Inland Flooding, an Awareness Guide.

Ohl, C., Tapsell, S., 2000. Flooding and human health: the dangers posed are not always obvious. Br. Med. J. 321 (7270), 1167–1168.

Peel, M.C., Blöschl, G., 2011. Hydrologic modelling in a changing world. Prog. Phys. Geog. 35, 249–261. http://dx.doi.org/10.1177/0309133311402550.

Petrow, T., Merz, B., 2009. Trends in flood magnitude, frequency and seasonality in Germany in the period 1951–2002. J. Hydrol. 371, 129–141. http://dx.doi.org/10.1016/j.jhydrol.2009.03.024.

Prudhomme, C., Wilby, R., Crooks, S., Kay, A., Reynard, N., 2010. Scenario-neutral approach to climate change impact studies: application to flood risk. J. Hydrol. 390, 198–209.

Rebora, N., et al., 2013. Extreme rainfall in the Mediterranean: what can we learn from observations? J. Hydrometeorol. 14, 906–922. http://dx.doi.org/10.1175/JHM-D-12-083.1.

Robinson, M., Cognard-Plancq, A.L., Cosandey, C., David, J., Durand, P., Führer, H.W., Hall, R., Hendriques, M., Marc, V., McCarthy, R.R., McDonnell, M., Martin, C., Nisbet, T., O'Dea, P., Rodgers, M., Zoller, A., 2003. Studies of the impact of forests on peak flows and baseflows: a European perspective. Forest Ecol. Manage. 186, 85–97. http://dx.doi.org/10.1016/S0378-1127(03)00238-X.

Rogger, M., Pirkl, H., Viglione, A., Komma, J., Kohl, B., Kirnbauer, R., Merz, R., Blöschl, G., 2012. Step changes in the flood frequency curve: process controls. Water Resour. Res. 48, W05544. http://dx.doi.org/10.1029/2011WR011187.

Rogger, M., Viglione, A., Derx, J., Blöschl, G., 2013. Quantifying effects of catchments storage thresholds on step changes in the flood frequency curve. Water Resour. Res. 49, 6946–6958. http://dx.doi.org/10.1002/wrcr.20553.

Salazar, S., Francés, F., Komma, J., Blume, T., Francke, T., Bronstert, A., Blöschl, G., 2012. A comparative analysis of the effectiveness of flood management measures based on the concept of "retaining water in the landscape" in different European hydro-climatic regions. Nat. Hazards Earth Syst. Sci. 12, 3287–3306. http://dx.doi.org/10.5194/nhess-12-3287-2012.

Salinas, J.L., Laaha, G., Rogger, M., Parajka, J., Viglione, A., Sivapalan, M., Blöschl, G., 2013. Comparative assessment of predictions in ungauged basins – Part 2: flood and low flow studies. Hydrol. Earth Syst. Sci. 17, 2637–2652. http://dx.doi.org/10.5194/hess-17-2637-2013.

Samuel, J.M., Sivapalan, M., 2008. Effects of multiscale rainfall variability on flood frequency: comparative multi-site analysis of dominant runoff processes. Water Resour. Res. 44 (9), W09423. http://dx.doi.org/10.1029/2008WR006928.

Schmocker-Fackel, P., Naef, F., 2010. More frequent flooding?, changes in flood frequency in Switzerland since 1850. J. Hydrol. 381, 1–8. http://dx.doi.org/10.1016/j.jhydrol.2009.09.022.

Si, Y., Quing, D., 1998. The world's most catastrophic dam failures: the august 1975 collapse of the Banqiao and Shimantan dams. In: Quing, D., Thibodeau, J.G., Williams, P.B. (Eds.), The River Dragon Has Come! the Three Gorges Dam and the Fate of China's Yangzi River and Its People. M.E Sharpe, Armonk, New York, pp. 25–38.

Silvestro, F., Gabellani, S., Giannoni, F., Parodi, A., Rebora, N., Rudari, R., Siccardi, F., 2012. A hydrological analysis of the 4 November 2011 event in Genoa. Nat. Hazards Earth Syst. Sci. 12, 2743–2752. http://dx.doi.org/10.5194/nhess-12-2743-2012.

Sivapalan, M., Wood, E.F., Beven, J., 1990. On hydrologic similarity. 3: a dimensionless flood frequency model using a generalized geomorphologic unit hydrograph and partial area runoff generation. Water Resour. Res. 26 (1), 43–58.

Sivapalan, M., Blöschl, G., Merz, R., Gutknecht, D., 2005. Linking flood frequency to long-term water balance: incorporating effects of seasonality. Water Resour. Res. 41 (6), 1–17. http://dx.doi.org/10.1029/2004WR003439.

Slager, K., 1992. De Ramp (The Disaster), second ed. Uitgeverij Atlas, Amsterdam. p. 557.

Szolgay, J., Danáèová, M., Jurèák, S., Spál, P., 2008. Multilinear flood routing using empirical wave-speed discharge relationships: case study on the Morava River. J. Hydrol. Hydromech. 56, 213–227.

Szolgayová, E., Laaha, G., Blöschl, G., Bucher, C., 2014. Factors influencing long range dependence in streamflow of European rivers. Hydrol. Processes 28 (4), 1573–1586. http://dx.doi.org/10.1002/hyp.9694.

Taleb, N.N., 2007. The Black Swan: the Impact of the Highly Improbable. Random House, New York, p. 366.

Viglione, A., Merz, R., Blöschl, G., 2009. On the role of the runoff coefficient in the mapping of rainfall to flood return periods. Hydrol. Earth Syst. Sci. 13, 577–593.

Viglione, A., Merz, R., Salinas, J.S., Blöschl, G., 2013. Flood frequency hydrology: 3. a Bayesian analysis. Water Resour. Res. 49, 675–692.

Virmani, J.I., Weisberg, R.H., 2006. The 2005 hurricane season: an echo of the past or a harbinger of the future? Geophys. Res. Lett. 33, L05707. http://dx.doi.org/10.1029/2005GL025517.

Vorogushyn, S., Merz, B., 2013. Flood trends along the Rhine: the role of river training. Hydrol. Earth Syst. Sci. 17 (10). http://dx.doi.org/10.5194/hess-17-3871-2013.

Vuichard, D., Zimmermann, M., 1986. The Langmoche flash-flood, khumbu himal, Nepal. Mt. Res. Dev. 6 (1), 90–94.

Vuichard, D., Zimmermann, M., 1987. The 1985 catastrophic drainage of a moraine-dammed lake, Khumbu Himal, Nepal: cause and consequences. Mt. Res. Dev., 91–110.

Wang, B., 2006. The Asian Monsoon. Springer, ISBN 978-3-540-37722-1, p. 787.

Wardekker, J.A., de Jong, A., Knoop, J.M., van der Sluijs, J.P., 2010. Operationalising a resilience approach to adapting an urban delta to uncertain climate changes. Technol. Forecast. Soc. Change 77 (6), 987–998. http://dx.doi.org/10.1016/j.techfore.2009.11.005.

Webster, P.J., Holland, G.J., Curry, J.A., Chang, H.-R., 2005. Changes in tropical cyclone number, duration, and intensity in a warming environment. Science 309, 1844–1846, 16 September, 2005.

Werritty, A., Paine, J., Macdonald, N., Rowan, J., McEwen, L., 2006. Use of multi-proxy flood records to improve estimates of flood risk: lower River Tay, Scotland. Catena 66, 107–119. http://dx.doi.org/10.1016/j.catena.2005.07.012.

Wheater, H., Evans, E., 2009. Land use, water management and future flood risk. Land Use Policy 26, S251–S264. http://dx.doi.org/10.1016/j.landusepol.2009.08.019.

Wood, E.F., 1976. An analysis of the effects of parameter uncertainty in deterministic hydrologic models. Water Resour. Res. 12 (5), 925–932. http://dx.doi.org/10.1029/WR012i005p00925.

Zhang, L.M., Xu, Y., Jia, J.S., 2009. Analysis of earth dam failures: a database approach. Georisk 3 (3), 184–189. http://dx.doi.org/10.1080/17499510902831759.

Chapter 2

Measuring and Mapping Flood Processes

Guy J.-P. Schumann [1], Paul D. Bates [2], Jeffrey C. Neal [2] and Konstantinos M. Andreadis [1]

[1] *Jet Propulsion Laboratory, California Institute of Technology, Pasadena, CA, USA,* [2] *School of Geographical Sciences, University of Bristol, Bristol, UK*

ABSTRACT

Floods are no doubt a major hazard and the risks they pose are increasing due to shifts in meteorological forcings, population pressures, as well as anthropogenic change to riverine landscapes. Flood waves and related processes are observed globally, through either river gauging networks or remote sensing acquisitions. River gauging stations are declining globally and although historical and current gauges are providing useful and frequent data in the developed world, the number of gauges in developing and emerging economies is very small and measurement stations are often very far apart and in remote locations thus making inference of processes and data collection difficult. Remote sensing, space-borne, and airborne can alleviate some of these limitations but has its own shortcomings. Hydrodynamic models can complement observations but the accuracy and complexity of the flood flow models used vary with both spatial scale at which they are applied and complexity of the topographic landscape. Furthermore, models are only as good as the data used to drive and calibrate them. In recent years, substantial efforts are being made to improve this complex situation of observing, mapping, and modeling flood processes, both in terms of flood model development and remote sensing, particularly satellite platforms. This chapter provides a detailed account of this complex interplay between models and data to observe, simulate, and understand flood processes on various scales and in different landscape settings.

2.1 INTRODUCTION

Flooding is a natural process and high flows in rivers and floodplain inundation are necessary to sustain life and ecosystem services all around the world. Water flow and storage also regulate the biogeochemical cycle. However, in the more populated areas in the world at certain times floods can have devastating impacts and thus monitoring and predicting floods is of great importance.

Hydro-Meteorological Hazards, Risks, and Disasters. http://dx.doi.org/10.1016/B978-0-12-394846-5.00002-3

In the developed world, dense river gauging networks, high quality channel survey data, as well as fine resolution, high precision, light detection and ranging (LiDAR) ground elevation data over floodplains exist to allow detailed long-term monitoring and modeling of flood events. In the developing world, however, the situation is very different. River-gauging stations are often sparse and are only operated in very large basins, channel survey data hardly exist and if so, they are regarded a state secret or are not shareable. Floodplain topography is most of the time only available as advanced spaceborne thermal emission and reflection radiometer (ASTER) or Shuttle Radar Topography Mission Digital Elevation Models (SRTM DEMs) with an accuracy of 6–9 m RMSE (root mean squared error) in the vertical for low-lying flat areas. Monitoring and predicting flood events or indeed building models is thus difficult to achieve in many areas around the world. In such areas, satellites are currently the only means of providing reliable data that can be used for mapping and modeling floods.

Remote sensing has a very long history in the collection of data and monitoring of processes in relation to flood inundation. With the advent of aerial photography and photogrammetry, stereo-photos have been used to construct DEMs that map floodplain topography, an essential boundary condition in floodplain inundation and flow processes and thus a prerequisite in any one-dimensional (1D) or two-dimensional (2D) hydrodynamic model like Hydrologic Engineering Center - River Analysis System (HEC-RAS) or LISFLOOD-FP. Aerial black and white and color photography is also used to map detailed flood inundation area and extent and, if fused with high resolution DEM data, can also yield accurate water stages (i.e., height above vertical datum) at the flood edge (Schumann et al., 2009b). In the past 30 years, satellite observations have been extensively used to complement aerial photography to map topography, flood extent, and area as well as retrieve either indirectly or directly water heights, and in a few cases even infer discharge.

Although remote sensing observations are of great value, they are in most cases only snapshot observations in the sense that the high temporal resolution is missing. This element is crucial for understanding the evolution of a flood process and monitoring of a river flood wave. Therefore most often remotely sensed flood information is integrated with flood models (Schumann et al., 2009b), through either simple data-model comparison to calibrate model parameters and validate model simulations or, more recently, data assimilation which can be very beneficial to flood forecasting.

Flooding affects societies, economies, and ecosystems worldwide and the current period is flood-rich compared with past records, so having a data-rich environment in terms of flood inundation observations and models is rather necessary. This data-rich environment will be further strengthened with new, upcoming, and potential satellite missions dedicated to flood hydrology (e.g., DLR's (German Aerospace Center) TerraSAR-X, ASI's

(Italian Space Agency) COSMO-SkyMed, CSA's (Canadian Space Agency) Radarsat-2). ESA's Sentinel-1 synthetic aperture radar (SAR) mission (set to launch 2014) is targeted at operational applications and the potential National Aeronautics and Space Administration/Centre National d'Etudes Spatiales (NASA/CNES) SWOT mission (with a tentative launch date of 2020) would measure water levels at 50 m resolution with cm accuracy and would potentially provide a global water database including river and lake area and width, water heights, water surface slopes, and even characterization of discharge. Table 2.1 lists existing and upcoming satellites operating in the microwave region, useful for measuring and mapping flood processes.

The scientific literature on remote sensing and modeling of flood inundation processes and rivers is vast and detailed reviews of all aspects in the field exist (e.g., Smith, 1997; Alsdorf et al., 2007; Marcus and Fonstad, 2008; Bates et al., 2014). The following sections outline the main (satellite) remote sensing observations of floods and rivers and also describe the models used to simulate flood inundation processes.

2.2 FLOODPLAIN TOPOGRAPHY

Elevation data in the form of a DEM are probably the most common remote sensing-derived product and are required for all types of environmental modeling. Table 2.2 provides a detailed overview and commonly reported accuracies of the different DEMs that exist.

Airborne photogrammetry and interferometric SAR technology from different viewing geometries allows the generation of fine spatial resolution, high accuracy (typically <50 cm vertical error) DEMs; however, errors rapidly increase up to several meters in the vertical when image pairs are acquired from satellites as in the case of SAR missions (e.g., ERS-2 SAR DEMs) or the ASTER DEM. It is clear that global coverage DEMs with low vertical accuracies are not at all useful for local to regional scale detailed floodplain inundation studies. Having said that, in low-lying areas, the SRTM-DEM at 90 m resolution for instance has been shown to be accurate to better than a couple of meters in the vertical in some floodplain areas (Sanders, 2007; Schumann et al., 2013). The improvement in vertical accuracy over low-lying, flat areas is generally the case and because riverine flooding occurs in those areas, many of the global coverage DEMs (e.g., 90 m SRTM-DEM) can be used for large scale flood inundation studies.

However, better results can only really be obtained with airborne LiDAR DEM at local to regional scales. It is noteworthy though that recent advances in hydrodynamic modeling are moving toward improved representation of physics in models that can credibly simulate flood processes at subgrid scale using coarse resolution and lower accuracy DEMs (e.g., LISFLOOD-FP subgrid channel (SGC) as developed by Neal et al. (2012).

TABLE 2.1 Recent and Planned Satellite Missions Dedicated to Hydrology Using Microwave Technology

Mission/Satellite	Launch Year	Band (GHz)	Revisit Time at Equator (days)	Spatial Resolution (m)	Primary Use for Flood Hydrology
ALOS	2006	L (1.3)	46	10–100	Flood mapping Soil moisture at small scale
TerraSAR-X (TerraSAR-X 2 planned)[a]	2007	X (9.6)	11	1–18	High resolution flood mapping
RADARSAT-2 (RADARSAT constellation mission—RCM—planned)[b]	2007	C (5.3)	24	3–100	High resolution flood mapping Soil moisture at small scale
COSMO-SkyMed	2007	X (9.6)	16 (4 to <1, with 4 satellites in constellation)	1–100	High resolution flood mapping

SMOS	2009	L (1.4) (passive)	2–3	30–50 km	Large scale flood mapping Soil moisture at large scale
Cryosat-2	2010	Ku (13.575) interferometric altimeter	369	~250 m imaging	Water level measurement Topography
TanDEM-X	2010	X (9.6) (interferometry)	11 (In tandem orbit with TerraSAR-X)	3 m	Water level dynamics Topography
Sentinel-1	2013	C (5.3)	12 (6, with 2 satellites in constellation)	5–100	High resolution flood mapping Soil moisture at small scale
TanDEM-L (potential)	2015	L (1–2) (interferometry)	8	5–100	Water level dynamics Topography
SWOT	2019	Ka (35) interferometric altimeter	22	<100 m imaging	Water level measurement Topography

[a] *Source: http://www.infoterra.de.*
[b] *Source: http://www.eoportal.org/.*
Source: After Schumann et al. (2012).

TABLE 2.2 Various Digital Elevation Model (DEM) Products and Their Resolution and Vertical Accuracy

DEM	Technique	Spatial Resolution	Vertical Accuracy	Coverage	End-User Cost (Source)
Photogrammetry	Aerial stereo–photography pairs	0.5–5 m	<20 cm	Local	High
LiDAR	Laser point cloud	<1–5 m	10–20 cm	Local	High
Airborne InSAR	SAR image pair interferometry	5 m	1–1.5 m	Regional/national	Moderate/high
Space-borne InSAR	SAR image pair interferometry	20–30 m	1.5–3 m	Regional/national	Moderate
TanDEM-X	Tandem satellite image pair interferometry	<12 m	<2 m	Regional/global	Low/free (DLR)
Stereoscopy	Satellite (SPOT, ASTER) *ortho*-stereo imaging	15–30	>10 m	Regional/national	Moderate
ASTER GDEM	Satellite *ortho*-stereo imaging	30 m	7–14 m	Global	Free (J-space systems)
SRTM	SAR dual antenna interferometry	90 m	7–16 m	Global	Free (NASA/CGIAR CSI)
ACE2 GDEM	Altimetry corrected global DEM	1 km	>10 m	Global	Free (EAPRS)
GTOPO30 (HYDRO1k)	Merged version of various DEM and vector height products	1 km	9–30 m	Global	Free (USGS)

Also important to note is that remote sensing acquisition of topographic height data means that undesirable features related to the data processing (voids and artificial hills) or related to the data acquisition process (e.g., vegetation canopy) need to be removed or corrected. For LiDAR processing such techniques are well advanced and lead to very accurate bare ground digital terrain models (DTM), typically less than 20 cm RMSE; however, in the case of SRTM for example, vegetation removal procedures are much less understood and only in a research phase meaning that the SRTM-DEM is currently not available as a bare ground DTM by default and vegetation needs to be removed before the DEM can be applied to flood inundation mapping or modeling. However, noteworthy is that the gradients in the SRTM-DEM at the river position can be used to infer reliable surface slopes of water, provided the reach over which gradients are taken is long enough to accommodate errors in SRTM height measurements (LeFavour and Alsdorf, 2005; Jung et al., 2010).

2.3 WATER AREA AND EXTENT MONITORING

Deriving area and extent information of permanent water bodies, flood inundation area, and shoreline extent from remote sensing is generally more straightforward than deriving information of other variables in hydrology. Information about the surface area of permanent water bodies and associated change can be used in a variety of applications, ranging from simple mapping and monitoring of water bodies to more complex water quality assessments of lakes and reservoirs. Data on inundation area and extent are commonly used to assess the magnitude and extent of a flood with the aim to support relief services and to calibrate and validate flood inundation (i.e., hydraulic) models. As alluded to earlier, although the mapping of permanent water bodies can be done with most satellite imaging platforms at almost any time, obtaining the area and extent of a flood is rather opportunistic and certain conditions on both the Earth surface and atmosphere during an event (such as emergent flooded vegetation and persistent cloud cover) may restrict suitable data acquisition technology only to a few remote sensing instruments, such as microwave sensors.

In general though, surface water area and extent can be measured with a variety of visible band sensors (e.g., Landsat, MODIS, and SPOT) (Marcus and Fonstad, 2008) with different repeat frequencies and by SAR imagery (e.g., RADARSAT, JERS-1, and ERS) (Schumann et al., 2012) with some success but their routine application is limited (Alsdorf et al., 2007).

Problems with optical imagery include cloud cover (and fire smoke), restriction to daylight operations, as well as low spatial resolution for sensors with high temporal coverage intervals (e.g., MODIS), although such sensors are applied to global monitoring of flood inundation area with success as demonstrated by the Dartmouth Flood Observatory (http://floodobservatory.colorado.edu) using MODIS data (Figure 2.1). This database currently represents the only global observed record of flood events. Also at the global scale using multisatellite observations of open water surfaces over a large spectral range, from visible (AVHRR) to microwave wavelength (passive (SSMI) and active (ERS scatterometer)), Prigent et al. (2012) produced a coarse resolution global wetland dynamics (classified inundated fraction at 0.25° spatial resolution) database from 1993 to 2007.

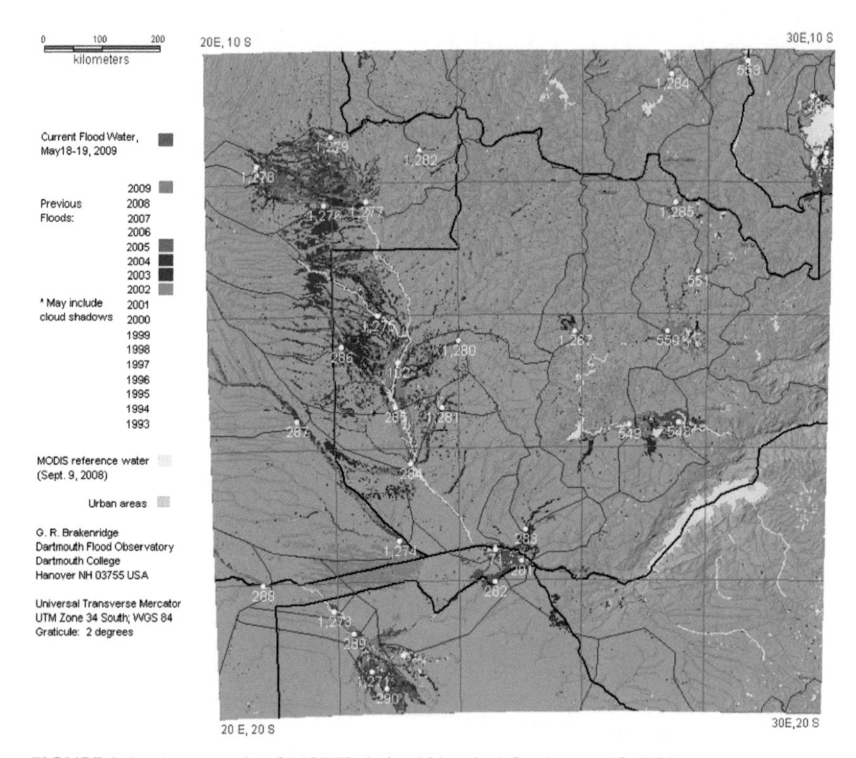

FIGURE 2.1 An example of MODIS-derived historical flood maps (© DFO).

To some extent passive radiometry can be used to map inundated area although the low resolution (about 25 × 25 km) limits the applicability of these sensors. Nevertheless, De Groeve (2010) showed that passive microwave-based flood extent corresponds well with gauged flood hydrographs when river overtopping occurs, although the signal-to-noise ratio is highly affected by variable local conditions, such as specific river bank geometry configuration that may prevent variations in width with rising water levels. Global Disaster Alert and Coordination System's (GDASC) experimental Global Flood Detection System (GFDS) monitors floods worldwide using near-real time satellite data. Surface water extent is observed using passive microwave remote sensing (Advanced Microwave Scanning Radiometer-Earth Observing System (AMSR-E) and Tropical Rainfall Measurement Mission (TRMM) sensors). When surface water increases significantly (anomalies with probability of less than 99.5 percent), the system flags it as a flood. Time series are calculated in more than 10,000 monitoring areas, along with small scale flood maps and animations.

Recently, there have been major advances in the field of flood area and extent mapping based on remote sensing technologies, particularly SAR (Matgen et al., 2007; Schumann et al., 2012). The ability of microwaves to penetrate clouds and light to moderate rain is highly advantageous because bad weather and cloudy conditions are most often associated with floods. The capability to acquire data during day and night and also to some extent map flooded vegetation (using L-band SAR, e.g., from the JERS or ALOS satellite images) are additional strengths of space-borne radar imaging of floods. Also, the specular (mirror-like) reflection of the SAR signal by smooth open water surfaces makes flood mapping from SAR imagery relatively simple and many image processing techniques exist to retrieve flooded area with varying degrees of algorithm complexity and accuracy. Schumann et al. (2009) have used this fuzziness in SAR flood area mapping to augment the information content of the final flood map, which increases flood model calibration and evaluation confidence (Di Baldassarre et al., 2009). Free SAR software tools, such as ESA's NEST (https://earth.esa.int/web/guest/software-tools) are available and can be used for SAR image preprocessing and well-established straightforward algorithms (e.g., based on image statistics or simple pixel value thresholding as used throughout the scientific literature) can be applied.

SAR-based estimates of surface water extent are confounded by difficulties with wind roughening of the water surface for the wavelengths used by most existing sensors and also by high rainfall intensities. Also in urban areas, SAR flood mapping is very limited, with the notable exception of very high resolution SAR imagery (e.g., TerraSAR-X as successfully demonstrated by Mason et al. (2010) and Giustarini et al. (2013)). This is because SAR signals are reflected off sharp corners such as buildings and lamp posts, etc. and only very few open areas can be successfully identified with most SAR sensor resolutions. Furthermore, given the rather "opportunistic" capturing of a (dynamic) flood with most currently available high resolution SAR satellites (at best ~11 day repeat orbit), routine flood mapping at the moment is only really feasible with low resolution, wide-swath mode SAR, or a constellation of SAR satellites, such as the COSMOSkyMed (e.g., Pulvirenti et al., 2011).

Multitemporal interferometric SAR coherence can also be used to delineate inundated surface area because the scattering characteristics of water surfaces continually change with waves, resulting in poor repeat-pass coherence over water (Smith and Alsdorf, 1998). This approach, however, requires the temporal coherence of the surrounding land surface to remain relatively greater, a condition not typically met for orbital repeat cycles greater than a few days (snow, rain, and wind between acquisitions alter dielectric properties of surrounding land and vegetation surfaces, resulting in poor coherence everywhere).

Rosenqvist and Birkett (2002) analyzed the utility of Japanese Earth Resources Satellite-SAR (JERS-SAR) mosaics to record the state of flooding in the Congo basin. The images were of low and high water portraying the Congo and its tributaries during their annual maximum and minimum water marks. They further complemented this information with river stage information derived from satellite radar altimeter (Topex/Poseidon) acquisitions supplemented by historical in situ records. Study results suggested that SAR mosaics may serve well to appraise the maximum extents of flooding in the Congo river basin but may be quite poor at assessing the dynamics and range of the variation.

Another example of surface water monitoring in the Congo basin is provided by Bwangoy et al. (2010) who improved the overall accuracy of Congo basin wetland maps by utilizing multisource satellite data in characterizing these wetlands. Passive optical (Landsat), active microwave (JERS-SAR) remotely sensed data, and derived topographic indices from the SRTM-DEM were used as inputs to create the final wetland map. Although the topographic indices were driving the overall characterization, Landsat water classification ranked as second most important, followed by the SAR high-water masking. Low and high water JERS imagery were also used in the Amazon basin to map floodplain inundation underneath the vegetation canopy (Hess et al., 2003). These maps have been extensively used to assess the performance of a hydraulic model simulating flow dynamics. Furthermore, Landsat fine resolution data (15 m panchromatic) were used to detect over a 1,000 floodplain channels (Trigg et al., 2012) to assess wetting and drying patterns with the aim to improve hydrodynamic modeling.

In a global context, Matgen et al. (2011) developed an automated flood mapping tool for different SAR image modes and resolutions based on a region-growing algorithm refined by change detection. The tool is hosted on European Space Agency's (ESA) G-POD system (http://gpod.eo.esa.int) and freely available to end-users who can query the ESA SAR database for a flood image and retrieve an automatically generated flood map. Like other flood mapping algorithms this tool was developed on SAR existing imagery (ESA's Envisat-ASAR image database) but is intended to work with SAR imagery of any spatial resolution and from future (ESA) SAR satellite missions, such as Sentinel-1. In a similar global context but for low, wide-swath resolution SAR imagery that can be provided in near real-time, Westerhoff et al. (2013) proposed a prototype automated technique that could also be embedded in an online service. Their tool classifies each SAR image tile based on the possibility of being dry or river-flooded based on training of an SAR image time series; DEM data can be integrated to correct for pixels misclassified as water based on information on slope magnitude away from a river. Figure 2.2 shows preliminary results from both tools developed for near real-time global SAR flood mapping.

2.4 MONITORING RIVER HYDRAULIC PARAMETERS

Monitoring river hydraulic parameters with remote sensing technology includes primarily retrieving water levels and stream network as well as river width and depth, and to a lesser extent (limited to a few research case studies) trying to infer discharge and velocities. In addition, networks of pervasive

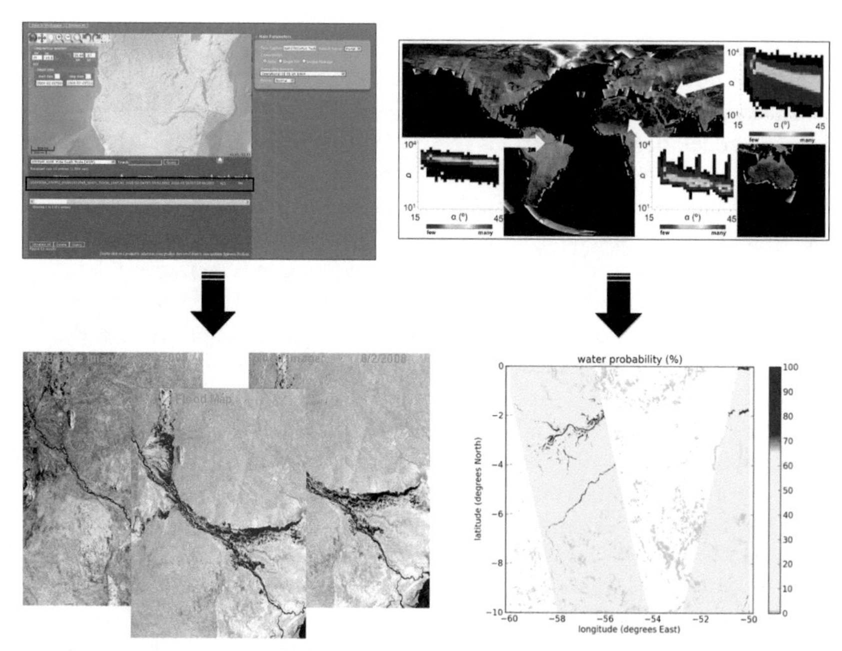

FIGURE 2.2 Preliminary results from proposed online tools for global flood mapping based on synthetic aperture radar (SAR) imagery. (Left) Results for a flood detected on the lower Zambezi River in 2008 using the tool on ESA G-POD developed by Matgen et al. (2011). The top panel shows the query results of the ESA catalog interface for the Zambezi area and the bottom panel shows the reference image and the flood image used to generate the final flood map (blue) without any user input. (Right) Test results from the prototype SAR mapping tool proposed by Westerhoff et al. (2013). The top panel shows the global ESA Envisat ASAR wide swath archive used to generate a series of image histograms that are used to depict flood possibility levels on an SAR image (0–100 percent) acquired for a given event as illustrated in the bottom panel. *Modified from Matgen et al. (2011) and Westerhoff et al. (2013).*

sensors (fixed or floating) and video imagery could represent promising technologies for observing river hydraulics. Hydraulic variables from remote sensing can be used to monitor river systems and associated change that is vital for effective water resources management, including river irrigation and navigation services, and also to build and assess hydraulic models that simulate river dynamics and water flow over land.

River (bankfull) width is an important hydraulic variable and, for many rivers around the world, can be obtained from high resolution optical (e.g., Landsat) or SAR imagery. If a river water mask can be successfully extracted, automated algorithms such as the *RivWidth* method (Pavelsky and Smith, 2008) may be applied to generate river width along an entire river network. With reference to large (ungauged) basins, river widths from Landsat have been generated for the middle reach of the Amazon (Trigg et al., 2012) and the Congo (O'Loughlin et al., 2013) rivers. Characterizing river and floodplain hydraulics of these two basins is crucial because they operate the world's two largest freshwater systems.

Water levels are typically obtained from either in-channel gauging stations or reflectorless total stations (e.g., Pasternack et al., 2008). Also, sonar boats that measure river depth (as employed by Trigg et al. (2012) on the Amazon River for accurate spatially distributed depth measurement) equipped with Global Positioning System (GPS) using real-time kinematic (RTK) satellite navigation to provide real-time corrections to a centimeter level of accuracy in water-level retrieval (Shields et al., 2003). A wireless sensor network (WSN) to measure changes in water levels or deployable pressure sensors linked to communication satellites (e.g., via GNSS—Global Navigation Satellite System) can provide accurate distributed (absolute) measurements of river water levels, water surface gradients, and provide velocity measurements. Even, surveys of water marks on objects such as buildings and lines of debris deposited at the flood edge, which are often referred to as trash lines or wrack marks, can be useful indicators of maximum flood extent and water surface elevation. These deposits can be surveyed after an event with a total station or differential GPS system and used to evaluate hydraulic models (e.g., Neal et al., 2009) or simply map maximum flood extents.

However the two most common form of remote sensing to measure water levels directly is via satellite profiling altimeters (Alsdorf et al., 2007; Frappart et al., 2006; Hall et al., 2012) to cover larger areas or SAR interferometry (Alsdorf et al., 2007). Rosenqvist and Birkett (2002) noted that the altimeter technique is a unique tool for measuring river dynamics in regions where in situ data are difficult to obtain, such as the Democratic Republic of Congo. Although the global average accuracy in measuring water levels with radar altimetry is around 0.5 m (accounting for the major sources of errors and uncertainties) and several global databases exist (e.g., the LEGOS database or the Lakes and Rivers database, Figure 2.3), the major limitation is the spatial resolution of current radar altimeters (>1 km footprints and 596 m along-track spacing), which limits its use to wide rivers. In some respects, the radar footprint shortcoming can be overcome using space-borne LiDAR (GLAS) onboard ICESat (2003–2009 with a follow-on ICESat-2 mission planned for 2016). In a proof-of-concept study Hall et al. (2012) demonstrated the applicability of ICESat's GLAS to river monitoring and showed that average vertical water level accuracies are in the order of 0.16 ± 0.73 m for the much higher 70 m footprint spacing and 172 m along-track spacing. Radar altimeters (Envisat, Topex/Poseidon, Jason) have been used extensively over the Amazon floodplain and channels to infer seasonal dynamics and water gradients (e.g., Frappart et al., 2006) and also to some extent over the Congo basin (Rosenqvist and Birkett, 2002; Lee et al., 2011). Hall et al. (2012) used ICESat over the Amazon River to adjust local gauge datums and obtain improved water-slope measurements. ICESat was also used by O'Loughlin et al. (2013)

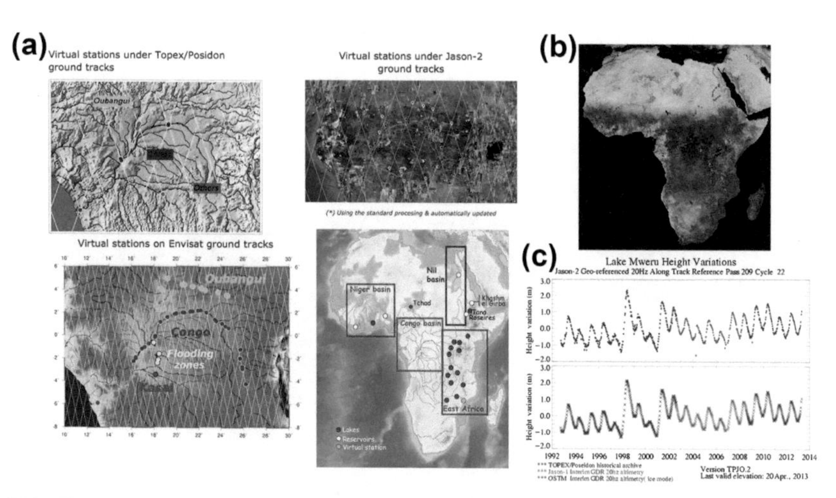

FIGURE 2.3 Different online resources for global radar altimetry data sets. *(From the Envisat, Jason and Topex/Poseidon satellite missions.)* (a) The database hosted by LEGOS (http://www.legos.obs-mip.fr/soa/hydrologie/hydroweb) provides radar altimetry data over lakes, reservoirs and also rivers as well as Gravity Recovery and Climate Experiment (GRACE) data to assess reservoir storage changes. (b) A snapshot of the locations of radar altimetry measurements provided by ESA's Rivers and Lakes project database (http://tethys.eaprs.cse.dmu.ac.uk/RiverLake). Similar to LEGOS, this database also provides radar altimetry data over lakes, reservoirs and rivers but processed with different methods. (c) Height variations over Lake Mweru (bordering Zambia and the DRC) from 1992 to 2013 obtained from the United States Department of Agriculture-Foreign Agricultural Service (USDA-FAS) database of reservoirs and lakes (http://www.pecad.fas.usda.gov/cropexplorer/global_reservoir).

over the Congo River to retrieve the spatial and temporal dynamics of water slopes and thus infer hydrodynamic processes.

For changes in water stage retrieval with InSAR technology, the specular reflection of smooth open water that causes most of the return signal to be reflected away from the antenna and the roughening of the surface (by e.g., wind or wavelength properties) results in the complete loss of temporal coherence between SAR images acquired at different times, rendering interferometric retrieval impossible. However, for emerging vegetation in inundated floodplains, Alsdorf et al. (2000, 2001) showed that it is possible to obtain reliable interferometric phase signatures of water stage changes (at centimeter scale) in the Amazon floodplain from the double bounced return signal of the repeat-pass L-HH-band Shuttle Imaging Radar. Jung et al. (2010) used the same procedure over the Amazon and Congo rivers to contrast the spatial and temporal dynamics of river and floodplain water level connectivities of both river systems (Figure 2.4).

Apart from direct measuring techniques from space, river stage can be estimated at the land-water interface, using high spatial resolution satellite or airborne imagery in combination with a DEM. Given that high resolution

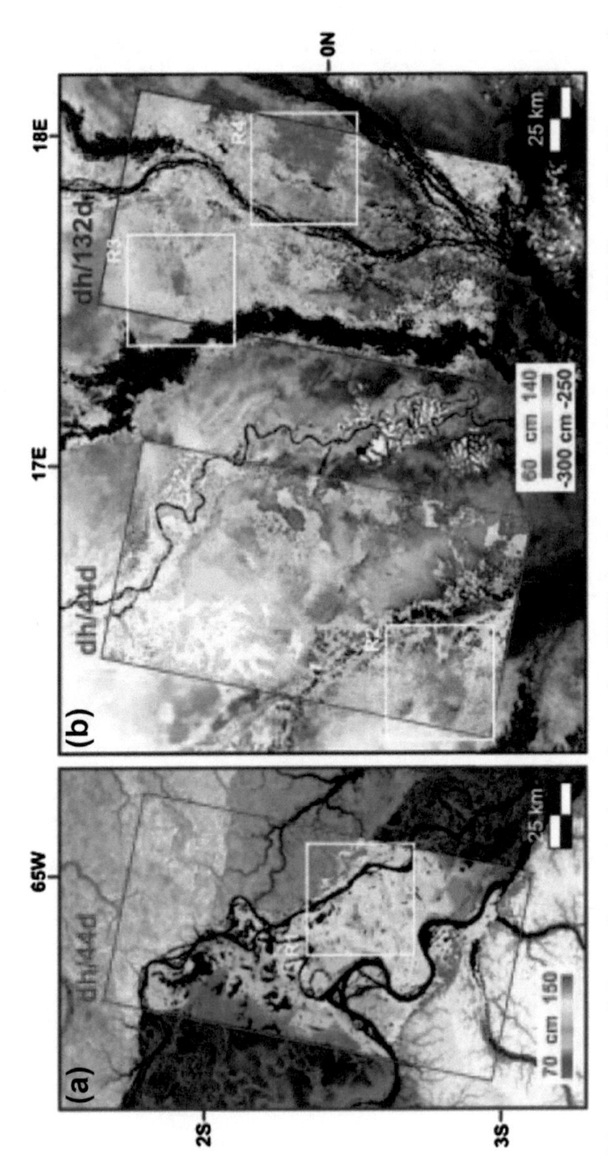

FIGURE 2.4 Measurements of changes in water level superimposed on SRTM elevation maps in (a) Amazon and (b) Congo. Spatial patterns of temporal water level changes are measured from repeat-pass interferometric JERS-SAR. Note that locations without interferometric measurements were not flooded during at least one of the overpasses. *Taken from Jung et al. (2010).*

DEMs are becoming more readily available, if flood boundaries can be adequately extracted from SAR, it is possible to map not only flood extent but also derive water stage at the shoreline. Inaccuracy is largely the result of a combination of two factors: (1) uncertain flood boundary position due to blurred signal response on the land-water contact zone and image positioning errors; (2) coarse resolution DEMs. Schumann et al. (2007) have shown that flood depth can be mapped with an RMS accuracy of better than 20 cm if a LiDAR DEM is available. Schumann et al. (2010) intersected a low resolution SAR flood map with the SRTM-DEM for a large flood on the Po River (Italy) to demonstrate the applicability to measure flood waves on large rivers. Table 2.3 summarizes different methodologies and their accuracies for retrieving water stages using DEMs.

It is evident that interferometry and altimetry constitute the only viable means to obtain up-to-date contemporary stage information in most data-poor regions of the world. A future instrument with a smaller spatial footprint than current altimeters but with even better retrieval accuracy would be highly desired, as it would mean a major step forward to improving our understanding of the role of major wetlands and river systems in a global context. Such an instrument, following the SRTM interferometry technology, is planned for the potential NASA/CNES Surface Water Ocean Topography (SWOT) mission that would measure terrestrial water levels at every 50 m pixels and associated reach water surface slopes with a centrimetric accuracy. Another SWOT mission goal would be to characterize discharge using an optimized inversion of Manning's equation (e.g., Durand et al., 2010) and data assimilation techniques. Although the potential SWOT mission might characterize discharge globally for rivers that are >100 m wide, obtaining discharge from remote sensing has only been explored to some degree in research and at present considerable limitations exist in providing reliable measurements operationally.

Since remote sensing cannot directly measure discharge, a procedure is needed to relate a measurable hydraulic variable, such as width or water level, to either modeled or gauged discharge. These relationships can then be extrapolated to predict discharge from the remotely sensed variable. River width from images can be used to build up power law relationships with discharge (Smith et al., 1996) or satellite altimetry-derived water levels can be used in a rating curve (stage-discharge relationship) approach or within an adapted formulation of Manning's equation to estimate discharge (e.g., Papa et al., 2010; see Michailovsky et al. (2012) for an application to the Zambezi River). Using microwave brightness temperature changes over rivers and associated inundated areas Brakenridge et al. (2007) demonstrated, through testing over different climatic regions of the world, that these changes can be related to modeled or gauged discharge thereby providing satellite-based river rating curves. Less explored methods for inferring flow or discharge using remote sensing technology include high resolution interferometry (e.g.,

TABLE 2.3 Reported Water Level Retrieval Techniques Based on Remotely Sensed Flood Extent and Digital Elevation Model (DEM) Fusion and Their Accuracies

Method	Error (RMSE)	Validation Data	Source
Landsat TM-derived flood extent superimposed on topographic contours for volume estimation	±21 percent	Field data	Gupta and Banerji (1985)
ERS SAR flood extent overlain on topographic contours	0.5–2 m	Field data	Oberstadler et al. (1997)
ERS SAR flood extent overlain on topographic contours	Up to 2 m	Model outputs	Brakenridge et al. (1998)
Intertidal area water from multiple ERS images superimposed onto simulated water heights	Mean error of 0.2–0.3 m	Field data	Mason et al. (2001)
Flooded vegetation maps from combined airborne Land C-SAR integrated with LiDAR vegetation height map	Around 0.1 m	Field data	Horritt et al. (2003)
Integration of high-resolution elevation data with event wrack lines	<0.2 m	Model outputs	Lane et al. (2003)
Fusion of RADARSAT-1 SAR flood edges with LiDAR	Correlation coefficient of 0.9	TELEMAC-2D model outputs	Mason et al. (2003)
Complex fusion of flood aerial photography and field based water stages from various floodplain structures	0.23 m	Mean between maximum and minimum estimation	Raclot (2006)
Fusion of ENVISAT ASAR flood edges with LiDAR and interpolation modeling	0.4–0.7 m	Field data	Matgen et al. (2007)
Fusion of ENVISAT ASAR flood edges with LiDAR and regression modeling	<0.2 m	Field data	Schumann et al. (2007)
Fusion of ENVISAT ASAR flood edges with LiDAR/topographic contours/SRTM and regression modeling	<0.35 m/ 0.7 m/1.07 m	1D model outputs	Schumann et al. (2008)

TABLE 2.3 Reported Water Level Retrieval Techniques Based on Remotely Sensed Flood Extent and Digital Elevation Model (DEM) Fusion and Their Accuracies—cont'd

Method	Error (RMSE)	Validation Data	Source
Fusion of hydraulically sensitive flood zones from ENVISAT ASAR imagery and LiDAR	0.3 m; 0.5 m uncertainty	Mean between maximum and minimum estimation	Hostache et al. (2009)
Fusion of ERS SAR flood edges from active contour modeling with LiDAR	Mean error of up to 0.5 m	Aerial photography	Mason et al. (2009)
Fusion of TerraSAR-X flood edge with LiDAR (rural area)	Error of 1.17 m (note that relative changes in levels from TerraSAR-X and aerial photography were mapped with an accuracy of 0.35 m compared to gauge data)	One field gauge	Zwenzner and Voigt (2009)
Fusion of uncertain ENVISAT ASAR WSM flood edges with SRTM heights	Error in median estimate of 0.8 m	LiDAR-derived water levels	Schumann et al. (2010)

Source: After Di Baldassarre et al. (2011).

TerraSAR-X or COSMO-SkyMed interferometry) (Nouvel et al., 2008) that can be applied to highly dynamic large rivers during high flows (and preferably high winds that generate waveforms) or sequences of satellite (or video) imagery techniques employed for making measurements of flow seeded with tracers in streams to estimate velocities and infer discharge (Muste et al., 2008; Chenand and Mied, 2013).

2.5 HYDRAULIC MODELS

This section focuses on hydraulic models that are applicable to large areas and can be run with a minimal amount of data given that such models (e.g., LISFLOOD-FP, CaMa-Flood, and HEC-RAS type 1D models to some extent) generally integrate better with currently available remotely sensed data related to flooding and flood inundation. However, in areas where fine resolution and high precision river bathymetric data and floodplain DEM data are available, state-of-the-art models, such as the Telemac-Mascaret modeling system (http://www.opentelemac.org), and commercial systems such as TUFLOW and MIKE, can be applied to study in high detail reach scale processes, for domains typically shorter than 50 km in length. Given the resolved details and the fine spatial resolutions at which such models operate, these types of studies also require high accuracy calibration/validation data in the form of flow parameters or/and fine resolution airborne or satellite imagery of flooding.

Hydraulic models (also termed hydrodynamic models or flood models) are computerized algorithms that simulate and predict a simplification of the dynamic process of water flow based on the shallow water equations. They are very complementary to the remote sensing-derived flood variables described in the previous sections in the sense that a model can provide continuous simulations of flood event variables thereby filling the gap linked to the time-continuity issue and sporadic data collection often encountered with observations of flood phenomena (mismatching orbital revisit times for most satellites, opportunistic remote sensing acquisitions, costly postevent point data collection, etc.).

Many different types of hydraulic models are used for a variety of spatial scales but they are all essentially applied to achieve the same objective. In simple terms, these models typically reproduce flooding by routing the flood wave over time through a series of river channels (most often represented by cross-sections based on ground surveying 3D data points) and propagation of water across the floodplain (represented by a high resolution DEM) in all directions using the governing flow equations. The continuity and momentum equations of the shallow water equations (St Venant equations) can be solved using both implicit or explicit schemes and may be discretized in space and time over 1D cross-sectional data, regular 2D grids or irregular 2D and 3D mesh structures. The St Venant equations themselves can be solved in a number of ways, but may also be simplified to provide higher speed efficiencies without significant loss of proper physics and accuracies for gradually varied flows (see e.g., Bates et al., 2010). Of course, water propagation is also influenced by external forces acting on the system, i.e., surface roughness or friction, often expressed by Manning's n, which exerts resistance to flow. It is worth noting that setting this parameter is difficult and input data (e.g., discharge) as well as the bathymetric and topographic boundary data are also uncertain. Nonetheless, hydraulic models are essential tools in any flood risk management plan and can be successfully applied to provide flood maps of certain flow return periods (important for producing flood risk maps) and for reanalysis as well as forecasting of flood events.

As noted before, a variety of hydraulic 1D, 1D–2D, or 2D models are available to simulate and predict river dynamics and floodplain flow, flow velocity, inundated area, and flood depth. In simple words, 1D models usually solve in-channel flow physics only at river cross-sections and floodplain inundation is simplified through either a DEM fill process or storage cell approaches. 1D–2D models are different in that these models simulate 2D flow processes operating in the floodplain and full 2D models combine 2D channel bathymetric information (from e.g., side-scan imaging sonar or interpolated cross-section surveys) with a floodplain DEM and simulate flow processes in 2D everywhere. The performance of all hydraulic models depends greatly on the accuracy of the underlying channel and floodplain topography that drives these models. This is of particular importance when modeling at large scales and in remote and data-poor areas where high quality and high resolution LiDAR elevation might not be available. All types of hydraulic models require a DEM to simulate flow processes. Other than elevation data, hydraulic models require input flow data (either from gauges or from hydrologic model outputs) and some sort of downstream boundary condition (either in the form of gauged time-varying water levels or a free boundary approximated by Manning's equation).

Some 1D models, such as HEC-RAS, solve the full St Venant shallow water equations to accurately reproduce in-channel flow dynamics. Since equations are solved in 1D and only at cross-sections (results are interpolated in-between cross-sections later), 1D models can be run over very long river reaches; however, these models require a priori knowledge about flow direction and in-channel bathymetry. Moreover, in the floodplain these models assume water height to be at level to that in the channel perpendicular to flow direction and therefore cannot reproduce diffusive floodplain flow dynamics correctly; although for most high flood inundation events this horizontal flow assumption is largely valid and can produce accurate results. Important to note here is that most river routing schemes employed in hydrologic models and also that within the EC-JRC/ECMWF GloFAS model for global ensemble streamflow forecasting and flood early warning (Alfieri et al., 2013) are simple 1D channel routing schemes.

2D models relevant to large area application include LISFLOOD-FP or LISFLOOD-FP SGC, CaMa-Flood, and the Modelo hidrológico para grandes bacias hidrográficas-Instituto de Pesquisas Hidráulicas (MGB-IPH) model). CaMa-Flood (Yamazaki et al., 2011) is an improved global river and floodplain routing model at 25 km resolution that explicitly parameterizes the subgrid scale topography of a floodplain (based on 500 m DEM data), thus describing floodplain inundation dynamics. Water flow is calculated using local inertia (Yamazaki et al., 2013), which realizes the backwater effect in shallow gradient river basins. A set of global-scale, river-flow simulations demonstrated an improved predictability of daily-scale river discharge in many major world rivers by incorporating the floodplain inundation dynamics. Detailed analysis of the simulated results for the Amazon River using satellite observations suggested that introduction of the diffusive wave equation is essential for simulating water surface elevation realistically. The MGB-IPH model (Paiva et al., 2013) is a large-scale, coupled hydrologic-hydrodynamic model. The hydrology model uses physical and conceptual equations to simulate land surface hydrologic processes and computed streamflow is routed through the channel by solving the 1D St Venant equations but uses a simple fill operation for the floodplain with prediction of storage volume only. Because the model lacks floodplain hydraulics, it cannot reproduce inundation area dynamically, which limits accuracy in floodplain inundation predictions.

A notable advance in large scale channel and floodplain flow simulations is evident in the recent update of the 2D LISFLOOD-FP model. This model operates on a regular grid structure and simulates water flow by solving the inertial momentum equation in 1D but on the floodplain integrates it in two dimensions (Bates et al., 2010). The resulting model is simple yet contains enough physics to describe processes adequately while requiring an order of magnitude fewer computational operations than a full shallow water model, and can thus be applied over large basins such as the Amazon or Congo rivers. A recent addition to LISFLOOD-FP, which is of particular importance to modeling in in situ data-poor regions, simulates river channel depth (i.e., bathymetry) in the model using a SGC routine based on hydraulic geometry that relates input discharge, width and channel depth via parameterization (LISFLOOD-FP SGC). This implementation does not require cross-section survey data and allows rivers with widths less than nominal model resolution be efficiently simulated without the need for a separate 1D channel model, thus making it extremely suitable for continental to global scale applications. LISFLOOD-FP SGC was developed and tested by Neal et al. (2012) on a large reach of the Niger River and was also coupled with the variable infiltration capacity (VIC) hydrology model for high resolution forecasting of flood inundation patterns in the lower Zambezi River basin (Schumann et al., 2013) (Figure 2.5).

The calibration and validation of flood models is an essential task to ensure flood inundation and flow patterns can be accurately simulated and agree with observations. Traditionally observations for model calibration and evaluation included gauge observations. In the last 30 years, however, gauge data have been complemented by satellite-derived flood area, extent and water levels. Satellite data are essential when simulating floods over large, ungauged rivers, such as the case studies on the Amazon, Niger, and Zambezi rivers. A variety of metrics exist to calibrate and validate the models using satellite observations. Flooded area observations (from satellites or air photography) have been used since the early 1990s to compare against model simulations using simple metrics, such as flood edge distance or spatial performance measures. The integration of uncertainty in observed flood maps has been shown to improve model calibration (e.g., Schumann et al., 2009a). Also, satellite altimetry measurements of water levels have been used in model calibration and validation (see Wilson et al. (2007) for model validation on the Amazon River and Schumann et al. (2013) for model calibration on the Zambezi River). Accuracy results depend on both the ability of the

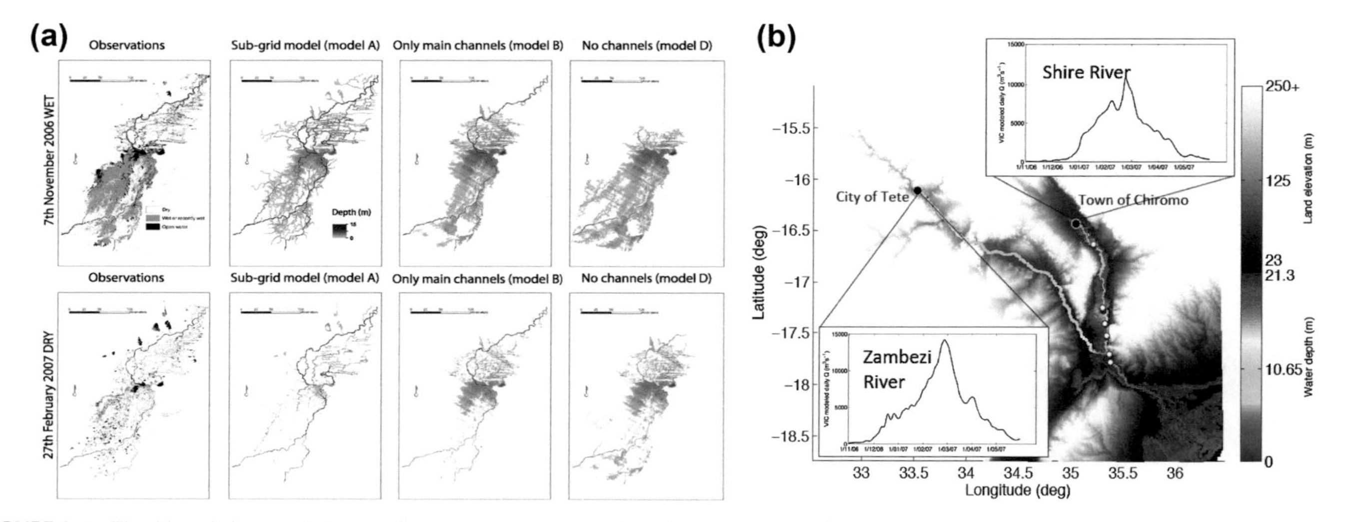

FIGURE 2.5 Flood inundation simulations performed with the LISFLOOD-FP subgrid channel (SGC) model. (a) Model development and testing on the Niger River; panels show Landsat-derived flood inundation maps compared with different LISFLOOD-FP model configurations (with SGCs, only main channels, no channels). (b) LISFLOOD-FP SGC model applied to the lower Zambezi River basin for flood inundation forecasting (circles depict ICESat altimeter data used to calibrate the model in nowcast mode). *Modified from Neal et al. (2012) and Schumann et al. (2013).*

model to simulate processes correctly and on the errors and uncertainty in the cal/val data used. Nevertheless, if uncertainties in the evaluation data can be approximated well, then model simulations can be verified even in the presence of low resolution and low accuracy (satellite) data. Generally, accuracies in water levels range from a couple of cm to 1–2 m during calibration but could easily exceed reasonable errors when validated on a different flood event using the same parameter values. Flooded area agreement ranges typically from about 40 percent to 90 percent, with values above 80 percent being less common.

More recently, assimilation of water levels or even flood area data has been applied. Assimilation (Ensemble Kalman Filter—EnKF, particle filter, 4D var, etc.) is a method that uses the error (uncertainty) characteristics in both observations and models to reach an improved performance in simulations. The method is widely used in numerical weather prediction and flood forecasting since it allows model states to be updated and therefore generate a much better forecast. Assimilation methods can be complex and applications in flood modeling (Andreadis et al., 2007) and forecasting are not well established yet, with only limited examples of successful application using nonsynthetic data (Hostache et al., 2010; Giustarini et al., 2011; Garca-Pintado et al., 2013).

2.6 INTEGRATION OF HYDRAULIC MODELS AND REMOTE SENSING DATA

Integration of hydraulic models and remote sensing can be a powerful approach to augment process understanding and prediction by merging significant information content from both. Bates et al. (1997) first provided an outline of how this integration can be done. Although this paper is now 15 years old, integration of flood models and remote sensing data to yield advances in understanding is still relatively unexplored. Nonetheless, substantial advances have been made in the past decade, of which the most significant ones are described in a thorough review by Schumann et al. (2009b). To provide illustrative examples, the following paragraphs briefly outline four different case studies that demonstrate very successfully the integration of both hydraulic models and remote sensing.

The first case study is the work by Bates et al. (2006) on a reach of the River Severn (UK). Multitemporal airborne SAR imagery was used to map river flood extent at four different times during a prolonged flood event. These high resolution (1.2 m) flood edges were subsequently fused with airborne LiDAR floodplain topography to yield dynamic changes in inundation area, total reach storage, rates of reach dewatering and new insights into floodplain drainage pathways. Coupling these data and different pieces of information with a simple LISFLOOD-FP hydraulic model of the same event, allowed for the first time, validation of the dynamic performance of a flood inundation model using observed extent maps (Figure 2.6(a)). It also highlighted structural weaknesses of the model and possible improvements to be made, for instance the incorporation of a better description of floodplain hydrological processes in the hydraulic model to better represent floodplain dewatering.

Using both fine- and coarse-resolution flood imagery, Schumann et al. (2011) retrieved similar information on the large Tewkesbury (UK) urban flood event of July 2007. A unique set of sequential aerial photography, a fine resolution TerraSAR-X image and several coarser resolution SAR images acquired at different times of the prolonged flood hydrograph allowed a complete remotely sensed reconstruction of the event. Dewatering rates of different urban as well as rural subregions were derived and compared to a very high resolution LISFLOOD-FP simulated reanalysis of the urban flooding. For most subregions, no significant differences in water-level dynamics could be found between model and observations; however, in parts of the urban area the comparison highlighted shortcomings in the model, such as missing inflows and drainage pathways not explicitly represented in the model. This study illustrates well that integration of data derived from different types of remotely sensed imagery, albeit with nontrivial errors and uncertainty, with hydraulic modeling leading to new insights that help better understand flood processes.

Even low-resolution, lower-accuracy remote sensing data can inform modeling efforts. In a study on the Po River (Italy), Schumann et al. (2010) fused flood-edge uncertainties from an SAR, wide-swath mode (75 m pixel resolution) with both an SRTM and a LiDAR DEM to derive flood event-specific, water-surface gradients. Results showed that if uncertainties are properly accounted for the SRTM-derived gradient is statistically similar to that derived from LiDAR. Furthermore, using an HEC-RAS model of the event with different parameterizations, it was demonstrated that the SRTM gradient contained enough information to discriminate between the different HEC-RAS parameterizations by rejecting unrealistic model realizations.

The obvious advantage of satellite remote sensing lies in obtaining meaningful information about processes in remote and difficult-to-access, large areas. A prime example that couples remote sensing with hydraulic models to advance process understanding of large river-floodplain systems is the Amazon case study. In 2007, Wilson et al. published the first hydraulic model (LISFLOOD-FP) of the middle reach of the Amazon River. This model was based on floodplain elevation data from SRTM and field-surveyed, channel cross-section data but showed substantial limitations, specifically at low water when compared to a JERS SAR-derived wetland inundation map and radar altimeter data in the channel. Since then, this model has been greatly improved with respect to both remote sensing and in situ observations by including diffusive channel wave propagation (Trigg et al., 2009), Landsat-detected floodplain channels (Trigg et al., 2012) and ICESat-leveled, water-height gauges (Hall et al., 2012) at the model boundaries.

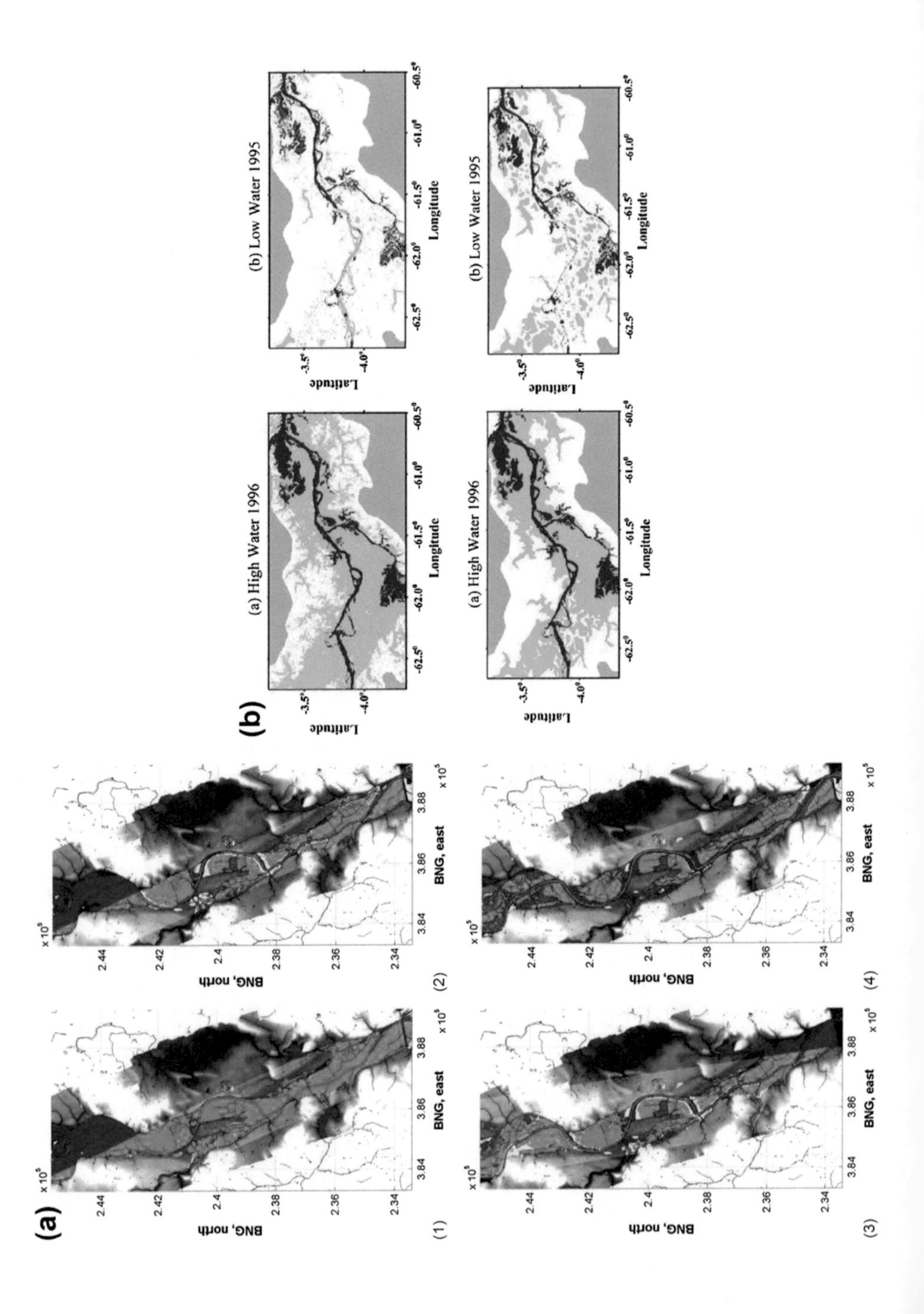
(a)
BNG, north
BNG, east
(1)
(2)
(3)
(4)
(b)
(a) High Water 1996
(b) Low Water 1995
(a) High Water 1996
(b) Low Water 1995
Latitude
Longitude

Wetting and drying abilities of this model have also been enhanced by improving vegetation removal from the SRTM floodplain DEM used to condition the model (Baugh et al., 2013) (Figure 2.6(b)).

2.7 CONCLUSIONS AND OUTLOOK

This chapter has described the utility of remote sensing and hydraulic models to monitor and map flood-inundation processes. Examples of applications in different landscape settings and at different spatial and temporal scales have been illustrated. In recent years, many satellite missions are collecting data that can inform directly or indirectly about flood inundation processes. Optical and thermal data collection from satellites is useful and can be used to infer information about rivers and floodplains but these types of data are very limited during flood events when bad weather conditions often prevail. Acquisitions in the microwave spectrum, however, can overcome this limitation but flooded vegetation, high winds and rain as well as urban areas create problems. Furthermore, long revisit times of most satellites and the opportunistic nature of capturing a flood wave traveling down a river or indeed a flood event make the use of remote sensing for flood inundation studies difficult.

Despite the many nontrivial issues, in the past 15 years, remote sensing has clearly shifted flood science and applications from a data-poor to a data-rich environment (Bates, 2004, 2012). This is also evident in the many engineering and scientific efforts that are currently made to further augment this data-rich environment by proposing the first satellite mission dedicated to hydrology. The proposed NASA/CNES Surface Water Ocean Topography mission would use Ka band radar that is switched on all the time. The proposed revisit time would be 21 days with sampling varying geographically. This novel technology built on the legacy of the SRTM would allow invaluable data gathering for hydrology. Information about water bodies and rivers (area

FIGURE 2.6 (a) Comparison of inundation extent predicted using LISFLOOD-FP with that predicted using the ASAR imagery: (1) November 8, 2000; (2) November 14, 2000; (3) November 15, 2000; (4) November 17, 2000. Light blue indicates areas predicted as flooded using both LISFLOOD-FP and the ASAR; dark blue represents areas predicted as flooded by the model but lying outside the limit of the ASAR swath; red indicates areas predicted as inundated in LISFLOOD-FP but not in the ASAR; yellow indicates areas predicted as inundated in the ASAR but not LISFLOOD-FP; the underlying gray scale represents the DEM used in the model, with the lighter band showing the extent and location of the ASAR swath for each overflight. The blue vectors show the drainage network. *(Taken from Bates et al. (2006).)* (b) Amazon floodplain inundation modeling/mapping. (Top) Comparison of inundation extents at (1) high water May 26, 1996 and (2) low water October 19, 1995 from the hydrodynamic simulation using the original SRTM and flood extent derived from JERS images using the threshold method of Hess et al. (2003). Blue: flooded in both, White: dry in both, Green: wet in JERS image only, Red: wet in original SRTM DEM only, and Gray: terra firma (upland areas never flooded (Hess et al., 2003)). (Bottom) Comparison of inundation extents at (1) high water May 26, 1996 and (2) low water October 19, 1995 from the hydrodynamic simulations using the original SRTM and 50 percent vegetation subtracted DEMs. *After Baugh et al. (2013).*

and extent) would be collected at high resolution at each satellite overpass. Measuring rivers with a width greater than 100 m and lakes with an area greater than 250 × 250 m would be the target. Water level and water-surface slopes would be measured with unprecedented accuracy and reach-scale discharge would be characterized.

Many of the existing and future satellite missions provide rich data with great potential for enhanced monitoring, measuring, and mapping of floods, improving hydraulic models through new data assimilation techniques and parameter scaling behavior, and ultimately for an exploration of the ways in which new data sources may reduce uncertainty in flood predictions (Bates, 2004; Schumann et al., 2012).

ACKNOWLEDGMENT

This chapter was written as part of the work at the Jet Propulsion Laboratory, California Institute of Technology, under a contract with the National Aeronautics and Space Administration.

REFERENCES

Alfieri, L., Burek, P., Dutra, E., Krzeminski, B., Muraro, D., Thielen, J., Pappenberger, F., 2013. GloFAS—global ensemble streamflow forecasting and flood early warning. Hydrol. Earth Syst. Sci. 17, 1161—1175. http://dx.doi.org/10.5194/hess-17-1161-2013.

Alsdorf, D.E., Melack, J.M., Dunne, T., Mertes, L.A.K., Hess, L.L., Smith, L.C., 2000. Interferometric radar measurements of water level changes on the Amazon floodplain. Lett. Nat. 404, 174—177.

Alsdorf, D.E., Smith, L.C., Melack, J.M., 2001. Amazon floodplain water level changes measured with interferometric SIR-C radar. IEEE Trans. Geosci. Remote Sens. 39 (2), 423—431.

Alsdorf, D.E., Rodrguez, E., Lettenmaier, D.P., 2007. Measuring surface water from space. Rev. Geophys. 45, RG2002. http://dx.doi.org/10.1029/2006RG000197.

Andreadis, K.M., Clark, E.A., Lettenmaier, D.P., Alsdorf, D.E., 2007. Prospects for river discharge and depth estimation through assimilation of swath—altimetry into a raster—based hydrodynamics model. Geophys. Res. Lett. 34 http://dx.doi.org/10.1029/2007GL029721.

Bates, P.D., Horritt, M.S., Smith, C.N., Mason, D., 1997. Integrating remote sensing observations of flood hydrology and hydraulic modelling. Hydrol. Processes 11 (14), 1777—1795.

Bates, P.D., 2004. Invited commentary: remote sensing and flood inundation modelling. Hydrol. Processes 18, 2593—2597.

Bates, P.D., Wilson, M.D., Horritt, M.S., Mason, D., Holden, N., Currie, A., 2006. Reach scale floodplain inundation dynamics observed using airborne synthetic aperture radar imagery: data analysis and modelling. J. Hydrol. 328, 306—318.

Bates, P.D., Horritt, M.S., Fewtrell, T.J., 2010. A simple inertial formulation of the shallow water equations for efficient two-dimensional flood inundation modelling. J. Hydrol. 387 (1—2), 33—45.

Bates, P.D., 2012. Invited Commentary: Integrating remote sensing data with flood inundation models: how far have we got? Hydrol. Processes 26 (16), 2515—2521.

Bates, P.D., Neal, J.C., Alsdorf, D., Schumann, G.J.P., 2014. Observing global surface water flood dynamics. Surv. Geophys. 35, 839—852. http://dx.doi.org/10.1007/s10712-013-9269-4.

Baugh, C.A., Bates, P.D., Schumann, G., Trigg, M.A., 2013. SRTM vegetation removal and hydrodynamic modeling accuracy. Water Resour. Res. 49, 5276–5289. http://dx.doi.org/10.1002/wrcr.20412.

Brakenridge, G.R., Tracy, B.T., Knox, J.C., 1998. Orbital SAR remote sensing of a river flood wave. Int. J. Remote Sens. 19 (7), 1439–1445.

Brakenridge, G.R., Nghiem, S.V., Anderson, E., Mic, R., 2007. Orbital microwave measurement of river discharge and ice status. Water Resour. Res. 43. http://dx.doi.org/10.1029/2006WR005238.

Bwangoy, J.-R.B., Hansen, M.C., Roy, D.P., De Grandi, G., Justice, C.O., 2010. Wetland mapping in the Congo basin using optical and radar remotely sensed data and derived topographical indices. Remote Sens. Environ. 114, 73–86.

Di Baldassarre, G., Schumann, G., Brandimarte, L., Bates, P., 2011. Timely low resolution SAR imagery to support floodplain modelling: A case study review. Surveys in Geophysics 32 (3), 255–269.

Di Baldassarre, G., Schumann, G., Bates, P.D., 2009. A technique for the calibration of hydraulic models using uncertain satellite observations of flood extent. J. Hydrol. 367, 276–282.

Chenand, W., Mied, R.P., 2013. River velocities from sequential multispectral remote sensing images. Water Resour. Res. 49, 3093–3103. http://dx.doi.org/10.1002/wrcr.20267.

Durand, M., Rodriguez, E., Alsdorf, D.E., Trigg, M., 2010. Estimating river depth from remote sensing swath interferometry measurements of river height, slope, and width. IEEE J. Sel. Top. Appl. Earth Obs. Remote Sens. 3, 20–31.

Frappart, F., Calmant, S., Cauhope, M., Seyler, F., Cazenave, A., 2006. Preliminary results of ENVISAT RA-2-derived water levels validation over the Amazon basin. Remote Sens. Environ. 100 (2), 252–264.

De Groeve, T., 2010. Flood monitoring and mapping using passive microwave remote. Geomatics, Nat. Hazards Risk 1 (1), 19–35.

Garca-Pintado, J., Neal, J.C., Mason, D.C., Dance, S.L., Bates, P.D., 2013. Scheduling satellite-based SAR acquisition for sequential assimilation of water level observations into flood modelling. J. Hydrol. 495, 252–266. http://dx.doi.org/10.1016/j.jhydrol.2013.03.050.

Giustarini, L., Matgen, P., Hostache, R., Montanari, M., Plaza, D., Pauwels, V.R.N., De Lannoy, G.J.M., De Keyser, R., Pfister, L., Hoffmann, L., Savenije, H.H.G., 2011. Assimilating SAR-derived water level data into a hydraulic model: a case study. Hydrol. Earth Syst. Sci. 15, 2349–2365.

Giustarini, L., Hostache, R., Matgen, P., Schumann, G.J.-P., Bates, P.D., Mason, D.C., 2013. A change detection approach to flood mapping in urban areas using TerraSAR-X. IEEE Trans. Geosci. Remote Sens. 51, 2417–2430.

Gupta, R.P., Banerji, S., 1985. Monitoring of reservoir volume using LANDSAT data. J. Hydrol. 77, 159–170.

Hall, A.C., Schumann, G.J.-P., Bamber, J.L., Bates, P.D., Trigg, M.A., 2012. Geodetic corrections to Amazon River water level gauges using ICESat altimetry. Water Resour. Res. 48. http://dx.doi.org/10.1029/2011WR010895.

Hess, L.L., Melack, J.M., Novo, E., Barbosa, C.C.F., Gastil, M., 2003. Dual-season mapping of wetland inundation and vegetation for the central Amazon basin. Remote Sens. Environ. 87 (4), 404–428.

Horritt, M.S., Mason, D.C., Cobby, D.M., Davenport, I.J., Bates, P.D., 2003. Waterline mapping in flooded vegetation from airborne SAR imagery. Remote Sens. Environ. 85 (3), 271–281.

Hostache, R., Matgen, P., Schumann, G., Puech, C., Hoffmann, L., Pfister, L., 2009. Water level estimation and reduction of hydraulic model calibration uncertainties using satellite SAR images of floods. IEEE Trans. Geosci. Remote Sens. 47, 431–441.

Hostache, R., Lai, X., Monnier, J., Puech, C., 2010. Assimilation of spatially distributed water levels into a shallow-water ood model. Part II: use of a remote sensing image of Mosel River. J. Hydrol. 390, 257–268.

Jung, H.C., Hamski, J., Durand, M., Alsdorf, D., Hossain, F., Lee, H., Hossain, A., Hasan, K., Khan, A.S., Hoque, Z., 2010. Characterization of complex uvial systems using remote sensing of spatial and temporal water level variations in the Amazon, Congo, and Brahmaputra rivers. Earth Surf. Processes Landforms 35, 294–304.

Lane, S.N., James, T.D., Pritchard, H., Saunders, M., 2003. Photogrammetric and laser altimetric reconstruction of water levels for extreme flood event analysis. Photogramm. Rec. 18 (104), 293–307.

Lee, H., Beighley, R.E., Alsdorf, D., Jung, H.C., Shum, C.K., Duan, J., Guo, J., Yamazaki, D., Andreadis, K., 2011. Characterization of terrestrial water dynamics in the Congo basin using GRACE and satellite radar altimetry. Remote Sens. Environ. 115, 3530–3538.

LeFavour, G., Alsdorf, D., 2005. Water slope and discharge in the Amazon river estimated using the shuttle radar topography mission digital elevation model. Geophys. Res. Lett. 32. http://dx.doi.org/10.1029/2005GL023836.

Marcus, W.A., Fonstad, M.A., 2008. Optical remote mapping of rivers at sub-meter resolutions and watershed extents. Earth Surf. Processes Landforms 33, 4–24.

Mason, D.C., Davenport, I.J., Flather, R.A., Gurney, C., Robinson, G.J., Smith, J.A., 2001. A sensitivity analysis of the waterline method of constructing a digital elevation model for intertidal areas in ERS SAR scene of Eastern England. Estuarine, Coastal Shelf Sci. 53, 759–778.

Mason, D.C., Cobby, D.M., Horritt, M.S., Bates, P.D., 2003. Floodplain friction parameterization in two-dimensional river flood models using vegetation heights derived from airborne scanning laser altimetry. Hydrol. Processes 17, 1711–1732.

Mason, D.C., Bates, P.D., Dall'Amico, J.T., 2009. Calibration of uncertain flood inundation models using remotely sensed water levels. J. Hydrol. 368, 224–236.

Mason, D.C., Speck, R., Devereux, B., Schumann, G.J.-P., Neal, J.C., Bates, P.D., 2010. Flood detection in urban areas using TerraSAR-X. IEEE Trans. Geosci. Remote Sens. 48 (2), 882–894.

Matgen, P., Schumann, G., Henry, J., Hoffmann, L., Pfister, L., 2007. Integration of SAR-derived inundation areas, high precision topographic data and a river flow model toward real-time flood management. Int. J. Appl. Earth Obs. Geoinf. 9 (3), 247–263.

Matgen, P., Hostache, R., Schumann, G., Pfister, L., Hoffmann, L., Savenije, H.H.G., 2011. Towards an automated SAR-based flood monitoring system: lessons learned from two case studies. Phys. Chem. Earth 36, 241–252.

Michailovsky, C.I., McEnnis, S., Berry, P.A.M., Smith, R., Bauer-Gottwein, P., 2012. River monitoring from satellite radar altimetry in the Zambezi River basin. Hydrol. Earth Syst. Sci. 16, 2181–2192.

Muste, M., Fujita, I., Hauet, A., 2008. Large-scale particle image velocimetry for measurements in riverine environments. Water Resour. Res. 44. http://dx.doi.org/10.1029/2008WR006950.

Neal, J., Schumann, G., Bates, P.D., Buytaert, W., Matgen, P., Pappenberger, F., 2009. An assimilation approach to discharge estimation from space. Hydrol. Processes 23, 3641–3649.

Neal, J.C., Schumann, G., Bates, P.D., 2012. A subgrid channel model for simulating river hydraulics and floodplain inundation over large and data sparse areas. Water Resour. Res. 48. http://dx.doi.org/10.1029/2012WR012514.

Nouvel, J.F., Dubois-Fernandez, P., Kosuth, P., Lasne, Y., July 7–11, 2008. Along-track interferometry on Rhone River. In: IEEE International Geoscience and Remote Sensing Symposium, 5, pp. 495–497. Boston, MA, USA.

Oberstadler, R., Hoensch, H., Huth, D., 1997. Assessment of the mapping capabilities of ERS-1 SAR data for flood mapping: a case study in Germany. Hydrol. Processes 10, 1415–1425.

O'Loughlin, F., Trigg, M., Schumann, G., Bates, P.D., 2013. Hydraulic characterization of the middle reach of the Congo River. Water Resour. Res. 49 (8), 5059–5070.

Paiva, R.C.D., Collischonn, W., Buarque, D.C., 2013. Validation of a full hydrodynamic model for large-scale hydrologic modelling in the Amazon. Hydrol. Processes 27, 333–346.

Papa, F., Durand, F., Rossow, W.B., Rahman, A., Balla, S.K., 2010. Satellite altimeter-derived monthly discharge of the Ganga-Brahmaputra River and its seasonal to interannual variations from 1993 to 2008. J. Geophys. Res. 115. http://dx.doi.org/10.1029/2009JC006075.

Pasternack, G.B., Bounrisavong, M.K., Parikh, K.K., 2008. Backwater control on riffle–pool hydraulics, fish habitat quality, and sediment transport regime in gravel-bed rivers. J. Hydrol. 357, 125–139.

Pavelsky, T.M., Smith, L.C., 2008. RivWidth: a software tool for the calculation of river widths from remotely sensed imagery. IEEE Geosci. Remote Sens. Lett. 5 (1), 70–73.

Prigent, C., Papa, F., Aires, F., Jimenez, C., Rossow, W.B., Matthews, E., 2012. Changes in land surface water dynamics since the 1990s and relation to population pressure. Geophys. Res. Lett. 39. http://dx.doi.org/10.1029/2012GL051276.

Pulvirenti, L., Chini, M., Pierdicca, N., Guerriero, L., Ferrazzoli, P., 2011. Flood monitoring using multi-temporal COSMO-SkyMed data: image segmentation and signature interpretation. Remote Sens. Environ. 115, 990–1002.

Raclot, D., 2006. Remote sensing of water levels on floodplains: a spatial approach guided by hydraulic functioning. Int. J. Remote Sens. 27 (12), 2553–2574.

Rosenqvist, A.A., Birkett, C.M., 2002. Evaluation of JERS-1 SAR mosaics for hydrological applications in the Congo river basin. Int. J. Remote Sens. 23, 1283–1302.

Sanders, B., 2007. Evaluation of online DEMs for flood inundation modeling. Adv. Water Resour. 30 (8), 1831–1843.

Schumann, G., Matgen, P., Pappenberger, F., Hostache, R., Puech, C., Hoffmann, L., Pfister, L., 2007. High-resolution 3D flood information from radar for effective flood hazard management. IEEE Trans. Geosci. Remote Sens. 45, 1715–1725.

Schumann, G., Matgen, P., Cutler, M.E.J., Black, A., Hoffmann, L., Pfister, L., 2008. Comparison of remotely sensed water stages from lidar, topographic contours and SRTM. ISPRS J. Photogramm. Remote Sens. 63, 283–296.

Schumann, G., Di Baldassarre, G., Bates, P.D., 2009a. The utility of space-borne radar to render flood inundation maps based on multi-algorithm ensembles. IEEE Trans. Geosci. Remote Sens. 47, 2801–2806.

Schumann, G., Bates, P.D., Horritt, M.S., Matgen, P., Pappenberger, F., 2009b. Progress in integration of remote sensing–derived flood extent and stage data and hydraulic models. Rev. Geophys. 47, RG4001. http://dx.doi.org/10.1029/2008RG000274.

Schumann, G., Di Baldassarre, G., Alsdorf, D., Bates, P.D., 2010. Near real-time flood wave approximation on large rivers from space: application to the River Po, Northern Italy. Water Resour. Res. 46. http://dx.doi.org/10.1029/2008WR007672.

Schumann, G.J.-P., Neal, J.C., Mason, D.C., Bates, P.D., 2011. The accuracy of sequential aerial photography and SAR data for observing urban flood dynamics, a case study of the UK summer 2007 floods. Remote Sens. Environ. 115 (10), 2536–2546.

Schumann, G.J.-P., Bates, P.D., Di Baldassarre, G., Mason, D.C., 2012. The use of radar imagery in riverine flood inundation studies. In: Carbonneau, P.E., Pi_egay, H. (Eds.), Fluvial Remote Sensing for Science and Management. Wiley-Blackwell, Chichester, UK, pp. 115–140 p. 458.

Schumann, G.J.-P., Neal, J.C., Voisin, N., Andreadis, K.M., Pappenberger, F., Phanthuwongpakdee, N., Hall, A.C., Bates, P.D., 2013. A first large scale flood inundation forecasting model. Water Resour. Res. 49 (10), 6248–6257.

Shields, F.D., Knight, S.S., Testa, S., Cooper, C.M., 2003. Use of acoustic doppler current profilers to describe velocity distributions at the reach scale. J. Am. Water Resour. Assoc. 39, 1397–1408.

Smith, L.C., Isacks, B.L., Bloom, A.L., Murray, A.B., 1996. Estimation of discharge from three braided rivers using synthetic aperture radar satellite imagery: potential application to ungauged basins. Water Resour. Res. 32, 2021–2034.

Smith, L.C., 1997. Satellite remote sensing of river inundation area, stage, and discharge: a review. Hydrol. Processes 11, 1427–1439.

Smith, L.C., Alsdorf, D.E., 1998. Control on sediment and organic carbon delivery to the arctic ocean revealed with spaceborne synthetic aperture radar: Ob River, Siberia. Geology 26, 395–398.

Trigg, M.A., Wilson, M.D., Bates, P.D., Horritt, M.S., Alsdorf, D.E., Forsberg, B.R., Vega, M.C., 2009. Amazon flood wave hydraulics. J. Hydrol. 374 (1–2), 92–105.

Trigg, M.A., Bates, P.D., Wilson, M.D., Schumann, G., Baugh, C., 2012. Floodplain channel morphology and networks of the middle Amazon River. Water Resour. Res. 48. http://dx.doi.org/10.1029/2012WR011888.

Westerhoff, R.S., Kleuskens, M.P.H., Winsemius, H.C., Huizinga, H.J., Brakenridge, G.R., Bishop, C., 2013. Automated global water mapping based on wide-swath orbital synthetic-aperture radar. Hydrol. Earth Syst. Sci. 17, 651–663.

Wilson, M., Bates, P., Alsdorf, D., Forsberg, B., Horritt, M., Melack, J., Frappart, F., Famiglietti, J., 2007. Modeling large-scale inundation of Amazonian seasonally flooded wetlands. Geophys. Res. Lett. 34. http://dx.doi.org/10.1029/2007GL030156.

Yamazaki, D., De Almeida, G.A.M., Bates, P.D., 2013. Improving computational efficiency in global river models by implementing the local inertial flow equation and a vector-based river network map. Water Resour. Res. 49 (11), 7221–7235.

Yamazaki, D., Kanae, S., Kim, H., Oki, T., 2011. A physically based description of floodplain inundation dynamics in a global river routing model. Water Resour. Res. 47. http://dx.doi.org/10.1029/2010WR009726.

Zwenzner, H., Voigt, S., 2009. Improved estimation of flood parameters by combining space based SAR data with very high resolution digital elevation data. Hydrol. Earth Syst. Sci. 13, 567–576.

Chapter 3

Palaeoflood Hydrology: Reconstructing Rare Events and Extreme Flood Discharges

Gerardo Benito [1] and Andrés Díez-Herrero [2]

[1] *Museo Nacional de Ciencias Naturales, Spanish Research Council (CSIC), Madrid, Spain,*
[2] *Instituto Geológico y Minero de España, (IGME), Geological Survey of Spain, Madrid, Spain*

ABSTRACT

The estimation of rare, large magnitude floods is problematic due to short gauging station records and their limited spatial distribution. The instrumental record can be lengthened by hundreds or thousands of years by estimating discharges of past floods using geological and botanical evidence (palaeostage indicators) left by flood waters. In the former, stratigraphic sequences of sand and silt deposited in slackwater and eddy sedimentary environments are described and dated by geochronological methods (radiocarbon and luminescence techniques). In the later, flood impacts on trees producing scars and other damages (e.g., candelabrum trees) are identified and dated using tree-ring counting. These palaeostage indicators enable to calculate flood discharges using hydraulic modelling, and both flood ages and magnitudes are the input data necessary for improving flood frequency analysis. The scientific and technological interest of these studies is evident for design purposes of critical structures (dams, bridges), risk planning, and for understanding the response of flood patterns to climate change.

3.1 INTRODUCTION

The availability of discharge records of sufficient length is always a major concern for the probabilistic analysis of flood hazards, particularly when the return intervals of floods exceed the length of systematic stream gauging. Longer term flood records can be obtained from human observations during the historical period, or from geological and botanical indicators of flood stage (e.g., flood sediments and damage to vegetation). Flood information obtained outside stream gauging is considered nonsystematic, although, in

Hydro-Meteorological Hazards, Risks, and Disasters. http://dx.doi.org/10.1016/B978-0-12-394846-5.00003-5

many instances, preinstrumental records fulfil the requirements of homogeneous and continuous flood series (Baker, 2008; Brázdil et al., 2006). Often past flood evidence refers to exceptional or rare floods preserved in narrative written sources (annals or chronicles) or as geological—botanical proxy evidence. The reconstruction of the timing and magnitude of these large floods provides valuable data to improve substantially our understanding of hazardous floods.

Flood hydrologists have traditionally used indirect discharge measurements based on field evidence of high-water marks (HWM) (e.g., drift wood, leaves, foam lines) and hydraulic energy equations for computing discharge as a means to extend flood records into the past or to obtain additional discharge data after the passage of a flood (Benson and Dalrymple, 1967). However, woody debris evidence tends to disappear rapidly after the flood, particularly in humid regions and, moreover, differentiation of water marks from different floods is a complicated task after some weeks from the flood peak. Kochel and Baker (1982) observed that "sediments deposited in the backwaters of large floods may accumulate thick sequences in tributary mouths" which can be interpreted in terms of flood frequency based on radiocarbon dating. These authors coined the term "palaeoflood" applied to events that occurred prior to the systematic gauging record and their magnitude and timing are estimated by indirect hydraulic, geological, geomorphological, sedimentological, and botanical methods. Consequently, palaeoflood hydrology is the reconstruction of the magnitude and frequency of recent, past, or ancient floods using geological (physical) evidence (Baker et al., 2002). The term "palaeo" has sometimes contributed to the general misconception that palaeoflood techniques are only used for estimating very old floods (geological scales). However, most palaeoflood studies involve prehistoric (past 5,000 years), historic (past 2,000 years), and even modern flood analysis in ungauged basins (Benito and O'Connor, 2013). Consequently, it is not the timescale of flooding that defines palaeoflood hydrology but the fact that flood evidence derives from the lasting physical effects of floods on natural recording indicators (palaeostage indicators, PSI's). Other nonsystematic records include historical floods or floods recorded from human observations and documents, that is, a manuscript, a piece of printed matter (book, newspaper), a picture or an artefact (a flood mark or an inscription on a building), which refers to the stage or impacts of hydrological extremes (Brázdil et al., 2006).

Over the past 30 years, palaeoflood hydrology has achieved recognition as a new branch of geomorphology and hydrology, developing new tools and applications for the study of extreme events in relation to climate change, water resources, and flood hazard assessment (Baker and Kochel, 1988; Baker, 2008). In a broad sense, palaeoflood studies involve multidisciplinary research addressing geological palaeostage indicators (stratigraphy, chemical analysis, radiometric techniques) and biological flood markers (dendrochronology, palynology, biomineralogy, lichenometry). Flood reconstructions based on geological and biological indicators have been successfully applied in several regions of the world (Figure 3.1). Palaeoflood studies in bedrock rivers, based on sedimentological indicators, gained scientific acceptance in the early 1980s, mainly under the auspices of Prof. Victor R. Baker and his students, focusing largely on the southwestern United States (e.g., Kochel et al., 1982; Ely and Baker, 1985; Partridge and Baker, 1987; Webb et al., 1988; O'Connor et al., 1986, Jarrett, 1990; Knox, 1985, 1999), and later extended to other parts of the world (Figure 3.1). Pioneer research on dendrogeomorphological palaeoflood indicators was carried out by Sigafoos (1964) in the eastern USA, and later extended to other sites in North America (USA and Canada), and Europe (Figure 3.1).

Recent advances in palaeoflood hydrology are the result of new applications and progress in computing science and new techniques related to computational fluid dynamics, remote sensing, geochronological and isotopic methods, environmental tracers, geophysical data acquisition and analysis, among others (Woodward et al., 2010). This chapter aims to describe different techniques available for extending the flood record back in time, as well as some fields of application of these palaeoflood data to solve different scientific and engineering problems.

3.2 PALAEOFLOOD APPROACHES AND METHODOLOGY

The primary goal of palaeoflood hydrology is to extend flood records (magnitude and timing) over centuries or millennia (Baker, 2008). Long-term flood records are of interest for flood hazards, understanding climate change impacts and for the assessment of water resources. These different palaeoflood applications are carried out through a common methodological framework for palaeoflood data collection and further applications follow specific methodological procedures (Benito and Thorndycraft, 2005). In palaeoflood data collection the methodological approach is dictated by river channel type (bedrock vs alluvial rivers), channel slope, and by the characteristics of the available PSIs.

The geomorphological characteristics of the studied reach may limit the quantity of PSIs and therefore the quality of discharge estimations. Bedrock channels provide the most suitable conditions for extending flood records as sediment deposition is favored by the sharp energy−flow changes between expansion and constriction reaches. Moreover, discharge estimations from stage indicators are more accurate under fixed-bedrock channel geometry. Alluvial rivers contain rich stratigraphic archives from which we may reconstruct flood history but discharge estimations in erodible channels lose accuracy. Alluvial rivers are here defined as streams with channel(s) flowing over alluvium with adjacent floodplain areas and alluvial fans. Mountain streams typically flow on bedrock or cemented alluvium, although the high flow−energy characteristics and coarse sediment size limit palaeoflood

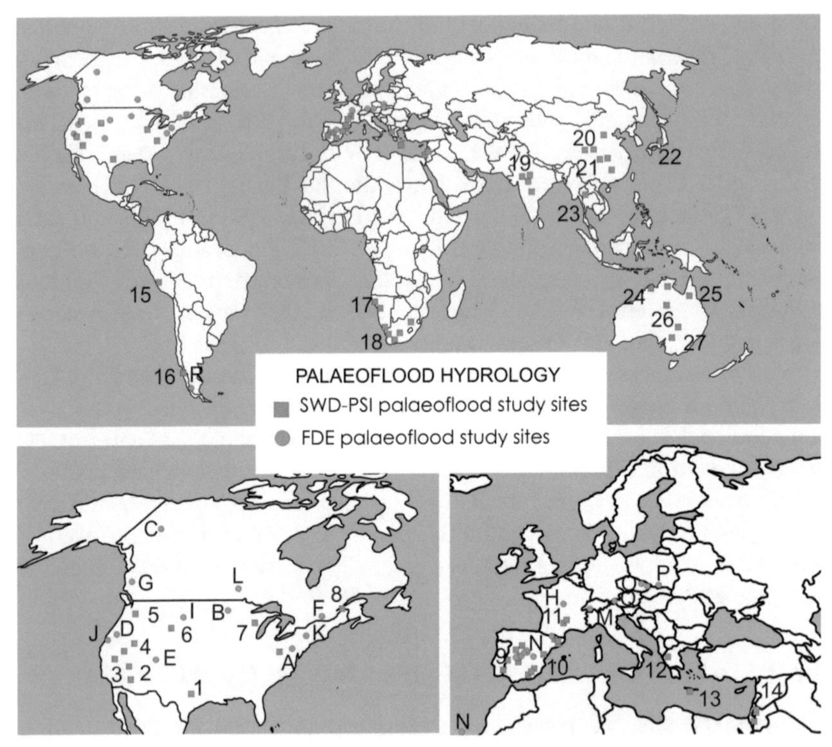

FIGURE 3.1 Global distribution of the most representative studies on palaeoflood hydrology that used slackwater flood deposits (SWD-PSI) and flood dendro evidence (FDE). Many more studies have been carried out in most of the areas than the indicated by the selected references. *SWD-PSI* (orange squares): 1. West and central Texas (Kochel et al., 1982); 2. Arizona and Utah (Ely and Baker, 1985; Partridge and Baker, 1987; Webb et al., 1988; Enzel et al., 1994; O'Connor et al., 1994; House et al., 2002; Webb et al., 2002); 3. California (Enzel, 1992); 4. Oregon and Idaho (Chatters and Hoover, 1994; Ostenaa et al., 2002; O'Connor et al., 2003); 5. Colorado (Jarrett and Tomlinson, 2000); 6. South Dakota (Harden et al., 2011); 7. Eastern USA (Springer and Kite, 1997; Kite et al., 2002); 8. Eastern Canada (Saint-Laurent et al., 2001); 9. Portugal (Ortega and Garzón, 2009); 10. Spain (Benito et al., 2003a, 2010; Thorndycraft et al., 2005a,b, 2006; García-García et al., 2013); 11. France (Sheffer et al., 2003, 2008); 12. Greece (Woodward et al., 2001); 13. Crete (Macklin et al., 2010); 14. Israel (Greenbaum et al., 1998, 2000, 2002); 15. Peru (Wells, 1990; Magilligan and Goldstein, 2001); 16. Chile (Dussaillant et al., 2010); 17. Nambia (Heine, 2004; Heine and Völkel, 2011; Grodek et al., 2013); 18. South Africa (Smith and Zawada, 1990; Smith, 1992; Zawada, 2000; Zawada and Hattingh, 1994; Benito et al., 2011a,b); 19. India (Ely et al., 1996; Kale et al., 1997, 2003; Kale and Baker, 2006); 20. Northeastern China (Zha et al., 2009, 2012; Huang et al., 2010, 2011; 2012a,b; Yang et al., 2000); 21. Central-Southeastern China (Zhang et al., 2013; Huang et al., 2013); 22. Japan (Jones et al., 2001; Grossman, 2001; Oguchi et al., 2001); 23. Thailand (Kidson et al., 2005a,b); 24. Northcentral Australia (Baker and Pickup, 1987; Sandercock and Wyrwoll (2005); 25. Northwestern Australia (Wohl et al., 1994); 26. Central Australia (Baker et al., 1983, 1985, 1987; Pickup et al., 1988; Pickup, 1991; Patton et al., 1993); 27. Southeastern Australia (Saynor and Erskine, 1993; Erskine et al., 2002). *FDEs* (green circles): A. Potomac River, Virginia and Maryland, USA (Sigafoos, 1964; Yanosky, 1982a, 1982b, 1983, 1984); B. North Dakota, USA (Harrison and Ried, 1967); C. Mackenzie Mountains, Northwest Territories, Canada (Butler, 1979); D. Mount Shasta, northern California, USA (Hupp, 1984;

techniques to botanical indicators (tilting and tree scars). Depending on the purpose of the research, some studies may require a complete palaeoflood record over some period of time, whereas others will focus on assessing the largest events possible within a catchment.

The methodological steps to conduct a standard palaeoflood study include (1) a preliminary inventory of potential sites using aerial photographs; (2) field visit and survey for the identification and selection of flood indicators (flood deposits, botanical indicators, and erosion marks); (3) at the selected sites, detailed descriptions of flood evidence, that is, stratigraphical description with emphasis on identifying flood units or tree coring for botanical evidence; (4) sample collection for age dating; (5) topographical survey of studied river reaches, with exact location and elevation of PSIs; (6) hydraulic modelling and discharge estimation; (7) comparison with available historical or systematic data; and (8) flood frequency analysis (FFA).

A preliminary inventory of potential areas for flood deposition and disturbed trees can be drawn up using aerial photograph interpretation. This preliminary identification of potential sites will limit the number of river reaches to be visited in the field. One of the most challenging tasks in palaeoflood studies is the field identification of flood evidence from which to estimate maximum flood stages for a particular river reach. This includes a close inspection to determine (1) the bedload being transported during a flood event (flood bars and large isolated boulders); (2) the identification of PSIs associated with flood events of different magnitudes (geological and botanical); and (3) the indication of maximum flood stage based on sedimentary, erosional, or pedological evidence. These flood stage indicators may be interpreted as follows: (1) close to peak flood stage (peak discharge level); (2) minimum flood stage; and (3) within a known flood level range. Hydraulic calculations based on rating curves (stage-discharge) will be used to assign a minimum discharge value to all/most flood indicators. FFA combines systematic and nonsystematic records, where palaeoflood data are treated as censured records over a stage-discharge threshold for a specific time period.

Hupp et al., 1987); E. Kanab Creek (Utah) and Arizona, USA (McCord, 1996); F. Upper St-Lawrence Estuary, Quebec, Canada (Begin et al., 1991); G. British Columbia (Gottesfeld, 1996); H. Saone River, France (Astrade and Bégin, 1997); I. Missouri River, Montana, USA (Scott et al., 1997); J. Eel River, California, USA (Sloan et al., 2001); K. Middlebury River Gorge, Vermont, USA (Coriell, 2002); L. Red River, Manitoba (Canada) and Red River of the North, North Dakota and Minnesota, USA (St. George et al., 2002; St. George, 2010); M. Ritigraben, Switzerland (Stoffel et al., 2003; Stoffel and Wilford, 2012); N. Navaluenga, Venero Claro, Guisando, Arenas de San Pedro, Segovia, Pajares de Pedraza, Peralejos de las Truchas and Valsaín (Central Spain) and Caldera de Taburiente (Canary Islands), Spain (Díez-Herrero et al., 2007, 2013b; Ballesteros Cánovas et al., 2010a,b, 2011a,b, 2013a; Ruiz-Villanueva et al., 2010, 2013); O. Bila Opava River and Moravskoslezsk, Beskydy Mts., Czech Republic (Malik and Matyja, 2008; Silhan, 2012); P. Tatra Mountains, Poland (Zielonka et al., 2008; Kundzewicz et al., in press); Q. Gratzental, Tyrol, Austria (Mayer et al., 2010); R Patagonian Andes, Argentina (Stoffel et al., 2012).

3.3 GEOLOGICAL AND BOTANICAL PALAEOFLOOD DATA

Geological and botanical flood indicators can be divided into two categories: (1) HWM and (2) PSIs. HWMs include mud, silt, seed lines, and flotsam (e.g., fine organic debris, grass, woody debris) representing the peak stage of a recent flood (Benson and Dalrymple, 1967). Preservation of fine organic debris is limited to weeks in humid climates and to a few decades in semiarid and arid climates (Williams and Costa, 1988).

PSIs are geological and botanical evidence left by flood waters (Kochel and Baker, 1988: Webb and Jarrett, 2002; Benito and O'Connor, 2013). The most common types of palaeostage evidence are the following:

Slackwater flood deposits (SWD) are fine sediments (sand and silt) transported as suspended load during floods, and deposited on slackwater flood areas (Kochel and Baker, 1982). SWDs are the most frequently used PSIs in bedrock river systems, typically placed at lower elevation than HWM. These sediments may be preserved for centuries or millennia when protected from post-flood erosion. At preserved sites, the accumulation of successive flood sediment layers provides excellent palaeoflood records.

Flood dendro evidence (FDE), including tree scars and damage to riparian trees (sprouting from tilting stems, eccentric ring growth), has been effectively used for documenting flood magnitude and the frequency of palaeofloods (Sigafoos, 1964). Recent studies of anatomical changes developed following flood damage open new research perspectives to detect the occurrence of extreme floods in trees with nonevident physical damage (Yanosky and Jarrett, 2002; Ballesteros Cánovas et al., 2010a,b).

Silt lines are subhorizontal linear deposits of silt- and clay-sized particles that line some portion of the bedrock canyon walls, rocks, and trees (Jarrett, 1985). These lines have been interpreted as derived from the suspended load of the flooded stream, left as the flood waters percolated into the bedrock, and serve as excellent PSIs (O'Connor et al., 1986).

Organic drift, including branches, leaves, and seeds, left stranded at the high-water line when water recedes. The finer organic drift (leaves, cornstalks, and seeds) are better flood stage makers than would be some scattered clumps of large drift (Benson and Dalrymple, 1967). Careful selection of proper organic drift is important to avoid affection by velocity head, as the marks upstream of obstacles (trees, buildings, rocks) will be overelevated and marks on the downstream side may be lower than normal. A mound (>20 cm) of fine organics at the very edge of the channel in slow velocities is considered an excellent flood stage indicator. The main problem is the poor preservation of organic drift.

Boulder deposits are PSIs in high-energy environments and/or high-slope streams (Jarrett and Malde, 1987). Boulder bars have been extensively used to estimate flood discharges in mountain streams together with flood-related scars on trees (Jarrett, 1990).

Scour marks and trimlines at high elevation on slope alluvium and soils are considered as evidence of the largest flood(s) occurring in a river reach, although their interpretation may be ambiguous and difficult to accurately decipher (Webb and Jarrett, 2002).

Some researchers have reported height differences of PSIs in relation to actual flood water depths (Kochel, 1980; Springer and Kite, 1997; House et al., 2002). In a systematic analysis of flood marks (PSI and HWM) in rivers in the United States, Jarrett and England (2002) concluded that, in short reaches, the elevation at the top of some flood sediments (PSIs) essentially equals the height of HWMs (with an average deviation of +15 mm in 192 observations), with almost no deviation when using deposits located at the channel margins. Nevertheless, further studies are needed to systematically evaluate the discrepancy between PSIs and actual water surface elevations for different river characteristics and climatic regions. The following section provides a detailed description of the most commonly used geological and botanical PSIs, namely slackwater flood deposits and dendrochronological analysis of tree scars.

3.3.1 Sedimentological Indicators

SWDs are composed of silts and sands with high settling velocities that accumulate relatively rapidly from suspension during major floods (Baker et al., 2002). The sedimentation is favored by sharp changes in flow energy conditions between the main flow in the channel and the valley margins. In these marginal areas and during flood stages, eddies, back-flooding and water stagnation occur, which significantly reduces flow velocities (<1 m s^{-1}) and favors deposition of suspended clay, silt, and sand (Figure 3.2).

Successful palaeoflood reconstruction over long periods of time depends upon (1) an appropriate sediment source or the presence of finegrained sediments within the catchment area and (2) the preservation conditions. Catchment geology dominated by granite and/or sandstones provides an abundant source of fine-grained material to be transported as suspended sediment during flood events. The deposition and preservation conditions are controlled by flood erosion/sedimentary processes and by post-flood erosion processes (slope and tributary erosion). The most common depositional environments include (1) rock shelters or caves in bedrock walls (Figure 3.2(a)); (2) channel expansions (Figure 3.2(b)); (3) channel bends; (4) back-flooded tributary mouths and valleys (Figure 3.2(c) and (d)); (5) obstacle shadows where flow separation causes eddies (Figure 3.2(e)); (6) channel widening (Figure 3.2(f)); and (7) on top of high alluvial or bedrock surfaces that flank the channel (Kochel et al., 1982; Ely and Baker, 1985; Baker and Kochel, 1988; Benito et al., 2003a; Sheffer et al., 2003; Thorndycraft et al., 2005a; Benito and Thorndycraft, 2005).

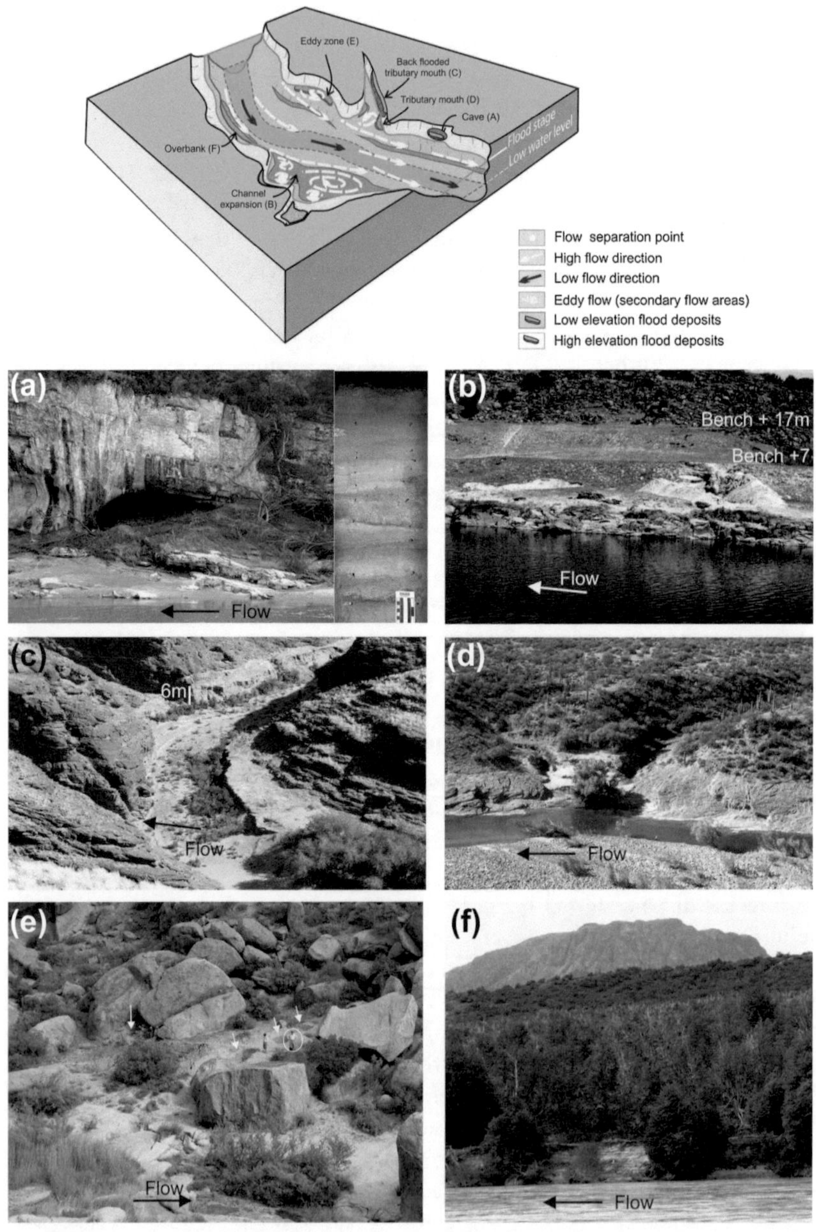

FIGURE 3.2 Upper part: Block diagram illustrating the location of sedimentary environments related to flood deposition (modified from Benito et al., 2003b). Photos: Field examples of palaeoflood stage indicators. (a) Rock shelter infilled with SWDs in the Gorge of the Gardon River (France) and stratigraphy of flood deposits within the cave (Sheffer et al., 2008). (b) Flood benches, at 17 and 10 m above the present channel bottom, deposited in an expansion reach of the Tagus River gorge (Central Spain; Benito et al., 2003a). (c) SWDs along a back flooded tributary

Flood deposition in narrow reaches is usually flushed away by subsequent flooding, although flood traces of the largest floods can be found. In narrow reaches, preservation of slack water palaeoflood deposits is optimized when they are deposited in rock shelters or rock overhangs (Figure 3.2(a)), protected from slope runoff and bioturbation. Typical slack water locations include areas upstream of tributary junctions (Figure 3.2(c) and (d)) and in transition reaches from channel expansion to constrictions (Figure 3.2(e)). In these settings, low-energy conditions favor the deposition of thick, high-standing flood deposits with bench morphology (Figure 3.2(b), (d) and (e)). Flood deposit benches are formed by vertical accretion of slackwater sediments deposited by successive floods. Deposition of a flood layer raises the minimum flood water level required for deposition of a new flood layer, meaning that a rising threshold or self-censoring level occurs over time (Kochel and Baker, 1982; House et al., 2002). Small floods may develop low elevation inset benches that are more susceptible to erosion during subsequent floods.

Individual flood beds commonly pinch out into the valley side as the layer rises in elevation. A transversal trench to the flood bench may expose complex disposition of bed contacts in response to the location and energy of recirculation eddies and sediment pulses delivered during floods. Flood benches may contain biased stratigraphic records with the higher flood bench containing a record of the largest palaeofloods over long periods of time, and the lowest benches recording more recent smaller floods.

The stratigraphical description of SWDs requires a major emphasis on interpreting the breaks and contacts between flood layers, and sedimentary structures (Baker and Kochel, 1988; Enzel et al., 1994; Benito et al., 2003a). The general criteria for identification of multiple flood layers are the following:

1. Identification of a distinct clay layer at the top of a flood unit, this representing the waning stage of a flood;
2. Deposition of a layer of sediments not deposited by a river flood that marks a clear boundary between two successive flood units. These may be colluvial, deposits falling from cave roof, or interbedded coarse tributary alluvium (couplets);
3. Bioturbation (plant and animal activity) that is indicative of an exposed sedimentary surface after the flood has passed;

of the Kuiseb River (Namibia; Grodek et al., 2013). The black arrow indicates the main flow direction through the Kuiseb River gorge. (d) Flood deposits at tributary mouth due to eddy circulation during flood stage in the Verde River, Central Arizona (Ely and Baker, 1985). (e) SWDs at eddy zone formed at the lee side of rock fall boulders in lower sector of the Buffels River (Namaqualand, South Africa; Benito et al., 2011a,b). (f) Flood deposits at overbank sedimentary environment in the Baker River, downstream of the confluence with the Colonia River (Patagonia region, Chile).

FIGURE 3.3 Examples of the most common types of dendrogeomorphological evidence useful in the study of past floods (FDEs): (a) candelabra growth ('Sigafoos' trees, Navaluenga, Central Spain); (b) scar and exposed floating roots, Taburiente (La Palma, Canary Islands); (c) bifurcations in tree stems, Navaluenga (Spain); (d) scars (stripped bark with callus marks) caused by sediment load or woody debris impacts, Venero Claro (Central Spain); (e) tilted and overturned tree, Pajares de Pedraza (Central Spain); (f) scars and exposed roots with stripped bark and erosion, Venero Claro, Spain; (g) bark erosion by abrasion, Taburiente (La Palma, Canary Islands); (h) scar and elbows and angles (sharp changes in the trunk growth direction; Taburiente, La Palma, Canary

4. An erosional boundary where the surface of an older flood unit has been eroded by a later flood event;
5. A change in the physical characteristics of the flood units, such as sediment color or particle size, that may be brought about by factors such as differing sediment source or differing energy conditions during separate flood events. This criterion is not always valid by itself and may need additional corroboration by other flood break indicators.
6. Development of palaeosols and changes in sediment hardness and mud cracks indicating soil-surface exposure.

Sedimentary structures developed within individual flood units provide valuable information on flow energy and sediment pulses during floods (McKee, 1938; Kochel and Baker, 1988; Benito et al., 2003b). The most systematic analysis of sedimentary structures associated with SWDs has been described by Benito et al. (2003b). In the flow separation zones textures are dominated by medium to coarse sand with sedimentary structures ranging from upper high-energy parallel lamination to large-scale planar or trough cross-bedding, ripple lamination, or massive intervals. In secondary flow areas toward the valley side, texture is dominated by silty sand and sandy silt, with weak stratigraphical breaks due to minor changes in texture and massive structure. Inside caves and rock shelters, characteristic sedimentary sequences are dominated by reverse flow structures (e.g., climbing ripples migrating upstream) due to eddies with a high sand concentration.

3.3.2 Botanical–Dendro Indicators

3.3.2.1 Dendrogeomorphology

Dendrogeomorphology uses information from flood damage and impacts recorded in tree trunks–stems, branches, and exposed roots of riparian forest communities (trees and bushes) located at river banks, bars, and floodplains. FDE includes more than 29 types of evidence produced by flooding. According to Díez-Herrero et al. (2007, 2013a), FDE can be classified by the spatial scale of the evidence (from macro- to microscopic, from kilometers to microns) and the scale of studied elements (from forest communities and vegetal formations to changes in cell structure and isotopic ratios). Examples of the most common FDE (impact scarring, candelabrum trees, exposed roots, adventitious branches, etc.) are shown in Figure 3.3.

Flood studies using FDE involve the following research protocol (Díez-Herrero et al., 2007, 2013a; Figure 3.4): (1) finding dendro evidence in

Islands); (i) internal scars and discontinuities, Arenas de San Pedro (Central Spain); (j) changes in tree-ring parameters (width, percent early wood, latewood, etc.); and (k) traumatic structures in wood tissues. See, for more details, Díez-Herrero et al. (2007, 2013a). All examples from Spanish study sites (Díez-Herrero et al., 2013b).

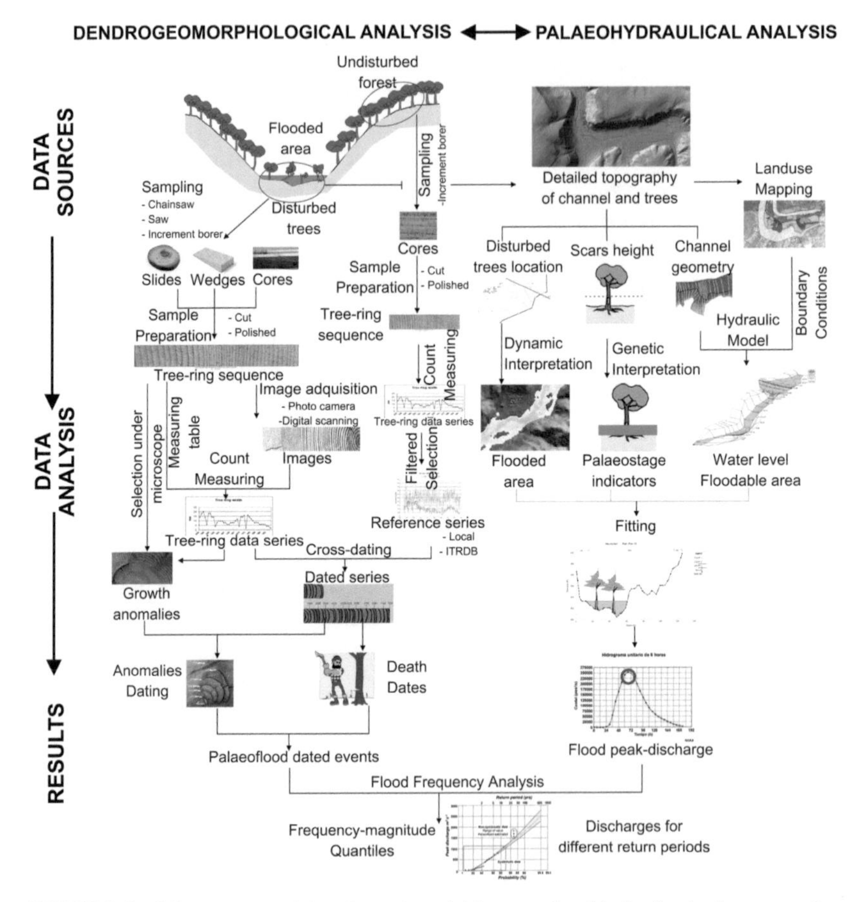

FIGURE 3.4 Scheme summarizing the main activities or tasks of both, the dendrogeomorphological and palaeohydraulic methodologies, from the data sources to the final expected results through analyses, and their mutual relations.

bottomland trees; (2) selection of tree species with homogeneous growth; (3) planned sampling of tree-ring sequences; (4) study of plant anatomy in search of characteristic signatures of past flood events; (5) synchronization of tree-ring sequences, by means of cross-dating, for past flood event dating; (6) hydraulic modelling of the river reach to estimate flow discharges using PSIs in trees; (7) assigning magnitudes to the estimated frequencies; and (8) combining palaeodischarges with systematic or modelled discharges to improve the frequency distribution function.

Dendrogeomorphological data have mainly been used for the characterization and dating of past floods (with annual precision) and the identification of time periods with different flood frequency (Coriell, 2002). These dendro-palaeostage indicators can be used to estimate the magnitude of past floods either by computing the peak flow using simple equations or by means of complex hydraulic models (Ballesteros Cánovas et al., 2011a). Palaeodischarges can be used for FFA using statistical functions (Ballesteros Cánovas et al., 2011a; Ruiz-Villanueva et al., 2013). Other palaeohydrological information derived from fluvial dendrogeomorphology includes the duration of the flood (hydrograph base time), meteorological origin (based on isotopic fractionation), and runoff genesis (Díez-Herrero et al., 2013c).

The dendrogeomorphologic indicators traditionally used (mainly scar evidence) and their use to infer flood frequency and magnitude have been restricted to a small, limited set of applications. New research opportunities in palaeoflood hydrology using dendrogeomorphologic data sources can be found in (1) the application of isotopic indicators ($^{18}O/^{16}O$ ratio) to interpret the meteorological origin of past floods (Díez-Herrero et al., 2013c); (2) the use of different dendrogeomorphic indicators (i.e., tilted trees, Ballesteros Cánovas et al., 2013b; or X-ray computed tomography techniques, Guardiola-Albert et al., 2012) to estimate peak flows with 2D and 3D hydraulic models; (3) improvements to the calibration of hydraulic model parameters (roughness); and (4) the application of statistics-based cost–benefit analysis to select optimal mitigation measures (Ballesteros Cánovas et al., 2013a).

The successful reconstruction of palaeoflood records from botanical evidence depends on the presence of bottomland trees of appropriate tree species for dendrochronological analysis, and the existence of external indicators of flood damage. Another limitation is the lack of statistical representativeness of the FDEs that depend on spatial distribution (i.e., number of replications and homogeneity along the river reach), and on the age distribution of available trees to determine the temporal length of the data series. Several sources of uncertainty can affect the dendrogeomorphological results applied to past flood characterisation: (1) mistakes in tree-ring dating due to cross-dating interference (double, false or missing tree-rings, multiple density bands, etc.); (2) inadequate interpretation of the observed height of the FDEs as PSIs; and (3) the complex reconstruction of hydraulic boundary conditions of past flood hydrodynamics around trees.

3.3.2.2 Lichenometry

Another group of palaeoflood techniques rooted in botany ecology are based on lichens and lichenometric dating of bedrock surfaces and deposits exposed to floods. The presence of lichens on boulders in the river channel can be used to date the last mobilization of the blocks (Gob et al., 2003), estimate flood competence (Gob et al., 2005, 2010), frequency (Macklin and Rumsby, 2007; Macklin et al., 2010), and magnitude based on palaeocompetence equations (Johnson and Warburton, 2002). By using size–frequency diagrams and regional growth curves calibrated with dated reference points, it is possible to determine the past flood event responsible for the last mobilization of each boulder covered with lichens (Gob et al., 2010). The first step is always to establish a lichenometric growth curve characteristic of each lichen species and subspecies used for the area studied (Jacob et al., 2002). The secular growth rate of the lichen colonies is obtained from tombstones and old monuments (buildings, rocky surfaces, megaliths). Lichens on canyon walls and terraces can also be used to date the incision phases or abrasion phases linked to floods and thereby date the floods and their boundaries, which are used in turn to infer magnitudes (Gregory, 1976).

The uncertainties in lichenometry studies are related to the data sources (lichen colonies) and the methodology itself. Errors in lichen measurements, their proper identification (species and subspecies), and the characterization of the growth curve bring uncertainties in flood dating. In addition, another source of error can come from the nonrenovation of lichen colonies after flood events, because of their varying level of susceptibility to stream power, abrasion, or the period of immersion required for mortality (Marsh and Timoney, 2005); and the complex patterns of overwashing and reworking (McEwen and Matthews, 2013). These uncertainties can be reduced by the simultaneous use of a combination of different data sources and palaeohydrological methods (Johnson and Warburton, 2002).

3.4 DATING PALAEOFLOOD EVIDENCE

Some flood layers within a stratigraphic profile should be dated to determine temporal flood behavior and recurrence. The methods applied to date palaeoflood sediments can be divided into three categories: (1) relative; (2) numerical correlated; and (3) numerical (Jacobson et al., 2003).

Relative methods allow a first diagnosis about the age of sediments and other flood indicators and help to decide the strategy and targets for other dating techniques. Common relative methods applied to alluvial methods include stratigraphic position, weathering characteristics, pedogenesis, and morphological position (inset or superimposed levels).

Numerical correlated are hybrid methods aiming to estimate numerical ages through presence or absence of diagnostic properties (e.g., weathering) or elements (e.g., pollen, tephra, macrofossils, and archaeological artifacts). These diagnostic features can be used for the correlation and dating of different stratigraphic profiles.

Numerical dating methods aim to provide an estimate of the time elapsed since sediment deposition. Radiocarbon dating is a standard absolute dating tool employed in palaeohydrologic work (e.g., Baker et al., 1985). Radiocarbon analysis measures the decay of the radioactive ^{14}C isotope in organic samples (animals and plants) since the time they died. Organic material such as wood, charcoal, seeds, and leaf fragments transported by floods are commonly deposited in conjunction with detrital sediment in slackwater sequences. The radiocarbon method is limited to the range ~200 years to 55,000 years BP. Radiocarbon dating can be carried out in two ways: conventional and accelerator mass spectrometry (AMS). The conventional radiocarbon method measures the remaining ^{14}C activity in a sample, and compares it to atmospheric ^{14}C, assuming that the level has not changed. This conventional method utilizes only a small number of atoms that decay during the experiment, and therefore a large sample (a few grams) is necessary. The AMS method utilizes every atom of the nuclide in the sample, allowing extremely small samples (a few milligrams) to be dated. Radiocarbon dates are given in radiocarbon years before present that need to be calibrated to calendar years to account for secular variations in radiocarbon production in the atmosphere (Stuiver, 1982). Precise curves of radiocarbon ages have been established for calibrating radiocarbon results (e.g., Reimer et al., 2013), although anthropogenic emissions into the atmosphere pose special problems in obtaining precise calibration of ages of less than 200 years.

In flood deposits, common sample material for radiocarbon analysis includes charcoal, seeds, wood, twigs, peat, shells, bones. The sample should be taken and sent in glass, plastic, or aluminum foil, preferably in dry and dark conditions. During collection special attention should be given to the sample location within the stratigraphic profile as well as the potential postflood impact of biological or human activities. In most studies, it is assumed that the radiocarbon age is close to the flood date, although strictly the radiometric date should be considered as a maximum limiting age. Dating of multiple samples along a stratigraphical profile can prove the validity of the assumption.

The optically stimulated luminescence (OSL) method (Aitken, 1998; Wintle and Murray, 2006) is a dating technique that indicates the burial time of deposits, principally quartz and feldspar minerals. In OSL dating, the time when the sediment was last exposed to direct sunlight (bleached) is determined. For the purposes of dating flood deposit sequences, the general assumption is that the sediment was last exposed to light during transport prior to deposition. Luminescence dating might be limited if the sediment to be measured has been heterogeneously exposed to light prior to being buried and shielded from daylight. In these cases, the measurement of multigrain aliquots would lead to overestimation of the burial dose and thus overestimation of the age. Developments in instrumentation are reducing the sample size to individual grains (Duller and Murray, 2000; Bøtter-Jensen et al., 2000) from which the heterogeneity within the grain population can be analyzed. This involves the use of statistical methods and the so-called minimum age models to estimate the true burial dose making the analysis more complex. Moreover, new analytical protocols have considerably improved the application of OSL dating (Murray and Wintle, 2000; Wintle and Murray, 2006), resulting in numerical ages known to 5–10 percent, even for young deposits (<300 years) (e.g., Ballarini et al., 2003; Duller, 2004; Medialdea et al., 2014). OSL dating can be an important tool, especially for deposits: (1) containing little or no organic material; (2) older than 40,000 years, the range of radiocarbon dating; or (3) younger than 500 years old, so that radiocarbon dating cannot

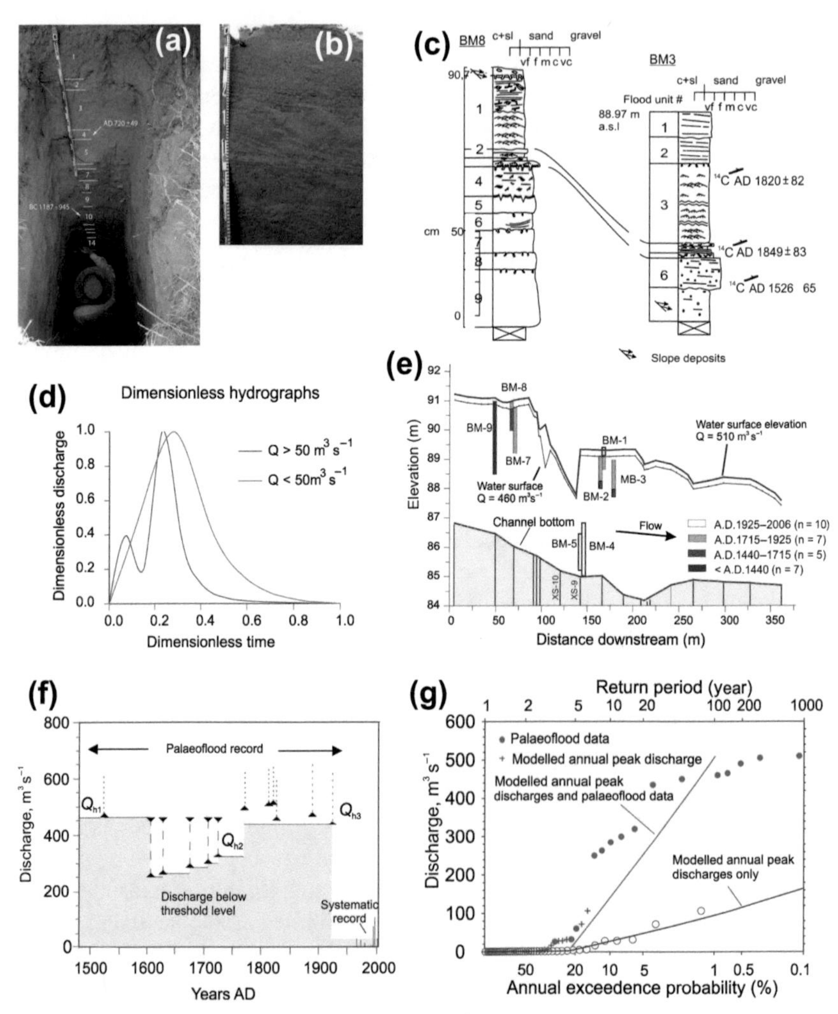

FIGURE 3.5 Buffels River palaeoflood case study in Messelpad reach based on data after Benito et al., 2011a,b. The general view of the Messelpad study reach and location of three profiles is shown in Figure 3.2(e). (a) View of the BM9 pit in Messelpad with indication of the stratigraphic units and radiocarbon dating results. (b) Lower part of slackwater unit 3 in BM8 profile (Messelpad) showing an organic detrital laminae and climbing ripples with both upstream and downstream flow direction (indicative of eddy circulation). (c) Two stratigraphic profiles of the Messelpad reach indicating dated samples (radiocarbon dates in years AD) and proposed correlations between sections. (d) The dimensionless hydrographs selected from a set of modelled hydrographs for small floods (<50 $m^3\,s^{-1}$) and large floods (<50 $m^3\,s^{-1}$). Probabilistic hydrographs were scaled to match the peak discharges obtained from the FFA (showed in F) (e) Longitudinal profile of the stream channel bed and water surface profiles obtained from HEC-RAS model for the highest palaeoflood deposits (510 $m^3\,s^{-1}$) and for a reference discharge of 460 $m^3\,s^{-1}$. Stratigraphic profiles are represented as vertical bars, with colors sketching sections of the columns with different age. (f) Palaeoflood and modelled discharges at Messelpad reach. Palaeofloods data are shown as triangles that show minimum discharge values above discharge

yield precise results. Samples are collected by hammering opaque PVC cylinders (internal diameter ~5 mm and length ~30 cm) into a cleaned vertical exposure until completely filled with sediment. Upon extrusion from the cliff face, these cylinders are sealed using thick duct tape. Extra sand subsamples for each sample are packed into airtight plastic containers holding approximately 200 g. These subsamples are to be used for determining the water content and the dose rate at the sample location. The cylinders of OSL samples are extruded under subdued red laboratory light and from them quartz grains with sizes of 180–250 μm are extracted using routine laboratory procedures (Porat, 2006).

Ideally, each flood unit should be dated, although in practice scarcity of datable material and high dating costs are the most common limiting factors on the number of samples. A first assessment based on relative age indicators such as degree of soil development, slope accumulation, and archaeological material may facilitate the decision-making process when choosing an age sampling strategy. It is advisable to separate the flood stratigraphy in sequences or sets and spend time and resources on dating the base and top of each sequence. The same applies to flood benches. Additional efforts should target the date of the largest floods and nonexcedence time intervals. Documentary and historical flood data is highly valuable to understand palaeoflood chronology (Benito et al., 2003b; Brázdil et al., 2006; Thorndycraft et al., 2006).

The problems of dating recent deposits (the past 150 years) with radiocarbon dating were solved using the analysis of modern radionuclides such as cesium-137 and lead-210 (Ely et al., 1992; Thorndycraft et al., 2005b). Cesium-137 is an artificial isotope that was introduced into the atmosphere during nuclear bomb testing in the 1950s. Since then, cesium-137 has been deposited on the land surface (including soils and sediments) from atmospheric fall-out. The presence of cesium-137 in flood sediments indicates a post-1950 age for the flood event. Sample techniques consist in collecting 100 gr of fresh sediment at 10–20 cm intervals down to a reasonable depth, informed by the results of other age dating controls. In the Llobregat flood deposits, samples were taken at the central portion of the flood units or at the upper and lower parts, depending on the unit thickness (Thorndycraft et al., 2005b).

3.5 PALAEOFLOOD DISCHARGE ESTIMATION

The water levels associated with the different flood units (Figure 3.5(a) and (b)) can be converted into discharge values (Lang et al., 2004), which is the

thresholds (shaded areas). Palaeoflood data within a discharge range are represented by the discontinuous lines between the normal and inverse triangles. The horizontal shaded areas represent the discharge threshold values used in the FFA. (g) Two component extreme value distribution fitted to annual series of modelled discharges and palaeoflood information (censored data) for Messelpad sites.

random variable used in the statistical analysis (Francés, 2004). The elevation of the uppermost contact of each flood unit represents the minimum water surface elevation during the flood event (Figure 3.5(b)). Cross-sections and flood deposit elevations, the input data for the hydraulic models, are surveyed along the study reaches. The discharge is obtained by trial and error using the appropriate hydraulic model by comparison between the observed water levels and the simulated ones (Figure 3.5(d)). A number of formulae and models exist to estimate past flood discharge from a known water surface elevation (O'Connor and Webb, 1988; Webb and Jarrett, 2002; Lang et al., 2004), ranging from simple hydraulic equations to one- or multidimensional hydraulic modelling. Most palaeoflood studies assume one-dimensional flow based on (1) uniform flow equations (Manning and Chézy equations); (2) critical flow conditions (King, 1954; Bodoque et al., 2011); (3) gradually varied flow models; and (4) one-dimensional St Venant equations. The selection of an appropriate approach for a particular site depends on local hydraulic conditions (Webb and Jarrett, 2002).

These models assume a fixed bed, and data required include slope, roughness (generally Manning's n), cross-sectional geometry and, for the step-backwater method, a boundary condition upstream or downstream depending on the flow type. The three principal sources of error are (1) the assumption that present channel geometry represents the channel conditions at the time of flooding and (2) an underestimate of the palaeodischarge due to the unknown level of the floodwaters above the palaeostage indicator (sediment, tree scar). The first error, that of cross-sectional stability, is substantially reduced by conducting palaeoflood studies in bedrock gorge reaches, where available. Bedrock channel geometry is significantly more stable than alluvial floodplain channels, for example, and will not have been substantially altered over the past centuries or millennia (Figure 3.2(e)). In case of uncertainty regarding the flow-boundary geometry, it is important to consider the plausible changes in channel geometry at the time of flooding and, consequently, provide a range of palaeoflood discharges. The unknown water depth above the flood sediments can be approached by a study of the sedimentology of the flood deposits (Benito et al., 2003a), which can enable interpretations regarding flow velocities and energy conditions at the site of deposition, thus allowing inferences to be made regarding the level of the water above the deposits.

The most common palaeoflood analysis situation is that of gradually varied flow at a steady state (constant discharge) for which depth varies with distance but not with time (O'Connor and Webb, 1988; Webb and Jarrett, 2002). The discharge of the different flood units/features is typically estimated by computing the water surface profiles for various hypothetical discharges using the step-backwater method (Chow, 1959), which solves the conservation of mass and energy equations in their one-dimensional forms (Figure 3.5(d)). Two-dimensional hydraulic models have been used for palaeodischarge estimation (Denlinger et al., 2002; Ballesteros Cánovas et al., 2011a,b). By comparing the model-generated profiles to the PSIs (e.g., slack water flood deposit elevations) minimum palaeodischarges are specified. Available public-domain computer routines, such as U.S. Army Corps of Engineers HEC-RAS (Hydraulic Engineering Center, 2010) software, allow for rapid calculation of water surface profiles for specific discharges, and energyloss coefficients.

3.6 FLOOD FREQUENCY ANALYSIS USING PALAEOFLOOD DATA

FFA with systematic data assumes that the distribution of the unknown magnitudes of the largest floods is well represented by the gauged record or that it can be obtained by statistical extrapolation from recorded floods (usually modest floods). However, the limit of credible statistical extrapolation relative to the typical length of gauged discharges (40–50 years) corresponds, at best, to a period of 200 years (England et al., 2006). The value of palaeoflood records is their potential for incorporating physical evidence of rare floods and limits to their largest magnitude (Figure 3.5(e)). The use of historical and palaeoflood information gives rise to two specific problems: (1) nonsystematic data (only the major floods remain known); and (c) nonhomogeneous data (hydroclimatically induced nonstationarity due to natural climatic variability within the past 1,000–10,000 years). These problems are discussed in detail by Redmond et al. (2002), Benito et al. (2004), Francés (2004).

In hydrology, flood observations reported as having occurred above some threshold are known as censored data sets (Leese, 1973). Palaeoflood information is considered data censored above a threshold (Figure 3.5(e)) and it is assumed that the number of k observations exceeding an arbitrary discharge threshold (X_T) in M years is known, similar to the partial duration series (Stedinger and Cohn, 1986; Francés et al., 1994). The value of the peak discharge for palaeofloods above X_T may be known or unknown. Palaeoflood data are organized according to different fixed threshold levels exceeded by flood waters over particular periods of time. Estimated flood discharges obtained from the minimum high-water palaeoflood indicators and maximum bounds (nonexceeded threshold sense; Levish et al., 1997) can be introduced as minimum and maximum discharge values (Figure 3.5(e)). Estimates of statistical parameters of flood distribution functions (e.g., Gumbel, LP3 or upper-bounded statistical models; Botero and Francés, 2010) are calculated using maximum likelihood estimators (Leese, 1973; Stedinger and Cohn, 1986; Figure 3.5(f)), the expected moment algorithm (Cohn et al., 1997), and a fully Bayesian approach (O'Connell et al., 2002; Reis and Stedinger, 2005), providing a practical framework for incorporating imprecise and categorical data as an alternative to the weighted moment method (U.S. Water Resources Council, 1982).

Before the statistical analysis is carried out, the general characteristics and stationarity of the flood series must be considered. The temporal changes in the trajectory and statistics of a variable may correspond to natural, low-frequency variations of the climate's hydrological system or to nonstationary dynamics related to anthropogenic changes in key parameters such as land use and atmospheric composition. Flood record stationarity from censored samples (systematic and/or nonsystematic) can be checked using Lang's test (Lang et al., 1999). This test assumes that the flood series can be described by a homogenous Poisson process. The 95 percent tolerance interval of the cumulative number of floods above a threshold, or censored, level is computed. Stationary flood series are those remaining within the 95 percent tolerance interval (Naulet et al., 2005). Recent advances in FFA have focused on modelling time series under nonstationary conditions, such as Generalized Additive Models for Location, Scale and Shape parameters (GAMLSS; Rigby and Stasinopoulos, 2005). The GAMLSS approach is able to describe the temporal variation of statistical parameters (mean, variance) in probability distribution functions (Gumbel, Lognormal, Weibull, Gamma). The statistical parameters may show increasing/decreasing trends that can be modelled using time as covariate (characterizing the trend or as a smooth function via cubic splines; Villarini et al., 2009), or they can be related to hydroclimatic covariates such as climatic indices that reflect low-frequency climatic variability (e.g., Pacific Decadal Oscillation, North Atlantic Oscillation, Arctic Oscillation; López and Francés, 2013). The application of these nonstationary models to palaeoflood hydrology requires a characterization of the occurrence rate (covariate) during the recorded period.

3.7 ESTIMATION OF PALAEOFLOOD VOLUME

Extreme flood hydrographs are needed for different engineering applications, such as risk analysis for dam safety (Nathan and Weinmann, 1999; Swain et al., 2006) or estimating floodwater recharge into alluvial aquifers (Greenbaum et al., 2002; Benito et al., 2011b). Extreme flood hydrographs are characterized by four major factors: peak, volume, duration, and shape (e.g., Yue et al., 2002). The few attempts to derive hydrographs from palaeoflood studies have used probabilistic hydrographs (England, 2003; Benito et al., 2011b). A probabilistic hydrograph is defined as one that preserves a peak discharge exceedance probability and dependence between volume and peak for a fixed duration (England, 2003). Probabilistic hydrographs are developed from scaling streamflow observations, or from rainfall–runoff models. This approach requires (England, 2003) (1) a peak discharge–probability analysis; (2) an extreme storm duration probability relationship; (3) correlation between peak discharge and maximum mean daily flow, and (4) observed hourly flow hydrograph(s) in natural regime.

Peak discharge probability relationships, hourly flow data and daily mean streamflow data, are used to construct probabilistic hydrographs. The peak discharge probability relationship uses data from gauge stations or from rainfall–runoff modelling combined with palaeoflood discharges (Figure 3.5(f)). Flood runoff duration can be obtained either from observed extreme storm duration (assuming a close similarity of storm and runoff duration) or based on the most common observed occurrences of peak duration at the study basin. The hydrograph shape is reproduced from averaged unit hydrographs (Cudworth, 1989) or based on mean dimensionless hydrographs (Craig and Rankl, 1978). The mean dimensionless hydrographs may be constructed by visually selecting a representative shape of observed or modelled hydrographs of large floods (Benito et al., 2011b, Figure 3.5(c)). The probabilistic hydrographs are then scaled to represent the estimated duration (from peak duration frequency distribution) and dimensionless hydrograph shape corresponding to the peak discharge calculated for different flood quantiles (Figure 3.4(c); Table 3.1).

3.8 APPLIED PALAEOFLOOD HYDROLOGY

Palaeoflood hydrology has been successfully applied to engineering and flood hazard studies (Benito and Thorndycraft, 2005), including (1) major improvements in flood risk assessment (House et al., 2002; Benito et al., 2004); (2) determination of the maximum limit of flood magnitude (Enzel et al., 1993) and nonexceedences to test the consistency of probable maximum flood (PMF), and safety risk analysis of critical facilities (e.g., dams, wastewater facilities, and power plants; Levish et al., 1996, 2003; Benito et al., 2006; Greenbaum, 2007); (3) understanding of long-term flood–climate relationships (Ely et al., 1993; Ely, 1997; Knox, 2000; Redmond et al., 2002; Thorndycraft and Benito, 2006; Benito et al., 2008); and (4) assessing the sustainability of water resources in dryland environments where floods are an important source of water for alluvial aquifer recharge (Greenbaum et al., 2002; Benito et al., 2011b).

3.8.1 Palaeofloods as an Analogue of Present and Future Flood Disasters

Palaeoflood hydrology has demonstrated in a number of cases that present catastrophic floods were already produced in the recent past (a few hundred years). In the Black Hills of South Dakota, on June 9–10th, 1972 an extreme convective rainfall of around 380 mm of rain in 6 h produced peak discharges of 1,461 $m^3 s^{-1}$ in Boxelder Creek (drainage area of 303 km^2) and 883 $m^3 s^{-1}$ in Rapid Creek (drainage area 135 km^2) upstream of Rapid City producing 238 casualties (Schwarz et al., 1975). This 1972 flood discharge represents a high outlier for the gauge records of the Black Hills region, but sedimentological

TABLE 3.1 Instrumental, Palaeoflood, the 10,000-year Return Period and Probable Maximum Flood (PMF) Discharges for Various River Basins in the USA and Spain. The 10,000-year Return Period Discharges Were Estimated Using the Palaeoflood Data in Each Basin (Benito and Thorndycraft, 2005)

Location	Drainage Basin Area (km^2)	Largest Instrumental Flood ($m^3 s^{-1}$)	Length of Palaeoflood Record (years)	Largest Palaeoflood Discharge ($m^3 s^{-1}$)	10,000-year Palaeoflood Discharge ($m^3 s^{-1}$)	PMF Discharge ($m^3 s^{-1}$)
Santa Ynez River, California (USA)[a]	1,080	2,265	2,900	2,550	2,690	13,060
Ochoco Creek, Oregon (USA)[b]	764	–	10,000–15,000	285	285	4,785
Crooked River, Oregon (USA)[c]	6,825	–	8,000–10,000	<1,100	1,100	7,225
South Fork Ogden River, Utah (USA)[c]	210	53	400	70	<215	3,075
Llobregat River (NE Spain)[d]	3,370	2500	2,800	4,680	–	18,985[e]
Caramel River (SE Spain)[d]	372	170	1,985	1,616	3,450	5,786

[a]*Data from Ostenaa et al., 1994 in Levish et al., 1996.*
[b]*Data from Ostenaa and Levish, 1996.*
[c]*Data from Levish et al., 1996.*
[d]*Data from Benito and Thorndycraft, 2005.*
[e]*Data from Francés (written communication, 2003).*

evidence of former floods deposited within rock shelters showed that similar magnitude floods have occurred at least five to seven times over the past 1,000 years on both creeks (Harden et al., 2011). In this case, as in others, palaeoflood analysis is a suitable deterministic means to provide a long-term perspective of the largest floods occurring at specific catchments.

Another further application of palaeoflood hydrology for the prevention of future flood disasters and the rational assessment of mitigation strategies is the incorporation of palaeoflood data in economic risk analysis. Few examples of these studies exist worldwide; one of them is the cost–benefit flood risk analysis made by Ballesteros Cánovas et al. (2013a) in the village of Navaluenga (Central Spain). Nonsystematic data derived from a dendrogeomorphological study of riparian trees were included in the FFA, improving the statistical function for flood estimation of medium and high recurrence periods. Flood damage was assessed by means of depth-damage functions, and the flooded urban areas were analyzed by applying a 2D hydraulic model. The best defense strategies were obtained via a cost–benefit procedure, where uncertainties derived from each analytical process were incorporated, based on a stochastic approach to estimating expected economic losses. The results showed that large structural solutions are not economically viable when compared with other smaller structural measures, presumably because the preestablished location of dams in the upper part of the basin is unable to laminate the flow generated in the headwaters. This analysis shows that uncertainties derived from nonsystematic data (i.e., tree-ring analysis) included in the flood frequency definition can be used in flood risk assessment.

3.8.2 Palaeoflood Hydrology Applied to Dam Safety

Dams are designed, constructed, and modified, when necessary, such that a catastrophic failure is prevented during a large flood. The design flood refers to the flood magnitude for which spillways and energy dissipating structures were designed, with a safety margin provided by the freeboard. Past experience indicates that overtopping represents more than 40 percent of dam failures, showing that extreme floods constitute an important risk for dam safety (ICOLD, 1995). The ICOLD (1995) recommends the estimation of a flood discharge return period of 1,000 years for the design of spillways, and 10,000 years for the safety of the dam structure. The estimation of such low frequency floods is based on either probabilistic or deterministic approaches. Probabilistic methods (statistical analysis and hydrometeorologic estimations) involve extrapolation of hydrological and meteorological instrumental records, and occasionally the use of historical data. The deterministic approach, particularly the estimation of the probable maximum flood (PMF), is nowadays a common practice in dam design. The rationale behind the PMF is the existence of a hydroclimatic limit to the supply of moisture to river basins through storms and precipitation, which has led to discussion on whether there is an upper limit of flood magnitude, duration, and volume that a specific drainage basin can generate (Wolman and Costa, 1984). The designed largest flood is then routed downstream by using watershed models (National Research Council, 1985). By definition, the PMF has no return period but arbitrarily it was assigned a return period of 10,000 to 1,000,000 years at the upper and lower confidence limits for FFA (National Research Council, 1985).

Palaeoflood studies, including those performed for dam safety purposes by the U.S. Bureau of Reclamation, show that the upper limit for palaeoflood magnitude is of an order of magnitude smaller than that implied by PMF calculations (Enzel et al., 1993; Levish et al., 1996). Studies performed in Spain show similar conclusions in the Llobregat and the Caramel river basins (Benito et al., 2006). In these studies the extrapolated discharges of the 10,000 year palaeoflood return period are between 5 and 20 percent of the calculated PMF for the United States (Levish et al., 1996) and around 60 percent for the Spanish case studies (Table 3.2), indicating that the calculated PMF discharges are very large overestimates. The value of PMF for dam safety studies is uncertain due to the lack of physical potential for these basins to generate the calculated peak discharges. Estimated discharges from the physical evidence left by floods over periods of thousands of years provide more realistic results, which can be combined with gauge station data using appropriate statistical tools and can subsequently be of great value for the planning of large-scale hydrological projects (e.g., Ostenaa et al., 1994; Ostenaa and Levish, 1996).

3.8.3 Flood Hazards in the Context of Climate Change

Climate affects flooding across a wide spectrum of spatial and temporal scales (Redmond et al., 2002). Temporal changes in flood magnitude and frequency have been described via instrumental (Robson et al., 1998), documentary/historical (Glaser et al., 2010), and palaeoflood records (Ely et al., 1993; Benito et al., 2008). Long-term palaeoflood records have revealed secular to multidecadal variability in flood patterns that may affect their frequency, magnitude, seasonality, and cause (e.g., Ely et al., 1993; Macklin et al., 2006; Benito et al., 2008; Glaser et al., 2010). A major driver for this flood variability is climate (Knox, 1999, 2000; Redmond et al., 2002), although land use changes could affect runoff generation conditions at basin scale, particularly (but not exclusively) during the nineteenth century (Greenbaum et al., 2000; Benito et al., 2010).

Periods with a higher frequency of large floods are interpreted as a result of changing atmospheric circulation patterns modulated by climate variability on decadal and secular timescales (Hirschboeck et al., 2000). In the Mediterranean region, palaeoflood records supported by documentary evidence indicate that periods of region-wide flooding can be identified in the second,

TABLE 3.2 Floodwater Recharge into the Spektakel Alluvial Aquifer in the Lower Buffels River (South Africa) Based on the Study by Benito et al., 2011b

Return Period, years	Peak Discharge, $m^3\ s^{-1}$	Duration, days	Hydrograph Volume, mm^3	Potential Infiltration, mm^3	Infiltration to Saturation[(a)] WT: 3 m, mm^3	Infiltration to Saturation[(b)], mm^3
5	23	6	4.2	4.2	4.2	4.20
10	140	12	27	11	9	11.83
25	287	12	54	17	9	13.60
50	397	12	76	18	9	13.60
100	505	12	97	21	9	13.60
500	752	12	144	23	9	13.60

First and second columns: flood quantiles for different return periods at messelpad reach (~5 km upstream of spektakel aquifer reach), obtained using a two-component extreme value (TCEV) distribution fitted to the combined modelled and censored palaeoflood data. Third and fourth columns: characteristics of the probabilistic hydrographs (duration and volume) based on dimensionless hydrographs and fix duration of 6 days to small floods ($<50\ m^3\ s^{-1}$) and 12 days for large floods ($<50\ m^3\ s^{-1}$). Fifth column: potential infiltration considering all the possible infiltration for the given probabilistic hydrograph assuming a limitless capacity of the aquifer. Sixth and seventh columns: infiltration to saturation estimated for two starting conditions assuming (a) a water table at 3 m below the surface and (b) a depleted aquifer conditions at the time of flooding. Floodwater recharge of this alluvial aquifer in Africa's dryland region may be enhanced keeping a low water table during wet years.

sixth–seventh, tenth, late fifteenth, and late eighteenth centuries AD, which all coincide with relatively wet and cold climatic conditions (Macklin et al., 2006; Thorndycraft and Benito, 2006; Benito et al., 2008; Luterbacher et al., 2012). Glaser and Stangl (2004) found that multidecadal (range of 30–100 years) increases or decreases in flood frequency have occurred several times in Central-western Europe. For instance, flood events in the Pegnitz and Main rivers in Southern Germany were more frequent in the second half of the sixteenth century than in the second half of the twentieth century. In the western United States, an increased frequency of high-magnitude palaeofloods coincides with periods of cool, wet climate, whereas warm intervals, such as the Medieval Climatic Anomaly, correspond to dramatic decreases in the number of large floods (Ely et al., 1993). In this region, a positive relationship between palaeofloods and long-term variations in the frequency of El Niño events is evident at least over the past 3,000 years (Ely, 1997).

The temporal change in flood frequency/magnitude may affect the assumption of stationarity of the statistical parametric models on which the random variable (flood discharge) is independently and identically distributed (Lang et al., 1999; Redmond et al., 2002). This assumption is being questioned because of the ways in which climate change and land use may alter flood hydrology (Milly et al., 2008). Palaeoflood and documentary records can be used to test this assumption of stationarity from the analysis of long-term flood records (Benito et al., 2011a; Grodek et al., 2013), their magnitude–frequency patterns, and their links to local and regional driving mechanisms (natural and anthropogenic).

3.8.4 Floodwater Recharge by Extreme Floods

In many arid environments, the availability of water resources for natural ecosystems and human societies depends on floodwater infiltrating into alluvial aquifers (indirect recharge). As floods are rare in the gauged records, limited data exist on which to base an evaluation of the long-term frequency of the recharging events, data that are of crucial importance for integrated water resource management. In recent years new methodologies have been developed, combining flood and palaeoflood records with data on modern transmission losses, to estimate shallow aquifer recharge processes. The approach was first successfully applied in the Negev Desert, with quantitative data provided on the volume of water lost during floods through transmission losses (Greenbaum et al., 2002). In the framework of the WADE Project, funded by the European Commission, instrumental records on the temporal variation of water content in the vadose zone during recharging floods (Dahan et al., 2003, 2008) were used to calibrate a model for long-term recharge estimation from palaeofloods (SWDs) linked to the isotopic composition of groundwater (Benito et al., 2011a,b). A major factor controlling the recharge is the infiltration rate of the channel bed. In alluvial channels with a limited infiltration rate (e.g., Kuiseb River, Namibia; 8.5–10 mm/h) flow duration and channel characteristics (length and width) become important factors in determining infiltration and total recharge volume (Morin et al., 2009; Grodek et al., 2013). In the Buffels river (South Africa; 10–60 mm/h) with a higher infiltration rate the effect of flood duration becomes less important and at the same time infiltration volume becomes more dependent on the magnitude of the flow and the size of the aquifer (Benito et al., 2011a,b). In the South African case, reconstructed palaeofloods were up to five times greater than the largest modelled peak discharges over the period 1965–2006. Dimensionless hydrographs scaled to the probabilistic flood discharge showed that small floods (return periods of 5–10 years) were able to fully saturate the alluvial aquifers (floods exceeding *ca* 120–140 $m^3 s^{-1}$ and 12 days duration; Benito et al., 2011b; Table 3.2).

3.9 CONCLUSIONS

Palaeoflood hydrology is a young and evolving discipline that has received criticism mostly because of misconceptions that persist in parallel disciplines, particularly in conventional hydrology. A review of the basic criticisms and their discussion can be found in Baker et al. (2002). Theoretical difficulties to be solved in the combined use of systematic and nonsystematic information stem from (1) methodological complexity and uncertainties associated with the reconstruction of the catalogue of palaeoflood data and (2) the influence and evidence of nonstationarity in long timescale flood records.

The methodological complexity associated with reconstructing past flood records can be solved by selecting the most appropriate settings, such as bedrock canyons for palaeoflood analysis, using SWDs. In terms of pure scientific research it is of interest to reconstruct the most complete catalogue of past floods and estimate the most accurate peak discharges possible so that issues such as the water depth above SWDs become of key importance. In practice, however, the critical issue for flood risk estimation is not the accurate compilation of the whole past flood record, but the frequency of floods that might have an impact on human activities. In this case, the number of floods exceeding or not exceeding a surface or altitudinal level during a time period, which represents stage and subsequently discharge limits, may result in a significant improvement in FFA.

A major effort must be focused on FFA for managing this nonsystematic and, even more importantly, nonstationary information. Different statistical tools, such as the maximum likelihood and Bayesian methodologies, have been successful in managing nonsystematic data, even data with a high degree of uncertainty regarding peak discharge values, which have been solved using censuring levels and thresholds. In the analysis of nonstationary flood series it is also critical to reach a better understanding of flood-producing mechanisms and flood–climate links during different time periods. A change in flood-generating mechanisms, or in flood frequency patterns, can be related to climatic variations and, therefore, a study of these variations, quantifying the frequency of climatic patterns responsible for flooding, is required. Another source of nonstationarity is related to land use and land cover changes (vegetation and soil moisture) that affect the response of a catchment to rainfall. As with other sedimentary records, a detailed analysis of geochemical and vegetation proxies (e.g., pollen and phytoliths) may provide insight into environmental changes in the studied basin at the time of flooding. Ultimately, nonstationarity in FFA is a problem with both systematic and/or nonsystematic records (Baker et al., 2002).

ACKNOWLEDGMENTS

The research conducted in this study was supported by the Spanish Inter-Ministry of Science and Technology (CICYT), through the projects CLARIES (CGL2011-29176) and MAS Dendro-Avenidas (CGL2010-19274); and the National Parks Research Program (MAGRAMA), through the project IDEA-GesPPNN (OAPN 163/2010).

REFERENCES

Aitken, M.J., 1998. An Introduction to Optical Dating. The Dating of Quaternary Sediments by the Use of Photon-stimulated Luminescence. Oxford University Press, Oxford, 280 pp.

Astrade, L., Bégin, Y., 1997. Tree-ring response of *Populus tremula* L. and *Quercus robur* L. to recent spring floods of the Sâone River, France. Ecoscience 4, 232–239.

Baker, V.R., Kochel, R.C., 1988. Flood sedimentation in bedrock fluvial systems. In: Baker, R.V., Kochel, R.C., Patton, P.C. (Eds.), Flood Geomorphology. John Wiley & Sons, New York, pp. 123–137.

Baker, V.R., Pickup, G., 1987. Flood geomorphology of the Katherine Gorge, Northern Territory, Australia. Geol. Soc. Am. Bull. 98, 635–646.

Baker, V.R., 2008. Paleoflood hydrology: origin, progress, prospects. Geomorphology 101 (1–2), 1–13.

Baker, V.R., Pickup, G., Polach, H.A., 1985. Radiocarbon dating of flood events, Katherine Gorge, Northern Territory, Australia. Geology 13, 344–347.

Baker, V.R., Pickup, G., Polach, H.A., 1983. Desert paleofloods in central Australia. Nature 301, 502–504.

Baker, V.R., Pickup, G., Webb, R.H., 1987. Paleoflood hydrologic analysis at ungaged sites, central and northern Australia. In: Singh, V.P. (Ed.), Regional Flood Frequency Analysis. D. Reidel, Boston, pp. 325–338.

Baker, V.R., Webb, R.H., House, P.K., 2002. The scientific and societal value of paleoflood hydrology. In: House, P.K., Webb, R.H., Baker, V.R., Levish, D.R. (Eds.), Ancient Floods, Modern Hazards: Principles and Applications of Paleoflood Hydrology, Water Science and Application Series, vol. 5. American Geophysical Union, Washington, DC, pp. 127–146.

Ballarini, M., Wallinga, J., Murray, A.S., van Heteren, S., Oost, A.P., Bos, A.J.J., van Eijk, C.W.E., 2003. Optical dating of young coastal dunes on a decadal time scale. Q. Sci. Rev. 22, 1011–1017.

Ballesteros Cánovas, J.A., Bodoque, J.M., Díez-Herrero, A., Sanchez-Silva, M., Stoffel, M., 2011a. Calibration of floodplain roughness and estimation of palaeoflood discharge based on tree-ring evidence and hydraulic modelling. J. Hydrol. 403, 103–115.

Ballesteros Cánovas, J.A., Eguibar, M., Bodoque, J.M., Díez-Herrero, A., Stoffel, M., Gutiérrez-Pérez, I., 2011b. Estimating flash flood discharge in an ungauged mountain catchment with 2D hydraulic models and dendrogeomorphic paleostage indicators. Hydrol. Process. 25, 970–979.

Ballesteros Cánovas, J.A., Stoffel, M., Bodoque, J.M., Bollschweiler, M., Hitz, O., Díez-Herrero, A., 2010a. Wood anatomy of *Pinus pinaster* Ait. following wounding by flash floods. Tree-Ring Res. 66 (2), 93–103.

Ballesteros Cánovas, J.A., Stoffel, M., Bollschweiler, M., Bodoque, J.M., Díez-Herrero, A., 2010b. Flash-flood impacts cause changes in wood anatomy of *Alnus glutinosa*, *Fraxinus angustifolia* and *Quercus pyrenaica*. Tree Physiol. 30, 773–781.

Ballesteros Cánovas, J.A., Sanchez-Silva, M., Bodoque, J.M., Díez-Herrero, A., 2013a. An example of integrated approach to flood risk management: the case of Navaluenga (Central Spain). Water Resour. Manage. 27 (8), 3051–3069.

Ballesteros Cánovas, J., Bodoque, J.M., Eguibar, M.A., Ruiz-Villanueva, V., Díez-Herrero, A., Stoffel, M., Sánchez-Silva, M., 2013b. Progress on the estimation of past flood discharge from dendromechanical analyses of tilted trees. In: 8th IAG International Conference on Geomorphology. Abstracts Volume, Paris, France, p. 374.

Begin, Y., Langlais, D., Cournoyer, L., 1991. A dendrogeomorphic estimate of shore erosion, Upper St-Lawrence Estuary, Quebec. J. Coastal Res. 7 (3), 607–615.

Benito, G., Thorndycraft, V.R., 2005. Palaeoflood hydrology and its role in applied hydrological sciences. J. Hydrol. 313, 3–15.

Benito G., O'Connor J.E., 2013. Quantitative paleoflood hydrology. In: Shroder, John F. (Editor-in-chief), Wohl, E. (Volume Ed.), Treatise on Geomorphology, vol. 9. Fluvial Geomorphology. Academic Press, San Diego. pp. 459–474.

Benito, G., Botero, B.A., Thorndycraft, V.R., Rico, M.T., Sánchez-Moya, Y., Sopeña, A., Machado, M.J., Dahan, O., 2011b. Rainfall-runoff modelling and palaeoflood hydrology applied to reconstruct centennial scale records of flooding and aquifer recharge in ungauged ephemeral rivers. Hydrol. Earth Syst. Sci. 15, 1185–1196. http://dx.doi.org/10.5194/hess-15-1-2011.

Benito, G., Lang, M., Barriendos, M., Llasat, M.C., Francés, F., Ouarda, T., Thorndycraft, V., Enzel, Y., Bardossy, A., Coeur, D., Bobée, B., 2004. Systematic, palaeoflood and historical data for the improvement of flood risk estimation. Nat. Hazards 31, 623–643.

Benito, G., Rico, M., Sánchez-Moya, Y., Sopeña, A., Thorndycraft, V.R., Barriendos, M., 2010. The impact of late Holocene climatic variability and land use change on the flood hydrology of the Guadalentín River, southeast Spain. Global Planet. Change 70, 53–63.

Benito, G., Rico, M., Thorndycraft, V.R., Sánchez-Moya, Y., Sopeña, A., Díez-Herrero, A., Jiménez, A., 2006. Palaeoflood records applied to assess dam safety in SE Spain. In: Ferreira, R., Alves, E., Leal, J., Cardoso, A. (Eds.), River Flow 2006. Taylor & Francis Group, London, pp. 2113–2120.

Benito, G., Sánchez, Y., Sopeña, A., 2003b. Sedimentology of high-stage flood deposits of the Tagus River, Central Spain. Sediment. Geol. 157, 107–132.

Benito, G., Sopeña, A., Sánchez, Y., Machado, M.J., Pérez González, A., 2003a. Palaeoflood record of the Tagus River (Central Spain) during the late Pleistocene and holocene. Q. Sci. Rev. 22, 1737–1756.

Benito, G., Thorndycraft, V.R., Rico, M., Sánchez-Moya, Y., Sopeña, A., 2008. Palaeoflood and floodplain records from Spain: evidence for long-term climate variability and environmental changes. Geomorphology 101, 68–77.

Benito, G., Thorndycraft, V.R., Rico, M.T., Sánchez-Moya, Y., Sopeña, A., Botero, B.A., Machado, M.J., Davis, M., Pérez-González, A., 2011a. Hydrological response of a dryland ephemeral river to southern African climatic variability during the last millennium. Quat. Res. 75, 471–482.

Benson, M.A., Dalrymple, T., 1967. General Field and Office Procedures for Indirect Measurements. Techniques of Water-Resources Investigations of the United States Geological Survey, Book 3, Washington, 30 pp.

Bodoque, J.M., Eguibar, M.A., Díez-Herrero, A., Gutiérrez-Pérez, I., Ruiz-Villanueva, V., 2011. Can the discharge of a hyperconcentrated flow be estimated from paleoflood evidence? Water Resour. Res. 47 (W12535), 1–14.

Botero, B.A., Francés, F., 2010. Estimation of high return period flood quantiles using additional non-systematic information with upper bounded statistical models. Hydrol. Earth Syst. Sci. 14, 2617–2628.

Bøtter-Jensen, L., Bulur, E., Duller, G.A.T., Murray, A.S., 2000. Advances in luminescence instrument systems. Radiat. Meas. 32, 523–528.

Brázdil, R., Kundzewicz, Z.W., Benito, G., 2006. Historical hydrology for studying flood risk in Europe. Hydrol. Sci. J. 51, 739–764.

Butler, D.R., 1979. Dendrogeomorphological analysis of flooding and mass movement, Ram Plateau, mackenzie mountains, northwest territories. Can. Geogr. -Geogr. Can. 23 (1), 62–65.

Chatters, J.C., Hoover, K.A., 1994. Response of the Columbia River fluvial system to Holocene climate change. Quat. Res. 37, 42–59.

Chow, V.T., 1959. Chow, Open-Channel Hydraulics, McGraw-Hill, New York, 680 pp.

Cohn, T.A., Lane, W.L., Baier, W.G., 1997. An algorithm for computing moments-based flood quantile estimates when historical flood information is available. Water Resour. Res. 33, 2089–2096.

Coriell, F., 2002. Reconstruction of a Paleoflood Chronology for the Middlebury River Gorge Using Tree Scars as Flood Stage Indicators (Unpublished senior thesis). Middlebury College, Department of Geology, Middlebury, VT, 45 pp.

Craig Jr, G.S., Rankl, J.G., 1978. Analysis of Runoff from Small Drainage Basins in Wyoming. U.S. Geological Survey Water-Supply Paper 2056, 70 pp.

Cudworth, A.G., 1989. Flood Hydrology Manual: A Water Resources Technical Publication. U.S. Department of the Interior, Bureau of Reclamation, Denver, CO, 243 pp.

Dahan, O., McDonald, E.V., Young, M.H., 2003. Flexible time domain reflectometry probe for deep vadose zone monitoring. Vadose Zone J. 2, 270–275. http://dx.doi.org/10.2113/2.2.270.

Dahan, O., Tatarsky, B., Enzel, Y., Külls, C., Seely, M., Benito, G., 2008. Dynamics of flood water infiltration and ground water recharge in hyperarid desert. Ground Water 46, 450–461. http://dx.doi.org/10.1111/j.1745-6584.2007.00414.x.

Denlinger, R.P., O'Connell, D.R.H., House, P.K., 2002. Robust determination of stage and discharge: an example from an extreme flood on the Verde River, Arizona. In: House, P.K., Webb, R.H., Baker, V.R., Levish, D.R. (Eds.), Ancient Floods, Modern Hazards: Principles and Applications of Paleoflood Hydrology, Water Science and Application Series, vol. 5. American Geophysical Union, Washington, DC, pp. 127–147.

Díez-Herrero, A., Ballesteros Cánovas, J.A., Bodoque, J.M., Eguíbar, M.A., Fernández, J.A., Génova, M., Laín, L., Llorente, M., Rubiales, J.M., Stoffel, M., 2007. Mejoras en la estimación de la frecuencia y magnitud de avenidas torrenciales mediante técnicas dendrogeomorfológicas. Bol. Geol. Min. 118 (4), 789–802.

Díez-Herrero, A., Ballesteros Cánovas, J.A., Bodoque, J.M., Ruiz-Villanueva, V., 2013a. A new methodological protocol for the use of dendrogeomorphological data in flood risk analysis. Hydrol. Res. 44.2, 234–247.

Díez-Herrero, A., Ballesteros Cánovas, J.A., Bodoque, J.M., Ruiz-Villanueva, V., 2013b. A review of dendrogeomorphological research applied to flood risk analysis in Spain. Geomorphology 196, 211–220.

Díez-Herrero, A., Ferrio, J.P., Ballesteros Cánovas, J.A., Voltas, J., Bodoque, J.M., Aguilera, M., Ruiz-Villanueva, V., Tarres, D., 2013c. Using tree-ring oxygen stable isotopes for studying the origin of past flood events: first results from the Iberian Peninsula. In: 8th IAG International Conference on Geomorphology. Abstracts Volume, Paris, France, p. 376.

Duller, G.A.T., 2004. Luminescence dating of quaternary sediments: recent advances. J. Quat. Sci. 19, 183–192.

Duller, G.A.T., Murray, A.S., 2000. Luminescence dating of sediments using individual mineral grains. Geologos 5, 88–106.

Dussaillant, A., Benito, G., Buytaert, W., Carling, P., Link, O., Espinoza, F., 2010. Repeated glacial-lake outburst floods in Patagonia: an increasing hazard? Nat. Hazards 54, 469–481.

Ely, L.L., 1997. Response of extreme floods in the southwestern United States to climatic variations in the late Holocene. Geomorphology 19, 175–201.

Ely, L.L., Baker, V.R., 1985. Reconstructing paleoflood hydrology with slackwater deposits: Verde river, Arizona. Phys. Geog. 6, 103–126.

Ely, L.L., Enzel, Y., Baker, V.R., Cayan, D.R., 1993. A 5000-year record of extreme floods and climate change in the southwestern United States. Science 262, 410–412.

Ely, L.L., Enzel, Y., Baker, V.R., Kale, V.S., Mishra, S., 1996. Changes in the magnitude and frequency of late Holocene monsoon floods on the Narmada River, central India. Geol. Soc. Am. Bull. 108, 1134–1148.

Ely, L.L., Webb, R.H., Enzel, Y., 1992. Accuracy of post-bomb 137Cs and 14C in dating fluvial deposits. Quat. Res. 38, 196–204.

England Jr., J.F., 2003. Probabilistic Extreme Flood Hydrographs that Use Paleoflood Data for Dam Safety Applications. Bureau of Reclamation, Denver, CO, p. 29.

England Jr., J.F., Klawon, J.E., Klinger, R.E., Bauer, T.R., 2006. Flood Hazard Study, Pueblo Dam, Colorado, Final Report, Bureau of Reclamation, Denver, CO, June, 160 pp.

Enzel, Y., 1992. Flood frequency of the Mojave River and the formation of late Holocene playa lakes, southern California. Holocene 2, 11–18.

Enzel, Y., Ely, L.L., Martinez-Goytre, J., Vivian, R.G., 1994. Paleofloods and a dam-failure flood on the Virgin River, Utah and Arizona. J. Hydrol. 153, 291–315.

Enzel, Y., Ely, L.L., House, P.K., Baker, V.R., 1993. Paleoflood evidence for a natural upper bound to flood magnitudes in the Colorado River basin. Water Resour. Res. 29, 2287–2297.

Erskine, W.D., Peacock, C.T., Dyer, F.J., Thoms, M.C., Olley, J.M., 2002. Late Holocene Flood Plain Development Following a Cataclysmic Flood. IAHS-AISH Publication No. 276, pp. 177–184.

Francés, F., 2004. Flood frequency analysis using systematic and non-systematic information. In: Benito, G., Thorndycraft, V.R. (Eds.), Systematic, Palaeoflood and Historical Data for the Improvement of Flood Risk Estimation. CSIC, Madrid, pp. 55–70.

Francés, F., Salas, J.D., Boes, D.C., 1994. Flood frequency analysis with systematic and historical or palaeoflood data based on the two-parameter general extreme value modes. Water Resour. Res. 30, 1653–1664.

García-García, F., Bohorquez, P., Martinez-Sanchez, C., Perez-Valera, F., Perez-Valera, L.A., Calero, J.A., Sanchez-Gomez, M., 2013. Stratigraphic architecture and alluvial geoarchaeology of an ephemeral fluvial infilling: climatic versus anthropogenic factors controlling the Holocene fluvial evolution in southeastern Spain drylands. Catena 104, 272–279.

Glaser, R., Stangl, H., 2004. Climate and floods in central Europe since AD 1000: data, methods, results and consequences. Surv. Geophys. 25, 485–510.

Glaser, R., Riemann, D., Schönbein, J., Barriendos, M., Brázdil, R., Bertolin, C., Camuffo, D., Deutsch, M., Dobrovolńy, P., Engelen, A., Enzi, S., Halíčková, M., Koenig, S., Kotyza, O., Limanówka, D., Macková, J., Sghedoni, M., Martin, B., Himmelsbach, I., 2010. The variability of European floods since AD 1500. Clim. Change 101, 235–256.

Gob, F., Bravard, J.P., Petit, F., 2010. The influence of sediment size, relative grain size and channel slope on initiation of sediment motion in boulder bed rivers. A lichenometric study. Earth Surf. Process. Landf. 35 (13), 1535–1547.

Gob, F., Jacob, N., Bravard, J.P., 2005. Determining the competence of mountainous mediterranean streams using lichenometric techniques. In: Geomorphological Processes and Human Impacts in River Basins, Proceedings of the International Conference hel at Solsona, Catalonia, Spain, May 2004, vol. 299. IAHS Publ., pp. 1–10.

Gob, F., Petit, F., Bravard, J.P., Ozer, A., Gob, A., 2003. Lichenometric application to historical and subrecent dynamics and sediment transport of a Corsican stream (Figarella River-France). Quat. Sci. Rev. 22, 2111–2124.

Gottsfeld, A.S., 1996. British Columbia flood scars: maximum flood stage indicators. Geomorphology 14, 319–325.

Greenbaum, N., 2007. Assessment of dam failure flood and a natural, high-magnitude flood in a hyperarid region using paleoflood hydrology. Water Resour. Res. 43 (W02401), 17. http://dx.doi.org/10.1029/2006WR004956.

Greenbaum, N., Margalit, A., Schick, A.P., Sharon, D., Baker, V.R., 1998. A high magnitude storm and flood in a hyperarid catchment, Nahal Zin, Negev Desert, Israel. Hydrol. Process. 12, 1–23.

Greenbaum, N., Schick, A.P., Baker, V.R., 2000. The palaeoflood record of a hyperarid catchment, Nahal Zin, Negev Desert, Israel. Earth Surf. Processes Landforms 25, 951–971.

Greenbaum, N., Schwartz, U., Schick, A.P., Enzel, Y., 2002. Palaeofloods and the estimation of long-term transmission losses and recharge to the Lower Nahal Zin alluvial aquifer, Negev Desert, Israel. In: House, P.K., Webb, R.H., Baker, V.R., Levish, D.R. (Eds.), Ancient Floods, Modern Hazards: Principles and Applications of Paleoflood Hydrology, Water Science and Application Series, vol. 5. American Geophysical Union, Washington, DC, pp. 311–328.

Gregory, K.J., 1976. Lichens and the determination of river channel capacity. Earth Surf. Process. 1, 276–285.

Grodek, T., Benito, G., Botero, B.A., Jacoby, Y., Porat, N., Haviv, I., Cloete, G., Enzel, Y., 2013. The last millennium largest floods in the hyperarid Kuiseb River basin, Namib Desert. J. Quat. Sci. 28, 258–270.

Grossman, M.J., 2001. Large floods and climatic change during the Holocene on the Ara River, central Japan. Geomorphology 39, 21–37.

Guardiola-Albert, C., Ballesteros Cánovas, J.A., Stoffel, M., Díez Herrero, A., 2012. Assessment of wood density structures of flood-damaged trees with X-ray computed tomography and variogram analyses. In: IX Conference on Geostatistics for Environmental Applications (GeoENV 2012), Valencia (Spain), pp. 133–134.

Harden, T.M., O'Connor, J.E., Driscoll, D.G., Stamm, J.F., 2011. Flood-Frequency Analyses from Paleoflood Investigations for Spring, Rapid, Boxelder, and Elk Creeks, Black Hills, Western South Dakota. U.S. Geological Survey Scientific Investigations Report 2011–5131, 136 p.

Harrison, S.S., Reid, J.R., 1967. A flood-frequency graph based on tree-scar data. Proc. N.D. Acad. Sci. 21, 23–33.

Heine, K., Völkel, J., 2011. Extreme floods around AD 1700 in the northern Namib Desert, Namibia, and in the orange river catchment, South Africa – were they forced by a decrease of solar irradiation during the little ice age? Geogr. Pol. 84, 61–80.

Heine, K., 2004. Flood reconstructions in the Namib Desert, Namibia and Little Ice Age climatic implications: evidence from slackwater deposits and desert soil sequences. J. Geol. Soc. India 64, 535–547.

Hirschboeck, K.K., Ely, L., Maddox, R.A., 2000. Hydroclimatology of meteorologic floods. In: Wohl, E. (Ed.), Inland Flood Hazards: Human Riparian and Aquatic Communities. Cambridge University Press, Cambridge and New York, pp. 39–72.

House, P.K., Pearthree, P.A., Klawon, J.E., 2002. Historical flood and paleoflood chronology of the Lower Verde River, Arizona: stratigraphic evidence and related uncertainties. In: House, P.K., Webb, R.H., Baker, V.R., Levish, D.R. (Eds.), Ancient Floods, Modern Hazards: Principles and Applications of Paleoflood Hydrology, Water Science and Application Series, vol. 5. American Geophysical Union, Washington, DC, pp. 267–293.

Huang, C.C., Pang, J., Zha, X., Su, H., Jia, Y., 2011. Extraordinary floods related to the climatic event at 4200 a BP on the Qishuihe River, middle reaches of the Yellow River, China. Quat. Sci. Rev. 30, 460–468.

Huang, C.C., Pang, J., Zha, X., Zhou, Y., Su, H., Li, Y., 2010. Extraordinary floods of 4100–4000 a BP recorded at the late Neolithic Ruins in the Jinghe river gorges, middle reach of the Yellow River, China. Palaeogeogr. Palaeoclimatol. Palaeoecol. 289, 1–9.

Huang, C.C., Pang, J., Zha, X., Zhou, Y., Su, H., Wan, H., Ge, B., 2012a. Sedimentary records of extraordinary floods at the ending of the mid-Holocene climatic optimum along the Upper Weihe River, China. Holocene 22, 675–686.

Huang, C.C., Pang, J., Zha, X., Zhou, Y., Su, H., Zhang, Y., Wang, H., Gu, H., 2012b. Holocene palaeoflood events recorded by slackwater deposits along the lower Jinghe River valley, middle Yellow River basin, China. J. Quat. Sci. 27, 485–493.

Huang, C.C., Pang, J., Zha, X., Zhou, Y., Yin, S., Su, H., Zhou, L., Yang, J., 2013. Extraordinary hydro-climatic events during the period AD 200-300 recorded by slackwater deposits in the upper Hanjiang River valley, China. Palaeogeogr. Palaeoclimatol. Palaeoecol. 374, 274–283.

Hupp, C.R., 1984. Dendrogeomorphic evidence of debris flow frequency and magnitude at Mount Shasta, California. Environ. Geol. Water Sci. 6 (2), 121–128.

Hupp, C.R., Osterkamp, W.R., Thornton, J.L., 1987. Dendrogeomorphic evidence and dating of recent debris flows on Mount Shasta, northern California. U.S. Geol. Surv. Prof. Pap. 1396-B, B1–B39.

Hydrologic Engineering Center, 2010. HEC-RAS, River Analysis System, Hydraulics Version 4.1. Reference Manual, (CPD-69). U.S. Army Corps of Engineers, Davis, 411 pp. http://www.hec.usace.army.mil/software/hec-ras/hecras-document.html.

ICOLD, 1995. Dam Failures. Statistical Analysis. Bulletin 99. ICOLD, Paris.

Jacob, N., Gob, F., Petit, F., Bravard, J.P., 2002. Croissance du lichen Rhizocarpon geographicum l.s. sur le pourtour nord-occidental de la Méditerranée: observations en vue d' une application à l'etude des lits fluviaux rocheux et caillouteux. Géomorphologie 4, 283–296.

Jacobson, R., O'Connor, J.E., Oguchi, T., 2003. Surficial geologic tools in fluvial geomorphology. In: Kondolf, M.G., Piégay, H. (Eds.), Tools in Fluvial Geomorphology. Wiley, Chichester, pp. 25–57.

Jarrett, R.D., Tomlinson, E.M., 2000. Regional interdisciplinary paleoflood approach to assess extreme flood potential. Water Resour. Res. 36, 2957–2984.

Jarrett, R.D., 1985. Determination of roughness coefficient for streams in colorado. U.S. Geol. Surv. Water Resour. Invest. Rep. 85–4004.

Jarrett, R.D., 1990. Hydrologic and hydraulic research in mountain rivers. Water Resour. Bull. 26, 419–429.

Jarrett, R.D., England Jr., J.F., 2002. Reliability of paleostage indicators for paleoflood studies. In: House, P.K., Webb, R.H., Baker, V.R., Levish, D.R. (Eds.), Ancient Floods, Modern Hazards: Principles and Applications of Paleoflood Hydrology, Water Science and Application Series, vol. 5. American Geophysical Union, Washington, DC, pp. 91–109.

Jarrett, R.D., Malde, H.E., 1987. Paleodischarge of the Late Pleistocene Bonneville flood, Snake River, Idaho, computed from new evidence. Geol. Soc. Am. Bull. 99, 127–134.

Johnson, R.M., Warburton, J., 2002. Flooding and geomorphic impacts in a mountain torrent: Raise Beck, Central Lake District, England. Earth Surf. Process. Landf. 27 (9), 945–969.

Jones, A.P., Shimazu, H., Oguchi, T., Okuno, M., Tokutake, M., 2001. Late holocene slackwater deposits on the Nakagawa river, Tochigi Prefecture, Japan. Geomorphology 39, 39–51.

Kale, V.S., Baker, V.R., 2006. An extraordinary period of low-magnitude floods coinciding with the Little Ice Age: palaeoflood evidence from Central and Western India. J. Geol. Soc. India 68, 477–483.

Kale, V.S., Mishra, S., Baker, V.R., 1997. A 2000-year palaeoflood record from Sakarghat on Narmada, central India. J. Geol. Soc. India 50, 283–288.

Kale, V.S., Mishra, S., Baker, V.R., 2003. Sedimentary records of palaeofloods in the bedrock gorges of the Tapi and Narmada Rivers, central India. Curr. Sci. 84, 1072–1079.

Kidson, R., Richards, K.S., Carling, P.A., 2005a. Reconstructing the ca. 100-year flood in Northem Thailand. Geomorphology 70, 279–295.

Kidson, R.L., Richards, K.S., Carling, P.A., 2005b. Hydraulic model calibration for extreme floods in bedrock-confined channels: case study from northern Thailand. Hydrol. Process. 20, 329–344.

King, H.W., 1954. Handbook of Hydraulics, fourth ed. McGraw-Hill, New York, NY.

Kite, J.S., Gebhardt, T.W., Springer, G.S., 2002. Slackwater deposits as paleostage indicators in Canyon reaches of the Central Appalachians: reevaluation after the 1996 Cheat River flood. In: House, P.K., Webb, R.H., Baker, V.R., Levish, D.R. (Eds.), Ancient Floods, Modern Hazards: Principles and Applications of Paleoflood Hydrology, Water Science and Application Series, vol. 5. American Geophysical Union, Washington, DC, pp. 257–266.

Knox, J.C., 1999. Long-term episodic changes in magnitudes and frequencies of floods in the Upper Mississippi Valley. In: Brown, A.G., Quine, T.A. (Eds.), Fluvial Processes and Environmental Change. John Wiley & Sons, New York, pp. 255–282.

Knox, J.C., 1985. Responses of floods to Holocene climatic change in the upper Mississippi Valley. Quat. Res. 23, 287–300.

Knox, J.C., 2000. Sensitivity of modern and Holocene floods to climate change. Quat. Sci. Rev. 19, 439–457.

Kochel, R.C., Baker, V.R., 1982. Paleoflood hydrology. Science 215, 353–361.

Kochel, R.C., Baker, V.R., 1988. Paleoflood analysis using slack water deposits. In: Baker, R.V., Kochel, R.C., Patton, P.C. (Eds.), Flood Geomorphology. John Wiley & Sons, New York, pp. 357–376.

Kochel, R.C., 1980. Interpretation of Flood Paleohydrology Using Slackwater Deposits, Lower Pecos and Devils Rivers, Southwestern Texas (Ph.D. dissertation). University of Texas, Austin.

Kochel, R.C., Baker, V.R., Patton, P.C., 1982. Paleohydrology of southwestern Texas. Water Resour. Res. 18, 1165–1183.

Kundzewicz, Z.W., Stoffel, M., Kaczka, R., Wyżga, B., Niedźwiedź, T., Pińskwar, I., Ruiz-Villanueva, V., Łupikasza, E., Czajka, B., Ballesteros Cánovas, J.A., Małarzewski, L., Choryński, A. Floods at the northern foothills of the Tatra Mountains – a Polish-Swiss research project. Acta Geophys. Pol. 62, 620–641.

Lang, M., Fernandez Bono, J.F., Recking, A., Naulet, R., Grau Gimeno, P., 2004. Methodological guide for paleoflood and historical peak discharge estimation. In: Benito, G., Thorndycraft, V.R. (Eds.), Systematic, Palaeoflood and Historical Data for the Improvement of Flood Risk Estimation: Methodological Guidelines. CSIC Madrid, Spain, pp. 43–53.

Lang, M., Ouarda, T.B.M.J., Bobée, B., 1999. Towards operational guidelines for over-threshold modeling. J. Hydrol. 225, 103–117.

Leese, M.N., 1973. Use of censored data in the estimation of gumbel distribution parameters for annual maximum flood series. Water Resour. Res. 9, 1534–1542.

Levish, D., Ostenaa, D., O'Connell, D., 1997. Paleoflood hydrology and dam safety. In: Mahoney, D.J. (Ed.), Waterpower 97: Proceedings of the International Conference on Hydropower. Am. Soc. of Civ. Eng. New York, pp. 2205–2214.

Levish, D.R., England Jr, J.F., Klawon, J.E., O'Connell, D.R.H., November 2003. Flood Hazard Analysis for Seminoe and Glendo Dams, Kendrick and North Platte Projects, Wyoming. Final Report. Bureau of Reclamation, Denver, CO, 126 pp. and two appendices.

Levish, D.R., Ostenaa, D.A., O'Connell, R.H., 1996. Paleo-hydrologic bounds and the frequency of extreme floods on the Santa Ynez river, California. In: California Weather Symposium. A Prehistoric Look at California Rainfall and Floods, pp. 1–19.

López, J., Francés, F., 2013. Non-stationary flood frequency analysis in continental Spanish rivers, using climate and reservoir indices as external covariates. Hydrol. Earth Syst. Sci. 17, 3189–3203.

Luterbacher, J., García-Herrera, R., Akcer-On, S., Allan, R., Alvarez-Castro, M.C., Benito, G., Booth, J., Büntgen, U., Cagatay, N., Colombaroli, D., Davis, B., Esper, J., Felis, T., Fleitmann, D., Frank, D., Gallego, D., Garcia-Bustamante, E., Glaser, R., González-Rouco, J.F., Goosse, H., Kiefer, T., Macklin, M.G., Manning, S.W., Montagna, P., Newman, L., Power, M.J., Rath, V., Ribera, P., Riemann, D., Roberts, N., Silenzi, S., Tinner, W., Valero-Garces, B., Van Der Schrier, G., Tzedakis, C., Vannière, B., Vogt, S., Wanner, H., Werner, J.P., Willett, G., Williams, M.H., Xoplaki, E., Zerefos, C.S., Zorita, E., 2012. A review of 2000 years of paleoclimatic evidence in the Mediterranean. In: Lionello, P. (Ed.), The Climate of the Mediterranean Region: From the Past to the Future, pp. 87–185.

Macklin, M.G., Rumsby, B.T., 2007. Changing climate and extreme floods in the British uplands. Trans. Inst. Br. Geogr. 32 (2), 168–186.

Macklin, M.G., Tooth, S., Brewer, P.A., Noble, P.L., Duller, G.A.T., 2010. Holocene flooding and river development in a Mediterranean steepland catchment: the Anapodaris Gorge, south central Crete, Greece. Global Planet. Change 70 (1–4), 35–52.

Macklin, M.G., Benito, G., Gregory, K.J., Johnstone, E., Lewin, J., Michczynska, D.J., Soja, R., Starkel, L., Thomdycraft, V.R., 2006. Past hydrological events reflected in the Holocene fluvial record of Europe. Catena 66, 145–154.

Magilligan, F.J., Goldstein, P.S., 2001. El Nino floods and culture change: a late Holocene flood history for the Rio Moquegua, southern Peru. Geology 29, 431–434.

Malik, I., Matyja, M., 2008. Bank erosion history of a mountain stream determined by means of anatomical changes in exposed tree roots over the last 100 years (Bila Opava River – Czech Republic). Geomorphology 98 (1–2), 126–142.

Marsh, J.E., Timoney, K.P., 2005. How long must northern saxicolous lichens be immersed to form a waterbody trimline? Wetlands 25 (2), 495–499.

Mayer, B., Stoffel, M., Bollschweiler, M., Hubl, J., Rudolf-Miklau, F., 2010. Frequency and spread of debris floods on fans: a dendrogeomorphic case study from a dolomite catchment in the Austrian Alps. Geomorphology 118 (1–2), 199–206.

McCord, V.A.S., 1996. Fluvial process dendrogeomorphology: reconstruction of flood events from the Southwestern United States using flood-scarred trees. In: Dean, J.S., Meko, D.M., Swetnam, T.W. (Eds.), Tree Rings, Environment and Humanity. International Conference on Tree Rings, Environment and Humanity – Relationship and Processes. TUCSON, Arizona, USA, pp. 689–699.

McEwen, L.J., Matthews, J.A., 2013. Sensitivity, persistence and resolution of the geomorphological record of valley-floor floods in an alpine glacier-fed catchment, Leirdalen, Jotunheimen, southern Norway. Holocene 23 (7), 977.

McKee, E.D., 1938. Original structures in Colorado river flood deposits of Grand Canyon. J. Sediment. Petrol. 8, 77–83.

Medialdea, A., Thomsen, K.J., Murray, A.S., Benito, G. Reliability of equivalent-dose determination and age-models in the OSL dating of historical and modern palaeoflood sediments. Quat. Geochronol. 22, 11–24.

Milly, P.C.D., Betancourt, J., Falkenmark, M., Hirsch, R.M., Kundzewicz, Z.W., Lettenmaier, D.P., Stouffer, R.J., 2008. Stationarity is dead: whither water management? Science 319, 573–574.

Morin, E., Grodek, T., Dahan, O., Benito, G., Kulls, C., Jacoby, Y., Van Langenhove, G., Seely, M., Enzel, Y., 2009. Flood routing and alluvial aquifer recharge along the ephemeral arid Kuiseb River, Namibia. J. Hydrol. 368, 262–275. http://dx.doi.org/10.1016/j.jhydrol.2009.02.015.

Murray, A.S., Wintle, A.G., 2000. Luminescence dating of quartz using an improved single-aliquot regenerative-dose protocol. Radiat. Meas. 32, 57–73.

Nathan, R.J., Weinmann, P.E., 1999. Estimation of Large to Extreme Floods: Book VI in Australian Rainfall and Runoff. A Guide to Flood Estimation. The Institution of Engineers, Australia.

National Research Council, 1985. Safety of Dams, Flood and Earthquake Criteria. National Academy Press, Washington, DC, 321pp.

Naulet, R., Lang, M., Ouarda, T.B.M.J., Coeur, D., Bobée, B., Recking, A., Moussay, D., 2005. Flood frequency analysis of the Ardèche River using French documentary sources from the last two centuries. J. Hydrol. 313, 58–78.

O'Connell, D.R.H., Ostenaa, D.A., Levish, D.R., Klinger, R.E., 2002. Bayesian flood frequency analysis with paleohydrologic bound data. Water Resour. Res. 38, 1058. http://dx.doi.org/10.1029/2000WR000028.

O'Connor, J.E., Webb, R.H., 1988. Hydraulic modeling for palaeoflood analysis. In: Baker, R.V., Kochel, R.C., Patton, P.C. (Eds.), Flood Geomorphology. John Wiley & Sons, New York, pp. 393–403.

O'Connor, J.E., Ely, L.L., Wohl, E.E., Stevens, L.E., Melis, T.S., Kale, V.S., Baker, V.R., 1994. A 4500-year record of large floods on the Colorado River in the Grand-Canyon, Arizona. J. Geol. 102, 1–9.

O'Connor, J.E., Curran, J.H., Beebee, R.A., Grant, G.E., Sarna-Wojcicki, A., 2003. Quaternary geology and geomorphology of the lower Deschutes River Canyon, Oregon. In: Grant, G.E., O'Connor, J.E. (Eds.), A Peculiar River: Geology, Geomorphology, and Hydrology of the Deschutes River, Oregon, Water Science and Applications, vol. 7. American Geophysical Union, pp. 77–98.

O'Connor, J.E., Webb, R.H., Baker, V.R., 1986. Paleohydrology of pool-and-riffle pattern development: Boulder Creek, Utah. Bull. Geol. Soc. Am. 97, 410–420.

Oguchi, T., Saito, K., Kadomura, H., Grossman, M., 2001. Fluvial geomorphology and paleohydrology of Japan. Geomorphology 39, 3–19.

Ortega, J.A., Garzón, G., 2009. A contribution to improved flood magnitude estimation in base of palaeoflood record and climatic implications – Guadiana River (Iberian Peninsula). Nat. Hazards Earth Syst. Sci. 9, 229–239.

Ostenaa, D., Levish, D., 1996. Reconnaissance Paleoflood Study for Ochoco Dam, Crooked River Project, Oregon: Seismotectonic. Report 96–2. U.S. Bureau of Rec-lamation, Seismotectonic and Geophysics Group, Denver.

Ostenaa, D., Levish, D., O'Connell, D., 1994. Paleoflood Study for Bradbury Dam. Reclamation Review Draft, Seismotectonic. Report 94–3. U.S. Bureau of Reclamation, Technical Service Center, Denver (Unpublished report).

Ostenaa, D.A., O'Connell, D.R.H., Walters, R.A., Creed, R.J., 2002. Holocene paleoflood hydrology of the big lost river, western Idaho national engineering and environmental laboratory, Idaho. In: Geology, Hydrogeology, and Environmental Remediation. Idaho National Engineering and Environmental Laboratory, Eastern Snake River Plain, Idaho, pp. 91–110.

Partridge, J., Baker, V.R., 1987. Paleoflood hydrology of the Salt River, Arizona. Earth Surf. Processes Landforms 12, 109–125.

Patton, P.C., Pickup, G., Price, D.M., 1993. Holocene paleofloods of the Ross River, central Australia. Quat. Res. 40, 201–212.

Pickup, G., 1991. Event frequency and landscape stability on the floodplain systems of arid Central Australia. Quat. Sci. Rev. 10, 463–473.

Pickup, G., Allan, G., Baker, V.R., 1988. History, palaeochannels, and palaeofloods of the Finke River, central Australia. In: Warner, R.F. (Ed.), Fluvial Geomorphology of Australia. Academic Press, Sydney, pp. 177–200.

Porat, N., 2006. Use of magnetic separation for purifying quartz for luminescence dating. Ancient TL 24, 33–36.

Redmond, K.T., Enzel, Y., House, P.K., Biondi, F., 2002. Climate variability and flood frequency at decadal to millennial time scales. In: House, P.K., Webb, R.H., Baker, V.R., Levish, D.R. (Eds.), Ancient Floods, Modern Hazards: Principles and Applications of Paleoflood Hydrology, Water Science and Application Series, vol. 5. American Geophysical Union, Washington, DC, pp. 21–45.

Reimer, P.J., Bard, E., Bayliss, A., Beck, J.W., Blackwell, P.G., Bronk Ramsey, C., Buck, C.E., Cheng, H., Edwards, R.L., Friedrich, M., Grootes, P.M., Guilderson, T.P., Haflidason, H., Hajdas, I., Hatté, C., Heaton, T.J., Hoffmann, D.L., Hogg, A.G., Hughen, K.A., Kaiser, K.F., Kromer, B., Manning, S.W., Niu, M., Reimer, R.W., Richards, D.A., Scott, E.M., Southon, J.R., Staff, R.A., Turney, C.S.M., van der Plicht, J., 2013. IntCal13 and Marine13 radiocarbon age calibration curves 0–50,000 years cal BP. Radiocarbon 55 (4), 1869–1887.

Reis Jr, D.S., Stedinger, J.R., 2005. Bayesian MCMC flood frequency analysis with historical information. J. Hydrol. 313, 97–116.

Rigby, R.A., Stasinopoulos, D.M., 2005. Generalized additive models for location, scale and shape. J. Roy. Stat. Soc. C 54, 507–554.

Robson, A.J., Jones, T.K., Reed, D.W., Bayliss, A.C., 1998. A study of national trend and variation in the UK floods. Int. J. Climatol. 18, 165–182.

Ruiz-Villanueva, V., Díez-Herrero, A., Bodoque, J.M., Ballesteros Cánovas, J.A., Stoffel, M., 2013. Characterization of flash floods in small ungauged mountains basins of Central Spain using an integrated approach. Catena 110, 32–43.

Ruiz-Villanueva, V., Díez-Herrero, A., Stoffel, M., Bollschweiler, M., Bodoque, J.M., Ballesteros Cánovas, J.A., 2010. Dendrogeomorphic analysis of flash floods in a small ungauged mountain catchment (Central Spain). Geomorphology 118, 383–392.

Saint-Laurent, D., Couture, C., McNeil, E., 2001. Spatio-temporal analysis of floods of the Saint-Francois drainage basin, Quebec, Canada. Environments 29, 74–90.

Sandercock, P., Wyrwoll, K.H., 2005. The historical and palaeoflood record of Katherine river, northern Australia: evaluating the likelihood of extreme discharge events in the context of the 1998 flood. Hydrol. Process. 19, 4107–4120.

Saynor, M.J., Erskine, W.D., 1993. Characteristics and implications of high-level slackwater deposits in the Fairlight Gorge, Nepean River, Australia. Aust. J. Mar. Freshwater Res. 44, 735–747.

Schwarz, F.K., Hughes, L.A., Hansen, E.M., Petersen, M.S., Kelly, D.B., 1975. The Black Hills-Rapid City Flood of June 9–10, 1972—A description of the storm and flood. U.S. Geol. Surv. Prof. Pap. 877, 47.

Scott, M.L., Auble, G.T., Friedman, J.M., 1997. Flood dependency of cottonwood establishment along the Missouri River, Montana, USA. Ecol. Appl. 7 (2), 677–690.

Sheffer, N.A., Enzel, Y., Benito, G., Grodek, T., Poart, N., Lang, M., Naulet, R., Coeur, D., 2003. Historical and palaeofloods of the Ardeche river, France. Water Resour. Res. 39. ESG 7–1/7-13.

Sheffer, N.A., Rico, M., Enzel, Y., Benito, G., Grodek, T., 2008. The palaeoflood record of the Gardon River, France: a comparison with the extreme 2002 flood event. Geomorphology 98, 71–83.

Sigafoos, R.S., 1964. Botanical evidence of floods and flood-plain deposition. U.S. Geol. Surv. Prof. Pap. 485-A, 1–35.

Silhan, K., 2012. Frequency of fast geomorphological processes in high-gradient streams: case study from the Moravskoslezsk, Beskydy Mts (Czech Republic) using dendrogeomorphic methods. Geochronometria 39 (2), 122–132.

Sloan, J., Miller, J.R., Lancaster, N., 2001. Response and recovery of the Eel River, California, and its tributaries to floods in 1955, 1964, and 1997. Geomorphology 36 (3–4), 129–154.

Smith, A.M., Zawada, P.K., 1990. Paleoflood hydrology – a tool for South Africa – an example from the Crocodile river near Brits, Transvaal, South Africa. Water SA 16, 195–200.

Smith, A.M., 1992. Paleoflood hydrology of the lower Umgeni River from a reach south of the Inanda Dam, Natal. S. Afr. Geogr. J. 74, 63–68.

Springer, G.S., Kite, S.J., 1997. River-derived slackwsater sediments in caves along Cheat River, West Virginia. Geomorphology 18, 91–100.

St. George, S., 2010. Tree rings as paleoflood and paleostage indicators. In: Stoffel, M., Bollschweiler, M., Butler, D.R., Luckman, B.H. (Eds.), Tree-Rings and Natural Hazards. A State-of The-Art. Springer, Dordrecht, Heidelberg, London, New York, pp. 233–239.

St. George, S., Nielsen, E., Conciatori, F., Tardif, J., 2002. Trends in Quercus macrocarpa vessel areas and their implications for tree-ring paleoflood studies. Tree-Ring Res. 58 (1/2), 3–10.

Stedinger, J.R., Cohn, T.A., 1986. Flood frequency analysis with historical and paleoflood information. Water Resour. Res. 22, 785–793.

Stoffel, M., Casteller, A., Luckman, B.H., Villalba, R., 2012. Spatiotemporal analysis of channel wall erosion in ephemeral torrents using tree roots – an example from the Patagonian Andes. Geology 40 (3), 247–250.

Stoffel, M., Gartner, H., Lievre, I., Monbaron, M., 2003. Comparison of reconstructed debris-flow event years (Ritigraben, Switzerland) and existing flooding data in neighboring rivers. In: Rickenmann, D., Chen, C. (Eds.), Debris-Flow Hazards Mitigation: Mechanics, Prediction, and Assessment, vols. 1 and 2, pp. 243–253. Davos, Switzerland.

Stoffel, M., Wilford, D.J., 2012. Hydrogeomorphic processes and vegetation: disturbance, process histories, dependencies and interactions. Earth Surf. Process. Landf. 37, 9–22.

Stuiver, M., 1982. A high-precision calibration of the AD radiocarbon time scale. Radiocarbon 24, 1–26.

Swain, R.E., England Jr., J.F., Bullard, K.L., Raff, D.A., 2006. Guidelines for Evaluating Hydrologic Hazards. Bureau of Reclamation, Denver, CO, 83 pp.

Thorndycraft, V., Benito, G., Rico, M., Sopeña, A., Sánchez-Moya, Y., Casas, M., 2005a. A long-term flood discharge record derived from slackwater flood deposits of the Llobregat River, NE Spain. J. Hydrol. 313, 16–31.

Thorndycraft, V.R., Benito, G., 2006. The Holocene fluvial chronology of Spain: evidence from a newly compiled radiocarbon database. Quat. Sci. Rev. 25, 223–234.

Thorndycraft, V.R., Barriendos, M., Benito, G., Rico, M.T., Casas, A., 2006. The catastrophic floods of A.D.1617 in Catalonia (NE Spain) and their climatic context. Hydrol. Sci. J. 51, 899–912.

Thorndycraft, V.R., Benito, G., Walling, D.E., Sopeña, A., Sánchez-Moya, Y., Rico, M., Casas, A., 2005b. Caesium-137 dating applied to slackwater flood deposits of the Llobregat River, NE Spain. Catena 59, 305–318.

United States Water Resources Council Hydrology Committee (USWRC), 1982. Guidelines for Determining Flood Frequency. Bull. 17B (revised). US Government Printing Office, Washington, DC.

Villarini, G., Serinaldi, F., Smith, J.A., Krajewski, W.F., 2009. On the stationarity of annual flood peaks in the Continental United States during the 20th Century. Water Resour. Res. 45, W08417. http://dx.doi.org/10.1029/2008WR007645.W08417.

Webb, R.H., Jarrett, R.D., 2002. One-dimensional estimation techniques for discharges of paleofloods and historical floods. In: House, P.K., Webb, R.H., Baker, V.R., Levish, D.R. (Eds.), Ancient Floods, Modern Hazards: Principles and Applications of Paleoflood Hydrology, Water Science and Application Series, vol. 5. American Geophysical Union, Washington, DC, pp. 111–125.

Webb, R.H., Blainey, J.B., Hyndman, D.W., 2002. Paleoflood hydrology of the Paria river, southern Utah and northern Arizona, USA. In: House, P.K., Webb, R.H., Baker, V.R., Levish, D.R. (Eds.), Ancient Floods, Modern Hazards: Principles and Applications of Paleoflood Hydrology, Water Science and Application Series, vol. 5. American Geophysical Union, Washington, DC, pp. 295–310.

Webb, R.H., O'Connor, J.E., Baker, V.R., 1988. Paleohydrologic reconstruction of flood frequency on the Escalante River, South-Central Utah. In: Baker, R.V., Kochel, R.C., Patton, P.C. (Eds.), Flood Geomorphology. John Wiley & Sons, New York, pp. 403–418.

Wells, L.E., 1990. Holocene history of the El Nino phenomenon as recorded in flood sediments in northern Peru. Geology 18, 1134–1137.

Williams, G.P., Costa, J.E., 1988. Geomorphic measurements after a flood. In: Baker, R.V., Kochel, R.C., Patton, P.C. (Eds.), Flood Geomorphology. John Wiley & Sons, New York, pp. 65–77.

Wintle, A.G., Murray, A.S., 2006. A review of quartz optically stimulated luminescence characteristics and their relevance in single-aliquot regeneration dating protocols. Radiat. Meas. 41, 369–391.

Wohl, E.E., Fuertsch, S.J., Baker, V.R., 1994. Sedimentary records of Late Holocene floods along the Fitzroy and Margaret rivers, Western-Australia. Aust. J. Earth Sci. 41, 273–280.

Wolman, M.G., Costa, J.E., 1984. Envelope curves for extreme flood events: discussion. J. Hydraul. Eng. 110, 77–78.

Woodward, J.C., Hamlin, R.H.B., Macklin, M.G., Karkanas, P., Kotjabopoulou, E., 2001. Quantitative sourcing of slackwater deposits at Boila rockshelter: a record of lateglacial flooding and paleolithic settlement in the Pindus Mountains, Northwest Greece. Geoarchaeology 16, 501–536.

Woodward, J.C., Tooth, S., Brewer, P.A., Macklin, M.G., 2010. The 4th international palaeoflood workshop and trends in palaeoflood science. Global Planet. Change 70, 1–4.

Yang, D.Y., Yu, G., Xie, Y.B., Zhan, D.J., Li, Z.J., 2000. Sedimentary records of large Holocene floods from the middle reaches of the Yellow River, China. Geomorphology 33, 73–88.

Yanosky, T.M., 1982a. Effects of flooding upon woody vegetation along parts of the Potomac River floodplain. U.S. Geol. Surv. Prof. Pap. 1206, 1–21.

Yanosky, T.M., 1982b. Hydrologic inferences from ring widths of flood-damaged trees, Potomac River, Maryland. Environ. Geol. Water Sci. 4, 43–52.

Yanosky, T.M., 1983. Evidence of floods on the Potomac River from anatomical abnormalities in the wood of flood-plain trees. U.S. Geol. Surv. Prof. Pap. 1296, 1–51.

Yanosky, T.M., 1984. Documentation of high summer flows on the Potomac River from the wood anatomy of ash trees. Water Resour. Bull. 20, 241–250.

Yanosky, T.M., Jarrett, R.D., 2002. Dendrochronologic evidence for the frequency and magnitud of paleofloods. In: House, P.K., Webb, R.H., Baker, V.R., Levish, D.R. (Eds.), Ancient Floods, Modern Hazards: Principles and Applications of Paleoflood Hydrology, Water Science and Application Series, vol. 5. American Geophysical Union, Washington, DC, pp. 77–89.

Yue, S., Ouarda, T.B.M.J., Bobee, B., Legendre, P., Bruneau, P., 2002. Approach for describing statistical properties of flood hydrograph. J. Hydrol. Eng. ASCE 7, 147–153.

Zawada, P., Hattingh, J., 1994. Studies on the paleoflood hydrology of South-African rivers. S. Afr. J. Sci. 90, 567–568.

Zawada, P.K., 2000. Palaeoflood Hydrological Analysis of Selected South African Rivers. Memoir of the Council for Geoscience, South Africa. In: Memoir, vol. 87, 173.

Zha, X., Huang, C., Pang, J., 2009. Palaeofloods recorded by slackwater deposits on the Qishuihe River in the Middle Yellow River. J. Geogr. Sci. 19, 681–690.

Zha, X., Huang, C., Pang, J., Li, Y., 2012. Sedimentary and hydrological studies of the Holocene palaeofloods in the middle reaches of the Jinghe River. J. Geogr. Sci. 22, 470–478.

Zhang, Y., Huang, C.C., Pang, J., Zha, X., Zhou, Y., Gu, H., 2013. Holocene paleofloods related to climatic events in the upper reaches of the Hanjiang River valley, middle Yangtze River basin, China. Geomorphology 195, 1–12.

Zielonka, T., Holeksa, J., Ciapala, S., 2008. A reconstruction of flood events using scarred trees in the Tatra Mountains, Poland. Dendrochronologia 26, 173–183.

Chapter 4

Global and Low-Cost Topographic Data to Support Flood Studies

Kun Yan [1], **Jeffrey C. Neal** [2], **Dimitri P. Solomatine** [1,3] **and Giuliano Di Baldassarre** [1,4]

[1] *UNESCO-IHE, Institute for Water Education, Delft, the Netherlands,* [2] *School of Geographical Sciences, University of Bristol, Bristol, UK,* [3] *Water Resources Section, Delft University of Technology, Delft, the Netherlands,* [4] *Department of Earth Sciences, Uppsala University, Uppsala, Sweden*

ABSTRACT

This chapter provides an overview of global and low-cost topographic data to support flood studies, with a focus on usefulness of shuttle radar topography mission (SRTM) topography in supporting two-dimensional hydraulic modeling of floods. In particular, flood propagation and inundation modeling of a 10-km reach of the River Dee (United Kingdom) was performed by using LISFLOOD-FP to simulate the December 2006 flood event. Flood extent maps from satellite imagery (ERS-2 Synthetic Aperture Radar SAR) and hydrometric information (downstream water levels) were used as evaluation data. Uncertainty analysis was carried out within the generalized likelihood uncertainty estimation framework using the roughness coefficients and downstream water surface slope as free parameters. The results of this study showed: (1) the potentials and limitations of SRTM topographic data in flood inundation modeling; (2) the value of downstream water levels in constraining uncertainty in hydraulic model of floods; (3) the impact of setting a water surface slope as downstream boundary on the results of the hydraulic model (e.g., predictions of water stages and flood extent).

4.1 INTRODUCTION

4.1.1 Growing Availability of Global Earth Observation Data

The recent catastrophic flood events (e.g., Central Europe, June 2013) encouraged more efforts in flood risk-prevention measures to reduce human losses and economic damages. To this end, modeling and mapping flood-inundation processes using hydraulic modeling techniques has become an

Hydro-Meteorological Hazards, Risks, and Disasters. http://dx.doi.org/10.1016/B978-0-12-394846-5.00004-7

essential component (de Moel et al., 2009; Van Alphen et al., 2009). The growing availability of distributed remote sensing data has provided a great potential in building and testing flood inundation models in recent years (Bates, 2012; Di Baldassarre and Uhlenbrook, 2012). In addition to the high resolution digital elevation models (DEMs), which are highly precise but costly, global low cost products also provide topographic data, such as the DEM derived by the shuttle radar topography mission (SRTM). These topographic data may potentially offer new opportunities to implement flood-inundation modeling in data scarce/poor areas. However, SRTM suffers from random noises and radar speckles due to the fact that it utilizes radar-based interferometry technology, which involves the reception of a back-scattered radar signal by two antennae. Speckle adds waviness with amplitudes of ~1.0 m to SRTM (Falorni et al., 2005). Additionally, vertical accuracy is degraded by its space-borne altitude and its inability to penetrate water surface and dense vegetation (Falorni et al., 2005). Hence, the low accuracy of SRTM (Rabus et al., 2003; Rodríguez et al., 2006) together with all its drawbacks listed above seem that utilizing SRTM data in flood modeling is rather challenging. However, SRTM is characterized by errors in flat areas lower than the errors occurring in high slope areas (Rodríguez et al., 2006). This feature of the SRTM is beneficial for the potential use of this topographic data to support large-scale inundation modeling as floodplains are usually flat and with a mild slope. In addition, floodplain flow above small-scale topography features, which are usually misrepresented by the SRTM, do not play a dominate role in large-scale flood inundation processes (Bates, 2012).

4.1.2 Recent Progress on Evaluation of Global Topographic Data in Supporting Flood Modeling

The recent scientific efforts on exploring the potential usefulness of SRTM data in supporting floodplain monitoring at large scale are encouraging. For example, LeFavour and Alsdorf (2005) derived the water-surface slope of Amazon based on SRTM topography and found that accurate main stem discharge values can be estimated with this water-surface slope in this biggest river of the world. Schumann et al. (2010) compared the water surface gradient generated by intersecting SAR image and SRTM DEM to that derived from intersecting SAR image with a high resolution and quality Light Detection And Ranging LiDAR DEM on River Po. They found that there the two estimates are remarkably close to each other. Sanders (2007) evaluated diverse public DEMs (including interferometric synthetic aperture radar (IfSAR) and SRTM) for flood inundation modeling, and found that airborne IfSAR was not appropriate for flood simulation, while SRTM topography led to a 25 percent larger flood zone when compared with the high-resolution topography in a steady-flow Santa Clara River application. Schumann et al. (2012) calibrated the hydrodynamic model by using highly accurate water levels on the main channel from the ICESat (Ice, Cloud, and land Elevation Satellite) laser altimeter and validated using multiple satellite acquisitions of the flood area in the forecasting for the Lower Zambezi River in southeast Africa. Results showed that satisfactory parameter values and performance, as well as acceptable prediction skills can be achieved at a very large scale and using coarse grid resolutions. In a recent study (Yan et al., 2013), the design flood profiles derived from hydraulic models based on high resolution and accuracy topography and bathymetry (LiDAR) and hydraulic models based on SRTM data were compared considering all the other major sources of uncertainty that unavoidably affect any modeling exercise. It was found out that the differences between the high resolution topography-based model and the SRTM-based model were not negligible, but within the accuracy that is typically associated with large-scale flood studies. However, the flood event considered in Yan et al. (2013) was confined by the lateral embankments of the River Po, and therefore a one-dimensional (1D) hydraulic model was used in that study. Moreover, studies at different scales mentioned above yield quite different conclusions. Hence, the value of SRTM topography in supporting two-dimensional (2D) flood inundation modeling remains largely unexplored, particularly for medium-small sized (with width smaller than 100 m) rivers.

4.1.3 Uncertainties in Inundation Modeling and Probabilistic Flood Mapping

Many studies have described that there are several sources of uncertainties intrinsic to flood inundation modeling, such as model structure, topographic data, model parameter, and inflow etc. (e.g., Aronica et al., 2002; Pappenberger et al., 2006; Di Baldassarre and Montanari, 2009). Among those, topography uncertainty is considered to be one of the major sources of uncertainty. The flood inundation maps are characteristically produced by hydraulic models using deterministic or probabilistic approaches. The deterministic flood maps which are produced by using a fully 2D physically based best-fit model, are precise, but potentially wrong, due to the fact that they ignore the above-mentioned uncertainties in inundation modeling. The probabilistic flood maps that explicitly consider various sources of uncertainties are believed to be theoretically more appropriate for visualizing flood hazard (Di Baldassarre et al., 2010), even though their application in flood risk studies is still limited.

4.1.4 Different Type of Data in Constraining Uncertainty in Flood Modeling

Recent advances in airborne and satellite remote sensing allow the parameterization, calibration, and validation of flood inundation models in a distributed manner (Bates, 2004). Hydraulic models are usually tested on flood extent data (e.g., Matgen et al., 2007; Pappenberger et al., 2007; Neal et al., 2013) rather than water level or flow data at particular points as the models may not perform well at the locations away from the gauged points (Bates et al., 2004). As pointed out by Pappenberger et al. (2007), data used to constrain model parameter uncertainty should be consistent with the modeling purpose. For example, models are better to be conditioned on flood extent data if the goal is predicting flood-prone areas, whereas high water marks are preferable if the purpose is estimating design flood profiles (Brandimarte and Di Baldassarre, 2012).

Yet, the use of flood extent data can sometimes be difficult to distinguish between different model parameterization when the flood extent is not sensitive to changes in water level. In addition, flood extents from satellite flood images are usually difficult to obtain. As a matter of fact, the overpass frequency of the satellites which provides high resolution flood imagery is usually low (e.g., 35 days of repeat cycle for ERS2-SAR, Schumann et al., 2010) even though there are few products with low revisit time recently available (e.g., COSMO-SkyMed offers 12 and 24 h revisit time, García-Pintado et al., 2013). This implies that finding a satellite image at the time of flooding may be difficult as flood duration time in small-medium catchment is usually shorter than the revisit time of satellite data (Hunter et al., 2007; Schumann et al., 2010). Hydrometric data such as water stages are relatively easier to find. They have a high temporal frequency, but are unavoidably sparse in space (Di Baldassarre et al., 2011).

A few scientists have explored the use of different types of data sets to constrain uncertainty in inundation models. For example, Horritt and Bates (2002) tested three hydraulic codes on a 60-km reach of the River Severn, UK using independent hydrometric and satellite data for model calibration. They found all models are capable of reproducing inundation extent and flood-wave travel time to the similar level of accuracy at optimal simulation. However, the predictions of inundation extent are in some cases poor when hydrometric data are used for model calibration. Hunter et al. (2005) calibrated an inundation model against flood images, downstream stage, and discharge hydrographs on a 35-km reach of the River Meuse, the Netherlands. They found that the evaluation of internal predictions of stage also offer considerable potential for reducing uncertainty over effective parameter specification.

4.1.5 The Dilemma of Downstream Water Level in Hydraulic Modeling

In hydraulic modeling, the normal depth (calculated from the water surface slope) is often used as downstream boundary condition. The water surface slope is normally unknown and is often estimated as the average bed slope under the assumption of a Manning's type relationship between water stage and discharge at the downstream end of the river reach. The results of flood inundation models (e.g., water levels, inundation extent) are affected by this assumption, especially when backwater effects are significant. Samuels (1989) proved the practical use of Eqn (4.1) to calculate the backwater length, L, for engineering applications:

$$L = \frac{0.7D}{s_0} \tag{4.1}$$

where D is the bankfull depth of the channel and S_0 is the bed slope. Only a few studies (e.g., Wang et al., 2005; Schumann et al., 2008) have investigated the impact of assuming a certain water surface slope as downstream boundary conditions on the results of 2D hydraulic models, such as inundation extent and water stage. Those impacts can be substantially reduced by extending the model domain and placing the downstream water level sufficiently far away from the points of interest. However, this is not always possible.

In this context, the aim of this chapter is twofold: (1) explore the potentials and limitations of SRTM data in supporting the 2D hydraulic modeling of floods; (2) examine the sensitivity of 2D hydraulic models on the water surface slope used as downstream boundary as well as the associated value of downstream water levels in constraining uncertainty of flood extent prediction.

4.2 TEST SITE AND DATA AVAILABILITY

The study is carried out on a river system including: (1) the 10-km reach of the River Dee, between Farndon and Iron Bridge, two gauging stations of the Environment Agency of England and Wales (hereafter called the EA); and (2) the 8-km reach of the River Alyn, between the EA gauging station of Pont-y-Capel and the confluence to the River Dee (Figure 4.1). A high resolution (2 m) LiDAR DEM of this test site is derived by the EA. Surface artefacts such as vegetation and buildings are removed from the raw LiDAR data. The EA also conducts a channel bathymetry ground survey of 36 cross-sections that are incorporated with the LiDAR data on floodplain. Hereafter, this hybrid high resolution DEM is called LiDAR DEM.

Another DEM of the test site is derived from the SRTM data postprocessed by the Consortium for Spatial Information of the Consultative Group for International Agricultural Research, e.g., fills in the no-data holes in the raw SRTM data (Jarvis et al., 2008). The SRTM DEM of the study area is reprojected into 75 m resolution with no speckles and surface artefacts removed. The two DEMs are strongly different, not only in terms of resolution (2 m versus 75 m), but also in terms of accuracy: the vertical accuracy of LiDAR data was of around 10 cm, whereas that of SRTM in Europe was found around 6 m (Rodríguez et al., 2006).

In December 2006, the River Dee underwent a low magnitude flood event (with the return period about 2 years). In this period, a high-resolution satellite image (ERS-2 SAR, see in Figure 4.1) was acquired. The ERS-2 SAR image is characterized by a pixel size of 12.5 m and a ground resolution of

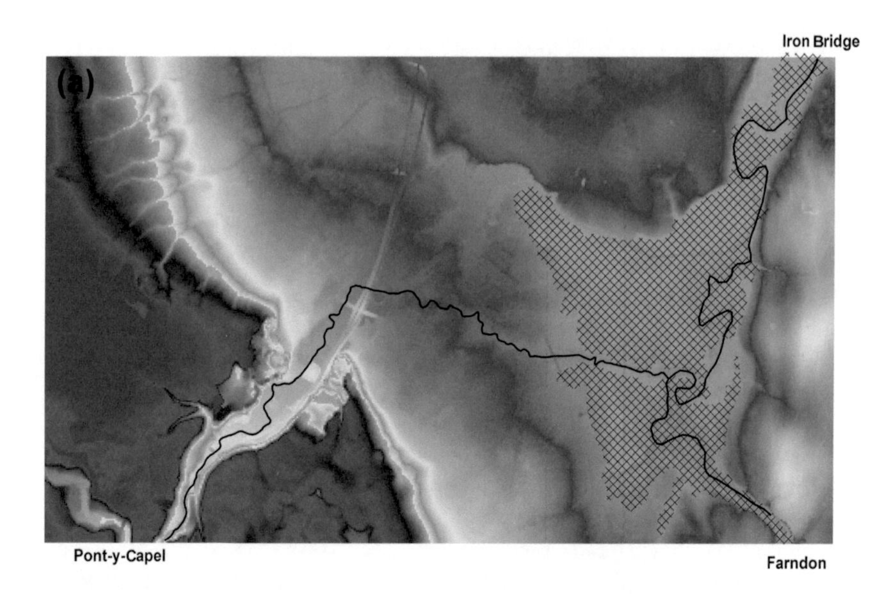

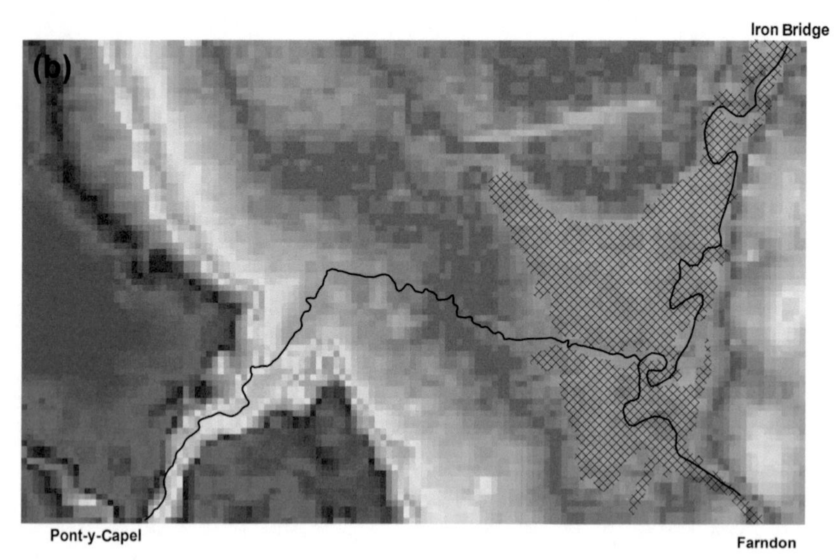

FIGURE 4.1 River Dee between Farndon and Iron Bridge and River Alyn from Pont-y-Capel (black lines); flood extent of 2006 event from ERS2-SAR flood image (crosshatch); (a) LiDAR digital elevation model (DEM) (grey scale); (b) shuttle radar topography mission DEM (grey scale).

approximately 25 m. The satellite image was processed by using visual interpretation procedure to derive a flood extent map (Schumann et al., 2009; Di Baldassarre et al., 2010). We reproject this flood extent map into the 20 and 75 m resolution for evaluating the LiDAR-based and SRTM-based models (see below).

4.3 INUNDATION MODELING

The LISFLOOD-FP (Bates and De Roo, 2000) raster-based hydraulic model is used to simulate the flood event in 2006. The main channel widths of River Dee and Alyn are on average 30 m and 12 m, respectively, which are much smaller than SRTM DEM cell resolution (i.e., 75 m). Therefore, the subgrid approach of LISFLOOD-FP (Neal et al., 2012), which can represent 1D channels with widths below the grid resolution, is applied for the SRTM-based model. The subgrid approach allows the modeler to specify channel width, channel depth as well as the bank elevation inside each cell so that it can better emulate flood propagation in the main channel for coarse resolution models. It was proved to change the floodplain inundation dynamics significantly and increase simulation accuracy in terms of water levels, wave propagation speed, and inundation extent compare with the pure 1D channel model or 2D floodplain model of LISFLOOD-FP (Neal et al., 2012). The subgrid approach uses the floodplain flow model of Bates et al. (2010), which introduced the local inertial term to the diffusive wave equation to significantly reduce the computation cost. However, computational cost can still be high for fine resolution (e.g., 1−10 m) grids. Therefore, the 2 m LiDAR DEM is aggregated into 20 m resolution to reduce the model computational time. The key topographic features such as embankments are manually identified in the aggregated DEM.

The channel-bed elevation of SRTM topography is found to be overall overestimated in the study area. This is due to the fact that radar wave cannot penetrate water surface to detect the channel-bed elevation and the channel is typically smaller than an SRTM pixel. Therefore, we improve the SRTM channel-bed elevation by using the boat survey data. However, the combination of boat-surveyed, channel-bed elevation and overestimated SRTM floodplain topography results in a very deep channel depth (around 8−10 m). As one of the main purposes of inundation modeling is to predict the flood extent correctly, we use the surveyed channel depth (bank elevation subtract bed elevation) to replace the SRTM channel depth rather than directly replacing SRTM bed elevation by the surveyed bed elevation.

Two hydraulic models (LiDAR and SRTM-based) are built to simulate the 2006-year flood event. The observed discharge hydrograph starting on December 6, 2006 at 11:00 h (around 144 h before the satellite overpass) is used as upstream boundary condition. A normal depth with the water surface slope (estimated as the average bed slope) is applied as the downstream boundary condition.

In this study, two types of observations are available for the model evaluation: (1) spatially distributed binary flood extent; (2) at-a-point time series of flood water stages (Iron Bridge). The simulated inundation areas are compared to the observed flood extent map (derived from the ERS2-SAR satellite imagery, Figure 4.1) using the performance measure, F (Aronica et al., 2002; Horritt et al., 2007):

$$F = \frac{A}{A + B + C} \tag{4.2}$$

where A is the number of cells correctly predicted by the model, B is the number of cells predicted as wet that is observed dry (overprediction), C is the number of cells predicted as dry that is observed wet (underprediction). F ranges from 0 to 1, the higher the better. As assumed in previous studies (e.g., Aronica et al., 2002; Pappenberger et al., 2007), only the cells with a simulated inundation depth greater than 20 cm are considered as flooded.

The evaluation of the simulated downstream water levels is conducted by using the observed time series of flood water levels at the downstream end of the river reach. The calibration focuses on the peak hours of the water stage hydrograph, starting on December 7, 2006 at 4:00 am and ending 127 h later, which is also the time of the satellite overpass. The root mean square error (RMSE) is used to evaluate model errors for both LiDAR and SRTM-based model.

4.4 THE EFFECT OF TOPOGRAPHY RESOLUTION ON INUNDATION MODELING

To better distinguish between the impact of the resolution and the accuracy of topographical input data, we first conduct a numerical experiment to isolate the resolution effect: the LiDAR DEM is aggregated into 80-m resolution DEM, which is similar to the SRTM resolution (i.e., 75 m). Then the subgrid LiDAR-based model (80 m of resolution) in which the channel has the same width, friction, and bed elevation to the LiDAR-based model (20 m of resolution) is built. The SRTM-based model (75 m of resolution) is used for comparison. The other model parameters among three models are identical.

The flood extents simulated by three models are compared to the flood extent derived from the ERS2-SAR image. The value of performance measure, F, is shown in Table 4.1. The coarse resolution LiDAR-based model performs slightly worse than the high resolution one (with 0.01 difference in terms of F), whereas the performance is much higher than the SRTM-based model (with 0.271 difference in terms of F). This shows that coarse resolution LiDAR-based model can simulate the flood extent equally well as the high resolution LiDAR-based model. The coarse resolution does not degrade the model performance whereas the vertical accuracy of floodplain cells might play an important role. Thus, we focus on the effect of DEM vertical accuracy on flood extent and downstream water level predictions in the following experiments.

TABLE 4.1 Comparison of Three Floodplain Models

	LiDAR-based model (80 m)	LiDAR-based model (20 m)	SRTM-based model (75 m)
F	0.781	0.791	0.510

SRTM, shuttle radar topography mission.

4.5 UNCERTAINTY ANALYSIS WITHIN A GENERALIZED LIKELIHOOD UNCERTAINTY ESTIMATION FRAMEWORK

To investigate the usefulness of SRTM data to support hydraulic modeling, the effects of topography uncertainty are evaluated within the generalized likelihood uncertainty estimation (generalized likelihood uncertainty estimation (GLUE), Beven and Binley, 1992) framework. GLUE is a simple and pragmatic methodology, which uses Monte Carlo simulations to produce parameter distributions and uncertainty bounds conditioned on available data. GLUE has been widely used in environmental modeling (e.g., Aronica et al., 1998; Romanowicz and Beven, 1998; Beven and Freer, 2001). It is worth noting that a number of authors (Montanari, 2005; Mantovan and Todini, 2006; Stedinger et al., 2008) showed that the GLUE methodology does not formally follow the Bayesian approach in estimating the posterior probabilities of parameters and of the output distribution. Also, a number of subjective decisions have to be made in GLUE; e.g., the priori distribution and feasible range of each parameter (generalized) likelihood function for model evaluation, the threshold between behavioral and nonbehavioral simulations. It is therefore necessary to clarify each decision to be transparent and unambiguous.

The assumed ranges of parameters can have an influence on resulting uncertainties (Aronica et al., 1998). Thus, large parameter ranges, which cover the extreme feasible values, were used to overcome the potential issue of subjective choice (e.g., Aronica et al., 1998). Therefore, we keep the roughness parameter range sufficiently large. Both Manning's channel and floodplain roughness coefficients are sampled randomly from the uniform distribution between 0.015 and 0.150 $m^{1/3}s^{-1}$ due to the lack of information regarding to the priori distribution as well as the feasible range of parameters. A similar parameter range was also used by Stephens et al. (2014) for the same study area. The range of average bed slopes is calculated from the two topographic data sets: the upper bound of the bed slope is calculated based on the reach from Fardon to Iron Bridge, whereas the lower bound is calculated based on the reach from the confluence to Iron Bridge for both LiDAR and SRTM-based model (Table 4.2).

The objective function (i.e., F and RMSE) values for each parameter set (i.e., Manning's coefficients in channel (n_{ch}), Manning's coefficients on floodplain (n_{fp}), and water surface slope (Sl)) are calculated to derive the generalized likelihoods, which are positive values with the summation of 1 (Wagener et al., 2001). The likelihood measure can be used to weight each model realization. The behavioral models can be selected by rejecting the simulations that underperform a user-defined threshold or a percentage of simulations. In this study, the best 10 percent of the realizations are assumed as behavioral models and then used to produce probabilistic inundation maps. Given the simulation results for the jth computational cell of $w_{ij} = 1$ for wet and $w_{ij} = 0$ for dry, the probabilistic inundation map is produced using Eqn (4.3):

$$M_j = \frac{\sum_i L_i w_{ij}}{\sum_i L_i} \tag{4.3}$$

where M_j indicates a weighted average flood state for the jth cell, L_i is the likelihood weight assigned for each simulation i. The posterior parameter distributions (PPD) of both two models are plotted as well.

4.6 RESULTS AND DISCUSSION

We conducted 1,000 simulations for LiDAR and SRTM-based models within the GLUE framework (see above). Two performance measures are evaluated according to two types of observations. The dotty plots are generated to show the parameter uncertainty given alternative performance measures and data sets for both LiDAR and SRTM-based model.

Figure 4.2(a) and (b) shows that the performance measure of LiDAR-based model increases as the n_{ch} and n_{fp} increase when conditioned on flood extent data for the small n_{ch} and n_{fp} values (below 0.05). The performance measure begins to remain unchanged for the larger n_{ch} and n_{fp} values. It is found out that there is a tendency to generally underestimate the flood extent for smaller Manning's coefficients. The simulated inundation extent is, as expected, increasing when n_{ch} and n_{fp} are increasing.

TABLE 4.2 Bed Slope Calculated from Two Topographic Data

	Farndon to Iron Bridge	Confluence to Iron Bridge
LiDAR-based model	0.0002	0.00002
SRTM-based model	0.0012	0.0002

SRTM, shuttle radar topography mission.

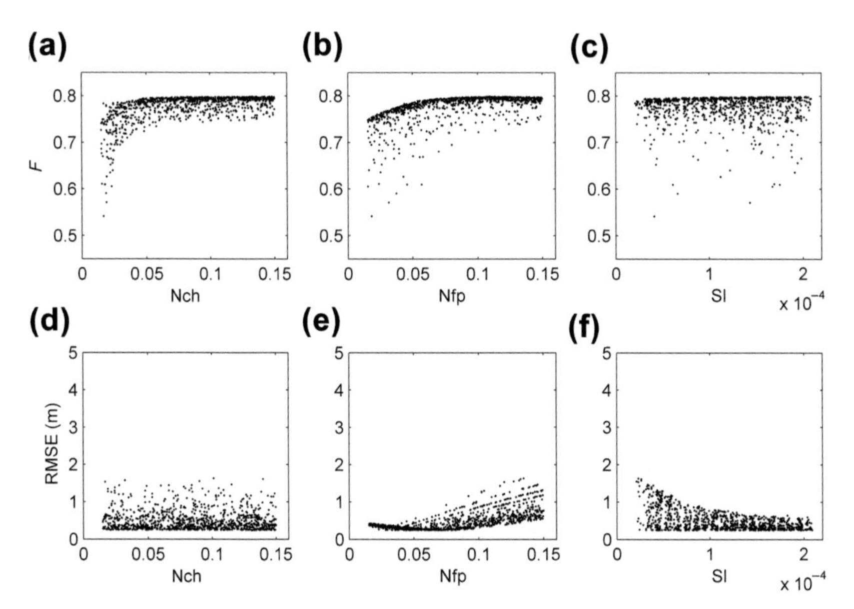

FIGURE 4.2 (a)–(c) Performance measure F for LiDAR-based model conditioned on ERS2-SAR flood image. (d)–(f) Performance measure root mean square error (RMSE) for LiDAR-based model conditioned on downstream water stage.

It is difficult to observe the optimal Sl when conditioned on flood-extent data as good simulations occur across the whole range of parameter values (Figure 4.2(c)). The flood extent is affected by the back-water effect and therefore related to the downstream water slope Sl. However, the influence of backwater to flood extent in this case is limited as the floodplain acts as a "valley-filling" case, whereby, once the valley is filled by flood water, increases of water depth do not lead to significant differences in flood extent (Hunter et al., 2005).

The sensitivity to parameters n_{fp} and Sl is assessed also by conditioning the model on downstream water levels (Figure 4.2(e) and (f)). The RMSE is increasing when n_{fp} is increasing with the optimal value around 0.05. The RMSE is decreasing when Sl is increasing (Figure 4.2(f)). The effects of the two parameters are compensated with each other (e.g., when Sl is increasing, one can keep almost the same water level but increase n_{fp}). The sensitivity of Sl is clearly visible as conditioned on water stage information, as expected. The predicted downstream water levels are strongly affected by the assumed water-surface slope.

In Figure 4.3(c), the average performance measure remains stationary with the change of Sl when the SRTM-based model is conditioned on flood-extent data. A clearly decreasing performance as n_{fp} is increasing (Figure 4.3(b)). A similar trend also occurs for n_{ch} (Figure 4.3(a)). The SRTM-based model essentially overestimates the flood extent (with very few underprediction cells),

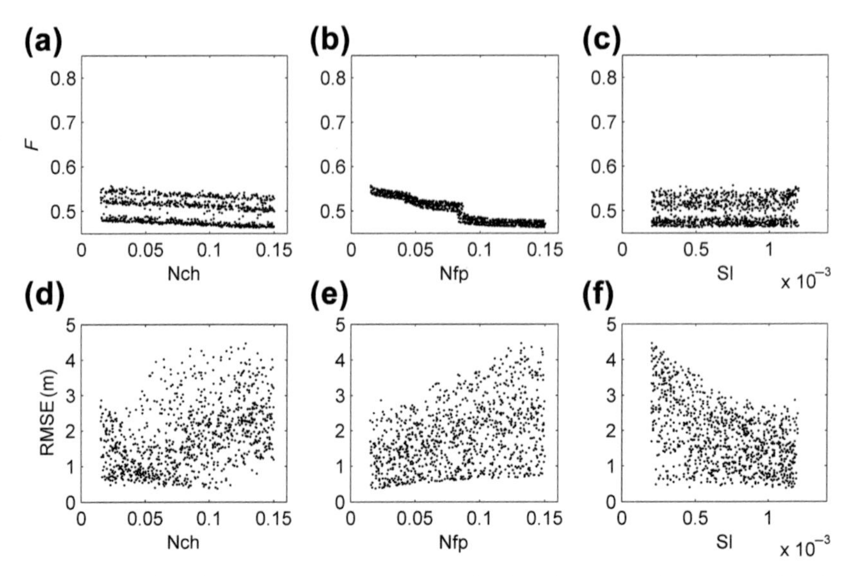

FIGURE 4.3 (a)–(c) Performance measure F for shuttle radar topography mission (SRTM)-based model conditioned on ERS2-SAR flood image. (d)–(f) Performance measure root mean square error (RMSE) for SRTM-based model conditioned on downstream water stage.

even with the optimal parameter sets. This is shown by the fact that the performance measure (F) keeps dropping while n_{ch} and n_{fp} are increasing due to the fact that the simulated inundation extent keeps increasing, which results in more overprediction.

The sensitivity of SRTM-based model conditioned on downstream water stage (Figure 4.3(d)–(f)) is overall larger than LiDAR-based model. The realizations with high performance (low RMSE) are more concentrated in an area with the small values of n_{ch} and n_{fp} rather than the high ones. Similarly to the LiDAR-based model, the SRTM-based model performs better with larger *Sl* values (Figures 4.2(f) and 4.3(f)).

Figure 4.4(b) and (e) shows that, when conditioned on downstream water stages, the best realizations are obtained with small Manning's floodplain roughness values (around 0.05), whereas high-performance realizations are found for higher Manning's floodplain values, when the model is conditioned on flood extent. The PPD for water surface slope, when conditioned on flood extent, and Manning's channel coefficients, when conditioned on water stage, are found nearly uniformly distributed for the LiDAR-based model (Figure 4.4(c) and (d)). If all the realizations are taken as behavioral, we might conclude that the simulations conditioned on hydrometric data (i.e., water-stage time series) may not predict the flood extent properly for LiDAR-based model (Figure 4.4(a)–(f)). However, the performances are very different after the rejection of nonbehavioral simulations. Figure 4.5 (upper

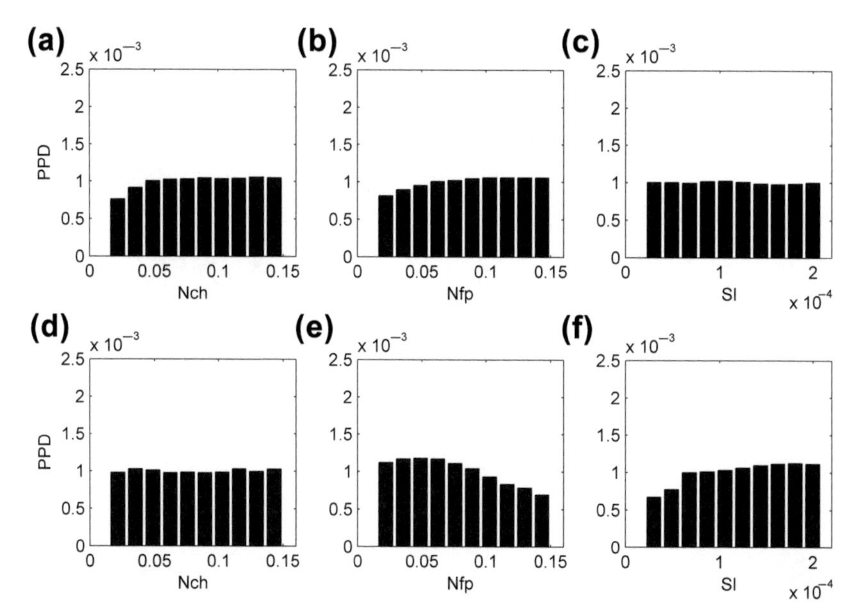

FIGURE 4.4 (a)−(c) Posterior parameter distribution (PPD) for LiDAR-based model conditioned on ERS2-SAR flood image. (d)−(f) PPD for LiDAR-based model conditioned on downstream water stage.

panel) shows the flood extent predicted by the best 10 percent simulations, conditioned on downstream water stages of which the average performance measure (*F*) is 0.771, given the best *F* among all 1,000 simulations is 0.799 (see Table 4.3).

In Figure 4.6, the PPD shows the performance of parameters for SRTM-based model is rather similar between the two performance measures (i.e., *F* and RMSE). This indicates that it might get relatively satisfactory predictions of flood extent when the SRTM-based model is conditioned on water stage data. It also shows SRTM-based model might be more flexible in conditioning on different data sets than LiDAR-based model. The model performances after the rejection of nonbehavioral simulations are also shown in Figure 4.5 (lower panel). The probabilistic inundation map of the best 10 percent simulations conditioned on water stage, of which the average performance measure (*F*) is 0.524, compare with the best *F* of 0.557 among all 1,000 simulations (Table 4.3).

Figures 4.2 and 4.3 show that the LiDAR-based model performs better than the SRTM-based model in predicting flood extent as well as the downstream water stage. This shows how the performance of hydraulic models can be affected by topographic errors. The prediction of downstream water stages shows a mean RMSE of 1.853 m for SRTM-based model and 0.504 m for the LiDAR-based model. Considering the predicted water stage is obviously

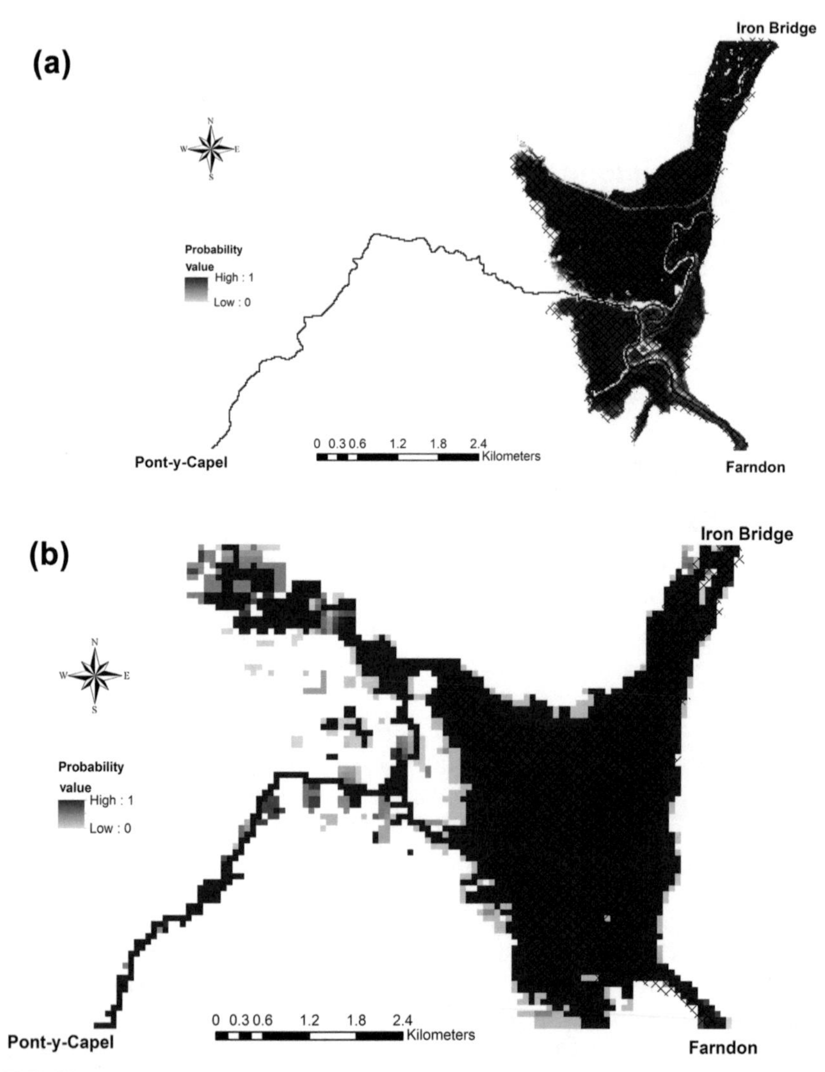

FIGURE 4.5 ERS2-SAR flood imagery (crosshatch) and probabilistic inundation map of 2006 event (from black, 1, to white, 0) with behavior simulations conditioned on downstream water stage: (a) LiDAR-based model; (b) shuttle radar topography mission-based model.

affected by the channel-bed elevation, the poor performance of the SRTM-based model is expected. On the other hand, the mean F of the best 10 percent realizations conditioned on ERS2-SAR is 0.543 for SRTM-based model and 0.797 for LiDAR-based model (Table 4.3). Despite this large difference, getting a performance above 50 percent in simulating flood extent is a reasonably good result for using SRTM topography to support the hydraulic modeling of a small-medium sized river.

TABLE 4.3 Performance Measure *F* of Two Models

	Performance Measure *F*		
	Average of Best 10% Simulations Conditioned on Water Stage	Average of Best 10% Simulations Conditioned on ERS2-SAR	Best of 1,000 Simulations
LiDAR-based model	0.771	0.797	0.799
SRTM-based model	0.524	0.543	0.557

SRTM, shuttle radar topography mission.

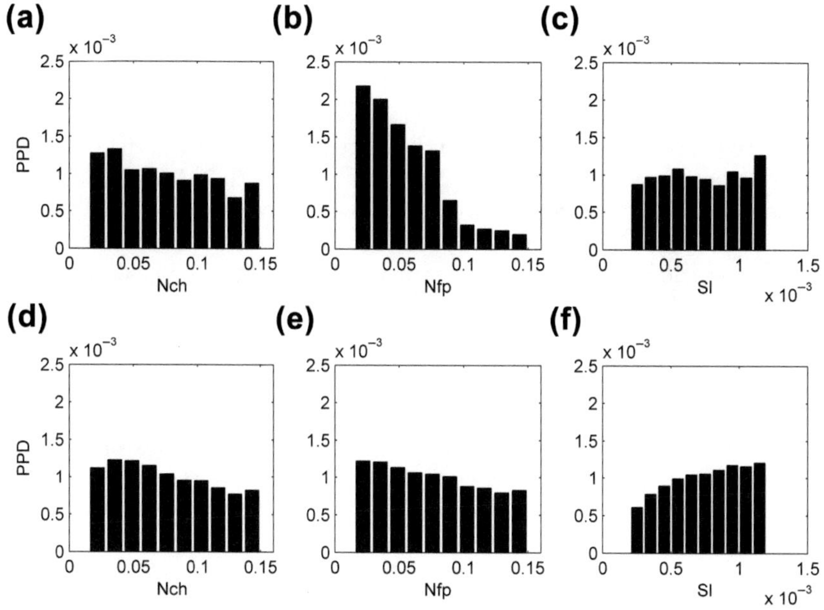

FIGURE 4.6 (a)−(c) Posterior parameter distribution (PPD) for shuttle radar topography mission (SRTM)-based model conditioned on ERS2-SAR flood image. (d)−(f) PPD for SRTM-based model conditioned on downstream water stage.

4.7 CONCLUSIONS

This chapter presents an evaluation of the potential usefulness of SRTM topography in supporting models predicting flood extent as well as downstream water stages, by taking into account parameter uncertainty within a GLUE framework. The topographic uncertainty is estimated by comparing the

SRTM-based model to a model based on high-resolution topography (i.e., LiDAR plus channel survey). The ERS2-SAR flood imagery and downstream time series of water stages of the 2006 flood event are used to constrain model uncertainty. Roughness coefficients in channel and floodplain as well as the water surface slope are sampled uniformly within their parameter space. The effect of water surface slope in affecting flood extent and downstream water stages is quantified. The ability of a 2D flood-inundation model conditioned on water stage to simulate flood extent is also evaluated.

The SRTM-based model performs poorly for the downstream water stage predictions, but it captures the majority of the inundation patterns. In addition, similar optimal parameters for the SRTM-based model conditioned on flood extent or water stage are encouraging. However, to generalize these findings, SRTM data should be tested on more case studies.

It is also shown that the optimal parameters are rather different when the LiDAR-based model is conditioned on either the flood extent or water stages. However, when behavioral simulations are conditioned on water stage, predictions of flood extent prediction are rather good. This is likely due to the fact that the differences in water levels do not imply changes in flood extent.

The water surface slope used as downstream boundary condition is found to have a negligible impact on flood extent predictions with the LiDAR-based model and a limited impact on flood-extent predictions with the SRTM-based model. In contrast, the downstream water surface slope is found to significantly affect water stage predictions of both models. This finding suggests that water-surface slope has to be selected with caution when one of the purposes of the hydraulic model is the prediction of downstream water stages and design flood profiles.

ACKNOWLEDGMENTS

The authors are extremely grateful to the to the European Space Agency (ESA) for allowing access to the flood images used in this study, the Environment Agency of England and Wales for the LiDAR data and other input data. The authors also acknowledge the EC FP7 research project KULTURisk which provided partial funding that made the preparation of this chapter possible.

REFERENCES

Aronica, G., Bates, P.D., Horritt, M.S., 2002. Assessing the uncertainty in distributed model predictions using observed binary pattern information within GLUE. Hydrol. Processes 16 (10), 2001–2016.

Aronica, G., Hankin, B., Beven, K., 1998. Uncertainty and equifinality in calibrating distributed roughness coefficients in a flood propagation model with limited data. Adv. Water Resour. 22 (4), 349–365.

Bates, P.D., 2004. Remote sensing and flood inundation modelling. Hydrol. Processes 18, 2593–2597.

Bates, P.D., 2012. Integrating remote sensing data with flood inundation models: how far have we got? Hydrol. Processes 26 (16), 2515–2521.

Bates, P.D., De Roo, A.P.J., 2000. A simple raster-based model for flood inundation simulation. J. Hydrol. 236 (1–2), 54–77.

Bates, P.D., Horritt, M.S., Aronica, G., Beven, K., 2004. Bayesian updating of flood inundation likelihoods conditioned on flood extent data. Hydrol. Processes 18 (17), 3347–3370.

Bates, P.D., Horritt, M.S., Fewtrell, T.J., 2010. A simple inertial formulation of the shallow water equations for efficient two-dimensional flood inundation modelling. J. Hydrol. 387 (1–2), 33–45.

Beven, K., Freer, J., 2001. Equifinality, data assimilation, and uncertainty estimation in mechanistic modelling of complex environmental systems using the GLUE methodology. J. Hydrol. 249 (1–4), 11–29.

Beven, K.J., Binley, A., 1992. The future of distributed models: model calibration and uncertainty prediction. Hydrol. Processes 6, 279–298.

Brandimarte, L., Di Baldassarre, G., 2012. Uncertainty in design flood profiles derived by hydraulic modelling. Hydrol. Res. 43 (6), 753–761.

Di Baldassarre, G., Montanari, A., 2009. Uncertainty in river discharge observations: a quantitative analysis. Hydrol. Earth Sys. Sci. 13 (6).

Di Baldassarre, G., Schumann, G., Bates, P.D., Freer, J.E., Beven, K.J., 2010. Flood-plain mapping: a critical discussion of deterministic and probabilistic approaches. Hydrol. Sci. J. 55 (3), 364–376.

Di Baldassarre, G., Schumann, G., Brandimarte, L., Bates, P., 2011. Timely low resolution SAR imagery to support floodplain modelling: a case study review. Surv. Geophys. 32 (3), 255–269.

Di Baldassarre, G., Uhlenbrook, S., 2012. Is the current flood of data enough? A treatise on research needs for the improvement of flood modelling. Hydrol. Processes 26 (1), 153–158.

Falorni, G., Teles, V., Vivoni, E.R., Bras, R.L., Amaratunga, K.S., 2005. Analysis and characterization of the vertical accuracy of digital elevation models from the Shuttle Radar Topography Mission. J. Geophys. Res. Earth Surf. 110 (F2), F02005.

García-Pintado, J., Neal, J.C., Mason, D.C., Dance, S.L., Bates, P.D., 2013. Scheduling satellite-based SAR acquisition for sequential assimilation of water level observations into flood modelling. J. Hydrol. 495 (0), 252–266.

Horritt, M.S., Bates, P.D., 2002. Evaluation of 1D and 2D numerical models for predicting river flood inundation. J. Hydrol. 268 (1–4), 87–99.

Horritt, M.S., Di Baldassarre, G., Bates, P.D., Brath, A., 2007. Comparing the performance of a 2-D finite element and a 2-D finite volume model of floodplain inundation using airborne SAR imagery. Hydrol. Processes. 21 (20), 2745–2759.

Hunter, N.M., Bates, P.D., Horritt, M.S., De Roo, A.P.J., Werner, M.G.F., 2005. Utility of different data types for calibrating flood inundation models within a GLUE framework. Hydrol. Earth Sys. Sci. 9 (4).

Hunter, N.M., Bates, P.D., Horritt, M.S., Wilson, M.D., 2007. Simple spatially-distributed models for predicting flood inundation: a review. Geomorphology 90 (3–4), 208–225.

Jarvis, A., Reuter, H.I., Nelson, A., Guevara, E., 2008. Hole-Filled SRTM for the Globe Version 4. International Centre for Tropical Agriculture (CIAT). Available from: http://srtm.csi.cgiar.org.

LeFavour, G., Alsdorf, D., 2005. Water slope and discharge in the Amazon River estimated using the shuttle radar topography mission digital elevation model. Geophys. Res. Lett. 32 (17), L17404.

Mantovan, P., Todini, E., 2006. Hydrological forecasting uncertainty assessment: incoherence of the GLUE methodology. J. Hydrol. 330 (1–2), 368–381.

Matgen, P., Schumann, G., Henry, J.B., Hoffmann, L., Pfister, L., 2007. Integration of SAR-derived river inundation areas, high-precision topographic data and a river flow model toward near real-time flood management. Int. J. Appl. Earth Obs. Geoinf. 9 (3), 247–263.

de Moel, H., van Alphen, J., Aerts, J.C.J.H., 2009. Flood maps in Europe – methods, availability and use. Nat. Hazards Earth Sys. Sci. 9 (2), 289–301.

Montanari, A., 2005. Large sample behaviors of the generalized likelihood uncertainty estimation (GLUE) in assessing the uncertainty of rainfall-runoff simulations. Water Resour. Res. 41 (8), W08406.

Neal, J., Keef, C., Bates, P., Beven, K., Leedal, D., 2013. Probabilistic flood risk mapping including spatial dependence. Hydrol. Processes 27 (9), 1349–1363.

Neal, J., Schumann, G., Bates, P.D., 2012. A subgrid channel model for simulating river hydraulics and floodplain inundation over large and data sparse areas. Water Resour. Res. 48, W11506.

Pappenberger, F., Frodsham, K., Beven, K., Romanowicz, R., Matgen, P., 2007. Fuzzy set approach to calibrating distributed flood inundation models using remote sensing observations. Hydrol. Earth Sys. Sci. 11 (2).

Pappenberger, F., Matgen, P., Beven, K.J., Henry, J.-B., Pfister, L., Fraipont de, P., 2006. Influence of uncertain boundary conditions and model structure on flood inundation predictions. Adv. Water Resour. 29 (10), 1430–1449.

Rabus, B., Eineder, M., Roth, A., Bamler, R., 2003. The shuttle radar topography mission—a new class of digital elevation models acquired by spaceborne radar. ISPRS J. Photogramm. Remote Sens. 57 (4), 241–262.

Rodríguez, E., Morris, C.S., Belz, J.E., 2006. A global assessment of the SRTM performance. Photogramm. Eng. Remote Sens. 72 (3), 249–260.

Romanowicz, R., Beven, K., 1998. Dynamic real-time prediction of flood inundation probabilities. Hydrol. Sci. J. 43 (2), 181–196.

Samuels, P., 1989. Backwater length in rivers. Proc. – Inst. Civ. Eng. 87, 571–582.

Sanders, B.F., 2007. Evaluation of on-line DEMs for flood inundation modeling. Adv. Water Resour. 30 (8), 1831–1843.

Schumann, G., Di Baldassarre, G., Alsdorf, D., Bates, P.D., 2010. Near real time flood wave approximation on large rivers from space: application to the River Po, Italy. Water Resour. Res. 46.

Schumann, G., Di Baldassarre, G., Bates, P.D., 2009. The utility of spaceborne radar to render flood inundation maps based on multialgorithm ensembles. Geosci. Remote Sens., IEEE Trans. 47 (8), 2801–2807.

Schumann, G., Matgen, P., Cutler, M.E.J., Black, A., Hoffmann, L., Pfister, L., 2008. Comparison of remotely sensed water stages from LiDAR, topographic contours and SRTM. ISPRS J. Photogramm. Remote Sens. 63 (3), 283–296.

Schumann, G., Neal, J.C., Phanthuwongpakdee, K., Voisin, N., Aspin, T., 2012. Assessing forecast skill of a large scale 2D inundation model of the Lower Zambezi River with multiple satellite data sets. In: XIX International Conference on Water Resources. University of Illinois at Urbana-Champaign.

Stedinger, J.R., Vogel, R.M., Lee, S.U., Batchelder, R., 2008. Appraisal of the generalized likelihood uncertainty estimation (GLUE) method. Water Resour. Res. 44, W00B06.

Stephens, E., Schumann, G., Bates, P.D., 2014. Problems with binary pattern measures for flood model evaluation. Hydrol. Processes. 28 (18), 4928–4937.

Van Alphen, J., Martini, F., Loat, R., Slomp, R., Passchier, R., 2009. Flood risk mapping in Europe, experiences and best practices. J. Flood Risk Manage. 2 (4), 285–292.

Wagener, T., Lees, M.J., Wheater, H.S., 2001. A toolkit for the development and application of parsimonious hydrological models. Math. Models Large Watershed Hydrol. 1, 87–136.

Wang, Y., Liao, M., Sun, G., Gong, J., 2005. Analysis of the water volume, length, total area and inundated area of the Three Gorges Reservoir, China using the SRTM DEM data. Int. J. Remote Sens. 26 (18), 4001–4012.

Yan, K., Di Baldassarre, G., Solomatine, D.P., 2013. Exploring the potential of SRTM topographic data for flood inundation modelling under uncertainty. J. Hydroinf. 15 (3), 849–861.

Chapter 5

Vulnerability and Exposure in Developed and Developing Countries: Large-Scale Assessments

S.F. Balica [3], **Q. Dinh** [2] **and I. Popescu** [1]

[1] *UNESCO-IHE, Institute for Water Education, Delft, The Netherlands,* [2] *Dipartimento di Elettronica, Informazione e Bioingegneria, Politecnico di Milano, Italy,* [3] *s.f.balica@gmail.com*

ABSTRACT

Natural hazards occur all over the world. The rising number of disasters is not only due to the increased number of natural hazards; but also (mainly) due to the fact that people prefer to live in the areas the most exposed to natural hazards, i.e., close to river banks and coastal areas. In order to decrease the number of causalities and affected people, populations should understand their level of exposure and vulnerability to such natural hazards. According to the Oxford dictionary "disaster" is "a sudden accident or a natural catastrophe that causes great damage or loss of life." Disaster is defined as being an impact of natural hazards which leads to human and economic losses. The list of natural hazards is long; however, the goal of the present chapter is to look at the flooding phenomena.

5.1 INTRODUCTION

Often broadcasting TVs are presenting breaking news about damages caused by natural hazards, which occur all over the world. The rising number of disasters is not only due to the increased number of natural hazards; but also (mainly) due to the fact that people prefer to live in the areas the most exposed to natural hazards, i.e., close to river banks and coastal areas (Di Baldassarre et al., 2010). In order to decrease the number of causalities and affected people, populations should understand their level of exposure and vulnerability to such natural hazards. According to the Oxford dictionary "disaster" is "a sudden accident or a natural catastrophe that causes great damage or loss of life." Disaster is defined as being an impact of natural hazards which leads to

Hydro-Meteorological Hazards, Risks, and Disasters. http://dx.doi.org/10.1016/B978-0-12-394846-5.00005-9

human and economic losses. The list of natural hazards is long; however, the goal of the present chapter is to look at the flooding phenomena.

The etymology of the word *flood* comes from Old English: Flod or Pleu. It generally means a sudden increase in the amount or level of a variable, which can be anything: rainfall, water level, water volume, etc.

Between 1993 and 2012, floods affected 2,480 million people, representing more than half of all people affected by all kinds of natural hazards. Costs over the last two decades of flood damage were estimated, at the US$ value of the year 2012, to be around 625 billion US$, occupying the fourth position after storms, earthquakes, and tsunamis. According to CRED (2013) in 2006 and 2007, more than 50 percent of all natural hazards were floods and, in spite of their decrease in occurrence after 2007, the number of floods still remained high (42 percent of all disasters from 2008 to 2011).

Floods are happening everywhere in the world; regardless of the development status of the country (i.e., Katrina Hurricane and Sandy, U.S; coastal floods in UK; or floods in Vietnam), however, the majority of flooding events are most often taking place in developing countries. The notion of "developing country" is defined by World Bank as "countries with a GNI of US$11,905 and less in 2010 are defined as developing," whereas countries with high gross domestic product (GDP) per capita would thus be described as developed countries (World Bank, 2012). The UN (2013), notes that the designations "developed" and "developing" are intended for statistical convenience and do not necessarily express a judgment about the socioeconomic conditions reached by a particular country or area in the development process.

This chapter of the book considers country classification as given by UNDP (2010) for the Human Development Index (HDI), which is a composite index of education, income, and life expectancy indicators. This index is used to classify countries into Very High, High, Medium and Low Human Development (UNDP, 2010). All the countries outside the Very High Human Development category are considered "developing" countries.

Having a vulnerable natural environment affects populations. The environment is as well, in turn, affected by several human activities; construction and operation of hydraulic structures (river bank protection, dykes, dams, etc.) or activities that involve use of water ways (shipping or travel). The environment provides tangible and direct economic benefits that prosper on the wealth of natural resources supply; for example, tourism or fisheries. Populations living close to particular risks, however, are in general more concerned about the flood risk, hence perceive vulnerability of the environment differently. Nevertheless, the system's vulnerability should worry everybody.

To assess flood exposure and vulnerability of the public, indices are developed that can show what is the level of this exposure and vulnerability. Flood vulnerability indices are based on social, economic, environmental, and physical indicators and help decision-makers in prioritizing natural hazard defense policies, measures, and activities (Dinh et al., 2012). In addition, "vulnerability can be measured directly through perceptions of those that are vulnerable" (Adger, 2006).

The number of people affected by floods and economic losses tends to diminish slowly over the years in Asia and more strongly in Europe. Though the level of flood damage is quite high because of countries' GDP, income, assets, and properties, the developing countries would still suffer more than the other ones, due to their inability to cope with natural hazards. This comes due to lack of the resources of developing countries that will allow recovery.

Mapping vulnerability to floods in coastal areas offers the opportunity to get a broader overview of the affected areas and give possible adaptation strategies to be applied on both developed and developing countries. Such kinds of maps could help in directing resources after identifying the most promising adaptation strategies. Such identification requires an in-depth investigation of the vulnerability maps.

This chapter examines vulnerability to coastal flooding of highly populated delta areas from the hydrogeological and socioeconomical point of view. The methodology involves developing large-scale maps of the flood exposure and vulnerability for two specific case studies; the Vietnam Mekong Delta (VMD) for the year 2000; and the Italian Po River Delta (PRD) for the year 2010. Vulnerability maps are created by applying The Coastal Cities Flood Vulnerability Index (CCFVI) as introduced by Balica et al. (2012).

The two considered case studies are based on the same kind of data availability and are covering developing and developed areas; the obtained maps could be helpful to the decision-makers who need to take measures on how to reduce and mitigate the coastal flood impact in the areas.

5.2 VULNERABILITY: DEFINITIONS AND COMPLEXITY

Although the concept of vulnerability is frequently used in disaster research, researchers' notions of vulnerability have changed over the past two decades, and consequently several attempts have been made to define and capture what is meant by the term. In the variety of definitions to vulnerability, the definitions of exposure to hazards differ for different fields of study (See Table 5.1). Similarly people's exposure to risk differs from one society to another.

The quantification of vulnerability can help in decision-making processes; hence parameters and indicators (indices) need to be designed to produce information for specific target areas. These parameters and indicators would provide enough information to properly prepare societies for different hazards they face. Because of the large number of people, economic activities and ecosystems that are impacted by the adverse effects of floods, they lately have gained a lot of attention and importance. In this context vulnerability can be determined by the economics and politics of the countries.

TABLE 5.1 Vulnerability Definitions

Year and Source	Vulnerability Definition
1979 United Nations	The degree of loss to a given element, or a set of such elements, at risk resulting from a flood of given magnitude and expressed on a scale from 0 (no damage) to 1 (total damage).
1980 Gabor and Griffith	Vulnerability is the risk context.
1981 Timmerman	The degree of a harm system at risk, the frequency of a hazardous incident.
1989 Chambers	"As a potential for loss," with two sides: The shocks and perturbations from outside exposure, and the ability or lack of ability from the internal side, its resilience.
1989 Ramade	The predisposition of goods, people, buildings, infrastructures, and activities to be damaged, offering low resistance.
1992 International Panel of Climate Change	The degree of incapability to cope with the consequences of climate change and sea-level rise.
1996 Cutter	A hazard which includes natural risks together with social response and action.
1997 International Panel of Climate Change	"As the extent to which climate change may damage or harm a system; it depends not only on a system's sensitivity but also on its ability to adapt to new climatic conditions."
1999 Lewis	"Vulnerability is the root cause of disasters" and elements at risk, exposure (damage potential), and (loss) susceptibility.
1999 Adger et al.	Social vulnerability to hazards is determined by their "existent state, that is, by their capacity"—or capacity to react and recover, and to deal with the everyday stresses.
2003 Sarewitz et al.	The potential damage or harm caused to a system by an extreme event or hazard.
2005 Cutter et al.	"Vulnerability as the potential for loss and involves a combination of factors that determine the degree to which a person's life or livelihood is put at risk by a particular event."
2004 Pratt et al.	A function of exposure, resilience, and resistance.
2005 Gheorghe	Vulnerability is a function of exposure, susceptibility, resilience, and state of knowledge.

TABLE 5.1 Vulnerability Definitions—cont'd

Year and Source	Vulnerability Definition
2006 Adger	Focus on "shocks and stressors" near "capacity for adaptive action," highlighting the vulnerability as the state of susceptibility to harm from exposure to stresses associated with environmental and social change and from the absence of capacity to adapt.
2001 IPCC	The vulnerability is related to exposure, sensitivity, and resilience (adaptive capacity).
2009 Balica et al.	The extent to which a system is susceptible to floods due to exposure, a perturbation, in conjunction with its ability (or inability) to cope, recover, or basically adapt.
2010 McEntire et al.	An evaluation of exposure to harmful occurrence and probability that individuals will lack the capacity to rebound.
2010 Quinn et al.	A geographical or territorial system as the result of different behavior and coping capacities in socially, economically, and technologically heterogeneous contexts.

All societies are vulnerable to floods; however, on a case-by-case basis they are very different, which make them somewhat unique. Understanding the distinctions for floods among societies may help to plan ahead and provide policy ideas to improve the quality of life for the people living in flooding situations on a frequent basis.

An overview of the exiting definitions of vulnerability is given in Table 5.1. The definition used in this chapter is the one given by Balica et al. (2009) "vulnerability is the extent to which a system is susceptible to floods due to exposure, a perturbation, in conjunction with its ability (or inability) to cope, recover, or basically adapt." Vulnerability is closely related to exposure, susceptibility, and resilience, which are detailed below.

In order to develop, societies need close access to water, which hinted to people to search for innovative ways to prosper, in spite of the limited resources as the population grows. This added pressure on the water resources. The results of the solutions for prosperity have been an important distinction on the development of societies, creating different problems for developed and developing countries.

Societies of the developed countries are well organized, however, their innovations harm the river system, i.e., most of them are heavily engineered (confined and leveed). Safety standards are basically sufficient to prevent floods.

Society's vulnerability to floods is mainly reflected by possible economic losses as development grows; the cities develop in flood-prone areas, leading to large economic growth and, thus increasing their vulnerability to floods.

Damages are extremely high when a flood defense structure fails, especially in urbanized areas, where the most important industries are located. For example, an interruption of electricity caused by flooding will disrupt the system from its normality, and the economic damages will be enormous. In developed countries the losses will be reflected mostly in the economy, whereas population is less affected economically. Moreover, the recovery process is faster due to governmental aids.

Developing countries on the other hand are characterized by general deficiency of necessary investments, high population density, high rates of unemployment, pressure on rural land, illiteracy, and an economy usually dominated by agriculture dependent on developed countries. The developing countries' vulnerability to floods can be reflected by the following factors:

- socioeconomic conditions, high poverty level and lack of development;
- most of the infrastructure, including dams is not multipurpose;
- during floods, often measures are taken that are not helping to reduce the impact of floods;
- countryside regions are heavily dependent on agriculture and being liable to agriculture will affect the economy more than the urban areas;
- lack of education and prevention;
- lack of nonstructural measures;
- a deficiency of adequate human and material resources to tackle the enormous flood disasters that happened in the past (Mirza, 2003).

Because of their vulnerability, often millions of people become homeless and hundreds of thousands are in need of food and medicines. Especially important during floods occurring in developing countries is the high number of persons that become infected with diseases. Houses, industries, infrastructure, and agriculture are highly vulnerable. In these societies the losses due to floods are mainly lives, cultural damage, agricultural fields, and cattle. The costs needed for reconstruction are high, and societies usually take a long time to recover, depending mainly on international aid (Davidson, 2004).

Coastal areas are one of the most densely populated areas in the world and consequently flood exposure is increasing in these areas (De Sherbinin et al., 2007) due to growing populations, properties, sea-level rise, and soil subsidence. Coastal flood vulnerability is increasing because exposure and susceptibility to floods are increasing, and also in cases that no adequate adaptation measures will be taken to expand flood resilience.

Taking into account how important the exposure and susceptibility to floods is, several definitions of them are presented in Tables 5.2 and 5.3, respectively.

The concept of *resilience* and the related concept of resistance, used in ecology, are used to describe a system's ability to deal with perturbations and to

TABLE 5.2 Exposure Definitions

Year and Source	Exposure Definitions
2010 UN	The values that are present at the location where floods can occur. These values can be goods, infrastructure, cultural heritage, agricultural fields, and people.
2003 Cardona	Physical fragility.
2004 UNDP	The extent to which humans and their homes are positioned in flood-risk areas.
2004 Karen O'Brien et al.	The degree of climate stress upon a particular unit of analysis. Climate stress can refer to long-term changes in climate conditions or to changes in climate variability and the magnitude and frequency of extreme events.
1977 Penning Rowsell et al.	The likelihood that humans and/or physical items (goods, infrastructure, cultural heritage, and agricultural fields) will be impacted by flooding.
2005 Hollenstein	The spatiotemporal distribution of elements at risk.
2005 Cardona	One element for measuring vulnerability.
2007 Balica	The predisposition of a system to be disrupted by a flooding event due to its location in the same area of influence.
2009 Fekete	The measure of susceptible elements within a region threatened by a hazard.

TABLE 5.3 Susceptibility Definitions

Year and Source	Susceptibility Definition
1977 Penning-Rowsell and Chatterton	Focuses on the relative damageability of property and materials during floods or other hazardous events.
1999 Klein and Nicholls	The current physical state, without taking into account temporal changes.
2001 IPCC	The affected system's degree, by climate-related stimuli.
2005 Veen & Logtmeijer	The probability and extent of flooding.
2006 Di Mauro	Integrates the probability of a hazardous event, the differential exposure and the possible sensitivity of an objective.
2012 Balica	The elements exposed within the system, which influence the probabilities of being harmed at times of hazardous floods.

TABLE 5.4 Resilience Definitions

Year and Source	Resilience Definition
1973 Holling	A measure of persistence of systems and their ability to absorb change and disturbance and still maintain the same relationships between populations and state variables.
1999 Pérez España and Arreguín Sánchez	The capacity of any system to re-gain its equilibrium after a reaction to a disturbance.
2003 Pelling	The capacity to adapt, to adjust to threats, and to mitigate or avoid harm.
2004 Walker	The capacity of a system to absorb disturbance and being reorganized while undergoing change, so as to still retain essentially the same function, structure, identity, and feedbacks.
2005 De Bruijn	The ability of the system to recover from floods.
2006 Galderisi et al.	The capacity of all systems, i.e., a society or community, potentially exposed to hazards to adapt to any change, by resisting or modifying itself, in order to maintain or to achieve an acceptable level of functioning and structure.
2012 Balica	The capacity of a system to endure any perturbation, such as floods, maintaining significant levels of efficiency in its social, economic, environmental, and physical components.

continue without huge irreversible changes in their most important characteristics. In the particular context of floods, resilience is defined as the ability of a system to prevent floods. Table 5.4 presents a series of resilience definitions.

5.3 APPROACHES TO VULNERABILITY

CRED (2013) shows that during the last few decades has occurred an increase in frequency intensity and economic effects due to meteorological-related events. The objective to develop indices is to provide decision-makers with tools to assess and analyze natural hazard events.

Vulnerability-related indices are given in Table 5.5, and the methodology for assessing flood exposure and vulnerability of delta areas is detailed below. Delta areas are selected as examples, because they are densely populated and as such more vulnerable.

TABLE 5.5 Screening through Water-Related Vulnerability Indices—Approaches of Assessing Vulnerability

Index's Name	Authors	Type of Natural Hazard	Number of Indicators Used	Scale	Index Formula
The Composite Vulnerability Index for Small Island States	Briguglio (2004)	Droughts, floods, cyclones, volcanoes, earthquakes, and windstorms	Four indices: • the vulnerability or susceptibility of the country in relation to natural disasters (*Vul*); • the economic exposure of the country, export dependence, • the average exports of goods • nonfactor services as a percentage of the GDP	Small Island Developing States	$CVISIS = 1.4142 + 0.0096\ Vul \times D + 0.0322\ Ex\text{-}Dep + 3.3442\ Div$
The Economic Vulnerability Index	Guillaumont (2008)	Droughts, floods, cyclones, volcanoes, earthquakes, and windstorms	Seven indices: • four shock indices; • three exposure indices	Small Island Developing States and Least Developed Countries	$EcVI\ (economic) = sqr\ [1 - (1 - EXP)(1 - SK)]$
Global Risk and Vulnerability Index Trends per Year (GRAVITY)	Peduzzi et al. (2001)	Any	Seven indicators: • urbanization indicator • indicator of corruption • human development index • population density • GDP/Capita • urban growth over last 3 years • population growth over last 3 years	Country	$Y = [victimsict]\ i = 1,\ldots,n$ $X = [x_1 i;\ x_{2i};\ \ldots\ ;\ x_{7i}]\ i = 1,\ldots,n$ where: x_1 - *popdct;* x_2 - *corupc*2000; x_3 - *hdic*1998; x_4 - *gdpcapct;* x_5 - *urbanct;* x_6 - *urbang3ct;* x_7 - *popg3ct*

Continued

TABLE 5.5 Screening through Water-Related Vulnerability Indices—Approaches of Assessing Vulnerability—cont'd

Index's Name	Authors	Type of Natural Hazard	Number of Indicators Used	Scale	Index Formula
Disaster Risk Index (DRI)	UNDP (2004)	Floods, earthquakes, cyclones, and droughts	26 indicators, included in: • *PhExp*—average number of people exposed to flood event; • *GDPcap*—the normalized Gross Domestic Product/capita (purchasing power parity); • *D*—population density	Country	$\ln(K) = 0.78 \ln(PhExp) - 0.45 \ln(GDP_{cap}) - 0.15 \ln(D) - 5.22$
The Environmental Vulnerability Index (EnVI)	South Pacific Applied Geosciences Commission Pratt et al. (2004)	Natural and anthropogenic hazards	50 "smart indicators" divided in 5 classes: • meteorological; • geological; • biological; • country characteristics; • anthropogenic	Small Island Developing States	$\text{EVI} = \sum X_i$ where $X_i =$ # of indicators
The Coastal Vulnerability Index	Gornitz (1990)	Physical changes due to sea-level rise	*Six indicators:* • geomorphology; • shoreline erosion/accretion rate; • coastal slope; • relative sea-level rise rate; • mean wave height; • mean tide range	Coastal Regions	$CVI = \sqrt{\frac{(a*b*c*d*e*f)}{6}}$

The Coastal Vulnerability Index	Balica et al. (2012)	Coastal floods	17 indicators: • sea-level rise, storm surge, foreshore slope, cyclones, coastal lines, river discharge, cultural heritage, shelters, % disable, population CL, Drain, recovery time, awareness, GCP, FHM	Coastal Cities, Coastal Districts	$Total_{CCFVI} = CCFVI_{Hydrogeological} + CCFVI_{Social} + CCFVI_{Economic} + CCFVI_{Political\text{-}Administrative}$
A multi-scale coastal vulnerability index	McLaughlin and Cooper (2010)	Coastal erosion	Three elements and 17 variables, 7—coastal characteristics: • (solid and drift geology, shoreline type, river mouths, elevation, orientation, inland buffer), Four coastal forcing: • (significant wave height, tidal range, difference in storm and modal wave height, storm, frequency); Six socioeconomic factors: • (population, cultural heritage, roads, land use, railways and conservation status)	Coastal Regions	$mCVI = \sum \frac{X_i}{3}$ where: X_i = the three components
"A simple and preliminary" coastal vulnerability index	Pethick and Crooks (2000)	Sensitivity of the landform	Four indicators: • Shoreline type • Event Frequency • Relaxation time • Vulnerability index	Coastal Regions	Narrative results
The Drought Vulnerability Index	Wilhelmi and Wilhite (2002)	Agricultural drought	Four factors • two biophysical, soil and climate • two social, land-use, and irrigation	Droughts in Nebraska	The 4 factors where combined in ERDAS Imagine GIS

Continued

TABLE 5.5 Screening through Water-Related Vulnerability Indices—Approaches of Assessing Vulnerability—cont'd

Index's Name	Authors	Type of Natural Hazard	Number of Indicators Used	Scale	Index Formula
The Climate Vulnerability Index	Sullivan (2002)	Climate change	The six major categories or components: • Resource (R), • Access (A), • Capacity (C), • Use (U), • Environment (E) • Geospatial (G)	Country	$CVI = \frac{w_r R + w_a A + w_c C + w_u U + w_e E + w_g G}{w_r + w_a + w_c + w_u + w_e + w_g}$
The Vulnerability to Climate Change Index	Yohe et al. (2006)	Climate change	• National average temperature (i.e., a rational-scaled variable) normalized index of national adaptive capacity	Country	$Vi(t) = \Delta Ti(t)/ACi(t)$
The Social Vulnerability Index	Cutter (2003)	Social vulnerability	Eleven indicators: • personal wealth, age, density of the built environment, single-sector economic dependence, housing stock and tenancy, race (Asian and African-American), ethnicity (Hispanic and native American), occupation, infrastructure dependence	US Counties	Algorithm: • first standardization of the indicators, • perform PCA, • selection of the components to be used (i.e., Kaiser Criterion), • rotate initial solution (i.e., varimax and quartimax), • interpretation and processing of the components, • the combination of those components (i.e., equal weights, first component only) • standardizing index values.

The Social Vulnerability to Climate Change for Africa Index	Adger (2006)	Climate change-induced changes in water availability	Five indicators: • economic wellbeing • stability demographic structure • global interconnectivity • institutional stability • wellbeing nature resource dependence	Country	$SVI = 0.2\ Iewb + 0.2Ids + 0.4Iis + 0.1Igi + 0.1Inrd$
Vulnerability assessments to aquifers	Neukum et al. (2008)	Karst vulnerability of the topmost aquifer	Three indicators: • the thickness of each layer in the unsaturated zone • the permeability of each stratum of the unsaturated zone • the amount of percolation water	Aquifers	$VSQ = X_i * Y_i * Z_i$
Water Poverty Index (WPI)	Sullivan (2002)	Water poverty	Five components: • access (population with access to safe water, with access to sanitation, fraction of land irrigated), • resources (groundwater, surface water, annual average precipitation), • capacity (GDP, mortality rate, education index), • environment (indices of water quality, water stress, informational capacity, regulation and management capacity and biodiversity) • use (domestic, industrial, agricultural).	Africa Regions	$WPI = \sum_{i=1}^{N} w_i * X_i$

Continued

TABLE 5.5 Screening through Water-Related Vulnerability Indices—Approaches of Assessing Vulnerability—cont'd

Index's Name	Authors	Type of Natural Hazard	Number of Indicators Used	Scale	Index Formula
The Flood Vulnerability Index	Connor and Hiroki (2005)	Floods	Eleven indicators: • frequency of heavy rainfall; average slope, urbanized area ration; TV penetration rate, literacy rate, population rate under poverty, years sustaining healthy life, population in flooded area, infant mortality rate; investment amount for structural measures, investment amount for nonstructural measures	River Basin	$FVI = \frac{w_c C + w_h H + w_s S}{w_m M}$
The Flood Vulnerability Index	Balica et al. (2009)	Floods	28 indicators divided in four components: • social, • economic, • environmental, • physical	River Basin Subcatchment Urban area	$FVI = \frac{E * S}{R}$ E = exposure S = susceptibility R = resilience
The Integrated Flood Vulnerability Index	Sebald, 2010	Floods	Four components: • social (total population, females, growing population, population density and unemployed), • economic (housing stock, dwellings and other units, vehicles registration and industrial commercial), • ecologic biological reserve/ protected areas and Natura 2000) • physical (flood extent 1999 and flood scenario 0.00 m > 4.00 m).	Local	$Vul = \sum_{i=0}^{n} (v_i * w_i)$

Indices are statistical concepts, in forms of a quantity or a state that are used for comparison. They are represented as numbers computed based on indicators (Sullivan, 2002). An indicator, is a number which quantitatively estimates the condition of a system. To create indices, scientists emphasize the need to use indicators that match reality. Vulnerability indicators are used to evaluate different risk-based assessment for social, economic, environmental, or engineering issues. Composite vulnerability indices are useful for decision-makers.

While identifying, understanding, and selecting the key vulnerability indicators to create composite vulnerability indices, limitation of the "high degree of heterogeneity in the way the data/indicators have to be collected" should be considered (Kaufmann et al., 1999).

5.4 METHODOLOGY

As mentioned previously, the CCFVI proposed by Balica et al., 2012, is used here as a demonstration for building a tool for decision-making, which could help in the process of directing investments in case of assessment of flood-mitigation strategies. The implementation of the proposed CCFVI could guide policy makers to analyze actions toward better coping with floods. The method allows a comprehensive interpretation of specific indicators and pinpoints actions to diminish focal spots of flood vulnerability.

The CCFVI is used to compute flood vulnerability based on a selected set of indicators belonging to the hydrogeological, social, and economic components of a coastal system. Coastal Flood Vulnerability assessment is done by gathering data and values for 15 indicators (see Table 5.6) for different spatial scales: districts, urban areas, and neighborhoods. The CCFVI involves converting each identified indicator into a normalized one, on a scale from 0 to 1, which is a dimensionless number using its minimum and maximum values as the normalization limits.

The CCFVI of each coastal component (hydrogeological, social, and economic) is computed based on the general flood vulnerability index (*FVI*) given by the Eqn (5.1):

$$FVI = \frac{E * S}{R} \tag{5.1}$$

The indicators of exposure and susceptibility are multiplied and then divided by the resilience indicators, because indicators representing exposure and susceptibility increase the flood vulnerability; they are therefore placed in the nominator, while the resilience indicators decreases flood vulnerability and thus are part of the denominator.

The total CCFVI is computed as a summation of the three CCFVIs components of the system (Eqn (5.2)):

$$\mathrm{Total}_{\mathrm{CCFVI}} = \mathrm{CCFVI}_{\mathrm{Hydrogeological}} + \mathrm{CCFVI}_{\mathrm{Social}} + \mathrm{CCFVI}_{\mathrm{Economic}} \tag{5.2}$$

TABLE 5.6 Coastal Cities Flood Vulnerability Index (CCFVI) Indicators

N°		Indicators	Abbreviation	Factor	Unit	Description	Functional Relationship with Vulnerability
1	Hydro-geological component	Sea level rise	SLR	Exposure	mm/yr	The increase of the sea level during one calendar year	The higher SLR, the higher the vulnerability
2		Storm surge	SS	Exposure	m	The rapid rise of the water level surface produced by onshore hurricane winds and falling barometric pressure.	The bigger increase in SS, the higher vulnerability
3		Number of cyclones	#Cyc	Exposure	#	Number of cyclones in the last 10 years	The higher # of cyclones, the higher vulnerability
4		River discharge	RD	Exposure	m^3/s	Maximum discharge recorded in the last 10 years	The higher RD, the higher vulnerability
5		Foreshore slope	FS	Exposure	%	Average slope of the foreshore beach	The lower FS, the higher vulnerability
6		Soil subsidence	Soil	Exposure	mm/yr	With how many mm/yr the level is decreasing?	The higher soil subsidence, the higher vulnerability
7		Coastal line	CL	Exposure	km	Kilometers of coastline along the districts or communes	The longer CL, the higher vulnerability
8	Social-economic component	Cultural heritage	CH	Exposure	#	Number of historical buildings, museums, etc., in danger if a coastal flood would occur	The higher # of CH, the higher vulnerability
9		Population close to coastline	PCL	Exposure	People	Number of people exposed to coastal hazard	The higher number of people, the higher vulnerability

10	Growing coastal population	GCP	Exposure	%	Percentage of population growth in urban areas in the last 10 years	The higher % of GCP, the higher vulnerability
11	Shelters	S	Susceptibility	#	Number of shelters, including hospitals, per square kilometer	The bigger # of S, the lower vulnerability
12	Percentage of disable persons (<14 & >65)	%Disab	Susceptibility	%	Percentage of population with any kind of disabilities, and population age range younger than 14 and older than 65	The higher % of disab, the higher vulnerability
13	Awareness and preparedness	A/P	Resilience	–	The level of awareness and preparedness of people, to hazard events such as floods. Did population in the area experienced any floods in the last 10 years? (Scaled see Annex 2a and b)	The higher # of past floods, more prepare/aware, the lower vulnerability
14	Recovery time	RT	Resilience	days	Time required by the city to recover o state of functional operation after the occurrence of a coastal flood event	The higher amount of time, the higher vulnerability
15	Kilometer of drainage	Drain	Resilience	km	Kilometer of drains and sewage existent in the city	The higher km, the lower vulnerability

CCFVI, Coastal Cities Flood Vulnerability Index.

TABLE 5.7 Vulnerability Zones

Vulnerability Zones	CCFVI
Very low	0–0.2
Low	0.2–0.4
Medium	0.4–0.6
High	0.6–0.8
Very high	0.8–1

CCFVI, Coastal Cities Flood Vulnerability Index.

In order to have comprehensive comparison, the final CCFVI results are normalized on a scale between 0 and 1; with 1 being the highest vulnerability found in the studied samples and 0 the lowest. The CCFVI values are grouped into five categories, as shown in Table 5.7, to represent five vulnerability zones.

Exemplification of the method is done by applying it to two deltas; one is the VMD, considered to be spread on developing countries; and the second one is the PRD in Italy, which is considered as representative for a developed country. Exposure and vulnerability maps are presented in Figures 5.1 and 5.2 in order to compare the exposure/vulnerability of a range of districts/communes under current conditions of both types of economies. The data used to assess flood exposure and vulnerability are based on h indicators as shown for few sample districts, in Table 5.8 and Table 5.9.

5.5 COPING WITH FLOOD VULNERABILITY IN DEVELOPING COUNTRIES

Assessments show that after 2000, flooding events affected nearly 949 million people all over the world. Generally, many of these people, 900 million, lived in countries of medium human development (IFRC, 2013). IPCC (2001) reports that the anticipated consequences of flooding are estimated to affect mostly developing countries.

According to the Munich Re, insurance company, flooding causes over one-third of the total estimated costs assigned to natural hazards and is responsible for two-thirds of people affected by natural hazards. In the period 1975–2001, 130,815 people were killed by floods in developing countries (UN-DESA, see Flood Losses). Asian countries are mainly exposed to frequent flooding; hence, the population living in these countries is more affected and more vulnerable to flooding than in other places. More than 90 percent of societies affected by natural hazards worldwide live in Asia (CRED Crunch 32, 2013). Considering the 100 times flood occurrence in Asia, in the

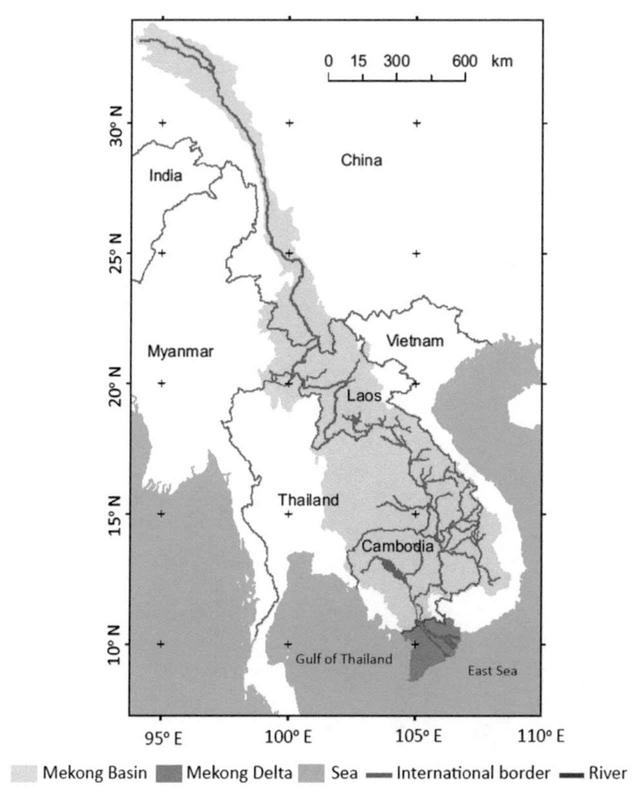

FIGURE 5.1 Location of the study area—Mekong River and Vietnam Mekong Delta (VMD).

last 20 years (CRED Crunch 32, 2013), it makes it the most exposed continent to flood hazards. Investments in developing countries are more focused on recovery from a disaster than on the creation of adaptive capacity.

Vietnam is considered to be Medium Human Development with an HDI of 0.617 in 2012 (0.539 HDI of education; 0.874 HDI of life expectancy; 0.501 HDI of income). Vietnam suffered an estimated annual economic loss equivalent to 1.3 percent of GDP or US$5.85 billion due to floods, in the period 1990–2009. In terms of destroyed and damaged houses, 32 percent are due to floods, being the second after storms with 36 percent. The majority of destroyed and damaged houses are situated in the VMD. In terms of loss of life due to floods, 48 percent of Vietnamese population was affected between 1989 and 2010, being, in the same time, the most reported hazard. In such natural hazards, 67 percent of the affected people died (Mekong River Commission, 2009).

Mekong River is one of the greatest rivers in Asia, among the first 12 longest rivers of the world (Mekong River Commission, 2009). The source of the Mekong River is located in the snow-capped Tibetan mountains. The Mekong River has a total length of 4,800 km, draining an area of 795,000 km^2

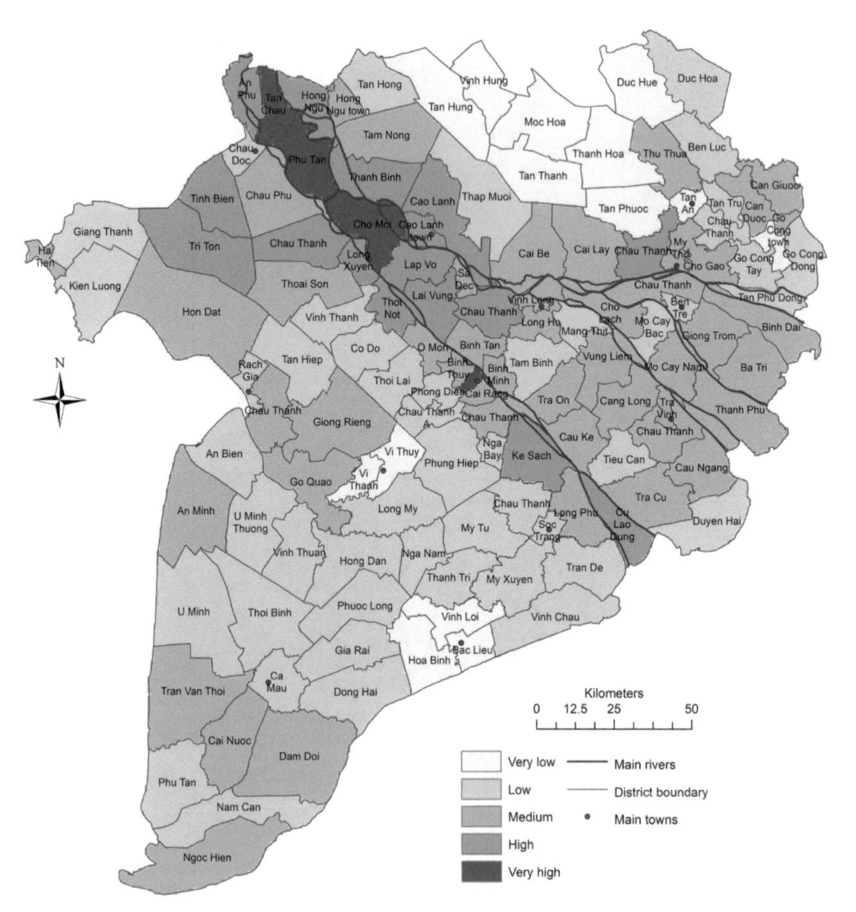

FIGURE 5.2 The Vietnam Mekong Delta (VMD) flood-exposure map for the year 2000. *After Balica et al. (2013).*

and an annual transported water volume of 135, 475 km[5.] The Mekong Delta begins in the city of Phnom Penh, where the river divides into its two main tributaries, the Mekong and the Bassac. The Mekong then divides into six main channels and the Bassac into three to form the "Nine Dragons" of the outer delta in Vietnam (Huy and Toan, 2010). The main delta is made up of a vast triangular plain which is on average 5 m above sea level, and consequently large areas of it are flooded every year. Two-third of the Mekong Delta is situated in the southern part of Vietnam (5.9 million ha) and one-third in Cambodia (1.6 million ha) (ASEAN Regional Center for Biodiversity Conservation, 2012) (see Figure 5.1).

VMD is one of the most productive and intensely cultivated areas in Asia. The VMD consists of 13 provinces with 105 districts and 21 urban areas, which cover an area of 39,352 km^2; constituting 12.1 percent of whole

TABLE 5.8 Vietnam Mekong Delta (VMD) Examples of Flood Exposure and Vulnerability Indicators Values

				District			
N°	Indicator	Factor	Unit	Long Xuyen	Chau Doc	Ninh Kieu	Source
1	SLR	Exposure	mm/year	2.0	2.0	2.0	MONRE (2009)
2	SS	Exposure	m	3.0	2.4	2.4	Mekong River Commission
3	#Cyc	Exposure	#	50.0	50.0	50.0	Mekong River Commission
4	RD	Exposure	m^3/s	15,459.2	8,143.8	23,432.9	ISIS-1D hydraulic model
5	FS	Exposure	%	0.0	0.1	0.0	Mekong River Commission
6	Soil	Exposure	mm/year	5.0	5.0	5.0	Syvitski et al. (2009)
7	CL	Exposure	km	15.0	7.0	8.6	Mekong River Commission
8	CH	Exposure	#	24.0	11.0	26.0	http://gis.chinhphu.vn/
9	PCL	Exposure	People	2,303.3	1,017.4	7,759.0	GSO (2009), GSO (2010) and GSO (2011)
10	GCP	Exposure	%	3.0	3.0	2.5	GSO (2009), GSO (2010) and GSO (2011)
11	S	Susceptibility	#	98.1	89.0	24.8	Mekong River Commission
12	%Disab	Susceptibility	%	41.6	41.6	41.6	GSO (2009), GSO (2010) and GSO (2011)
13	A/P	Resilience	–	6.0	6.0	6.0	Scaled, see Annex 2
14	RT	Resilience	days	300.0	300.0	300.0	Assumed based on GDP per capita and proximity to river/coastal line
15	Drain	Resilience	Km	212.4	273.5	73.4	Mekong River Commission

TABLE 5.9 Po River Delta (PRD) Examples of Flood Exposure and Vulnerability Indicators Values

				Commune			
N°	Indicator	Factor	Unit	Adria	Rovigo	Ferrara	Source
1	SLR	Exposure	mm/year	1.2	1.2	1.2	Brochier and Ramieri (2001)
2	SS	Exposure	m	0.6	0.4	0.5	Lionello et al. (2012)
3	#Cyc	Exposure	#	57.0	57.0	57.0	Lionello et al. (2012)
4	RD	Exposure	m^3/s	1,453.3	779.7	1,559.3	ARPAV (2010), ARPAV (2011)
5	Altitude	Exposure	%	4.0	7.0	9.0	ENEA
6	Soil	Exposure	mm/year	1.1	1.1	1.1	Simeon and Corbau (2008), Correggiari et al. (2005), Gambolati and Teatini (1998)
7	CL	Exposure	km	7.3	4.6	14.5	AdbPo-Autorità di bacino del fiume Po
8	CH	Exposure	#	6.0	15.0	12.0	AdbPo-Autorità di bacino del fiume Po
9	PCL	Exposure	People	180.1	486.3	334.8	http://www.comuni-italiani.it/029/index.html
10	GCP	Exposure	%	−2.3	−0.8	1.3	http://www.comuni-italiani.it/029/index.html
11	S	Susceptibility	#	96.5	76.0	343.7	AdbPo-Autorità di bacino del fiume Po
12	%Disab	Susceptibility	%	34.5	33.3	36.5	http://www.comuni-italiani.it/029/index.html
13	A/P	Resilience	–	7.0	5.0	7.0	Scaled, see Annex 2
14	RT	Resilience	days	250.0	90.0	250.0	Assumed based on GDP per capita and proximity to river/coastal line
15	Drain	Resilience	Km	1,486.6	987.6	620.8	http://idt.regione.veneto.it/app/metacatalog/index?deflevel=165, AdbPo (2006)

country's area (General Statistics Office of Vietnam (2010)). Vietnam is a major producer of rice and contributes 22 million tons (60 percent) to world rice production. The VMD is Vietnam's rice bowl and accounts for 90 percent of the country's rice exports (General Statistics Office of Vietnam (2010)). Within the delta which is dominated by rice, nowadays the farming system also includes activities related to aquaculture, such as fish, crustaceans, mollusks, and aquatic plants, but also rearing of animals, inland fisheries, cash crops, and fruit trees. Fresh and saline water shrimp are raised within the paddy rice fields. In 2008, the area of aquaculture product was 0.75 million ha, contributing to 71 percent of the total area of VMD and provides 59 percent of the demands of Vietnam (General Statistics Office of Vietnam (2010)).

The VMD experiences a monsoonal climate, with the seasonal precipitation as the primary source for river runoff. The wet season lasts from mid-May to October, and accounts for over 90 percent of annual precipitation in the area (General Statistics Office of Vietnam (2010)). Peak river flow at the head of the delta (Phnom Penh) usually occurs in September or October, with the high flow season extending from June to November. Annual minimum flows occur in March or April. Flooding is a major problem of the VMD and severely affects the region. Two types of flooding occur that are affecting the VMD: the first type causes floods from the upstream Mekong Delta, which are long-term flood inundations (from 2 to 6 months) (Mekong River Commission, 2010) and the second type is the tidally induced flood triggered by the tidal regimes in the East and West Sea. Large parts of the delta are flooded in the wet season, forming the Plain of Reeds. The discharge of the Mekong river at the most upstream point of the delta varies from 2,100 to 40,000 m^3/s occurring on April and December, with a peak in September (General Statistics Office of Vietnam (2010)).

5.5.1 Flood-Exposure Maps

The flood-exposure maps related to coastal floods vulnerabilities (see Figure 5.2) are constructed using ArcGIS tools based on the CCFVI results of the several districts, communes, and urban areas. The flood-exposure maps are constructed to visualize part of the flood vulnerability and therefore flood risk. The VMD flood-exposure results are based on adding up a series of indicators which are influencing the natural system, such as sea-level rise, storm surge, number of cyclones, river discharge, foreshore slope, soil subsidence, coastal line, cultural heritage, population close to coastal line, and growing coastal population.

5.5.1.1 VMD Flood-Exposure Results

From the flood-exposure map it can be seen that 15.8 percent of the whole VMD represent very high (901 km^2) and high (4,515 km^2) exposure to floods, mainly districts situated along the rivers and the coast; The most highly exposed to floods are the province of An Giang with Tan Chau, Phu Tan, Cho Moi districts; and the province Can Tho with the city of Ninh Kieu; followed by the high exposed districts An Phu, Hong Ngu, Thanh Binh, Cao Lahn, Cao Lahn town, Chau Thahn (Tien Giang), Tra Vinh, Chau Thahn (Dong Thap), Chau Thahn (Ben Tre), Chau Thahn (An Giang), Lap Vo, Lai Vung, Cu Lao Dung, Ke Sach, Thot Not, Long Xuyen, Chau Thahn (Tra Vinh), Tri Ton, and Tinh Bien. The high value of the exposure index is given by the high river discharge, high number of people living in the area, high number of cultural heritages exposed and also to the direct contact with the river or coast. The remaining majority of districts represent medium-exposure to floods (14,539 km^2, 36.9 percent of the total VMD). More than half of the VMD is under medium, high and very high exposure; these districts are situated along the Bassac and Mekong River banks and mainly on the Northern and North-eastern sides of the delta, but also on Ca Mau Peninsula. The 49.3 percent (19,397 km^2) of the whole VMD represents low and very low exposure to floods. The least exposed districts are the ones from Long An Province.

5.5.2 Flood Vulnerability Maps

Flood vulnerability maps are generally used by stakeholders and municipalities in order to develop and/or improve flood-risk management, public awareness raising and preparedness, and insurance companies. Mapping vulnerability can be used as a tool to picture the most significant vulnerable areas, where stakeholders, governments, and decision-makers should focus their investments (Figure 5.3).

5.5.2.1 VMD Flood Vulnerability Results

From the flood vulnerability map it can be seen that 10.2 percent of the whole VMD represent very high (209 km^2) and high (3,809 km^2) vulnerability to floods. Mainly the districts situated along the rivers and the coast are vulnerable. The remaining majority of districts represent medium vulnerability to floods (14,951 km^2, 38 percent of the total VMD). These districts are situated along the Bassac and Mekong River banks and on the Northern and Northeastern sides of the delta. The 51.8 percent (20,383 km^2) of the whole VMD represents low vulnerability to floods. The least vulnerable districts are the ones from Ca Mau Peninsula.

5.6 COPING WITH FLOOD VULNERABILITY IN DEVELOPED COUNTRIES

Flooding does not affect only the least developed countries, but also happens in the most of the developed and industrialized countries as well.

In Europe floods were at their highest level in 2000–2005, diminishing afterward, however still reaching an important peak in 2010. The year 2013 brought Central Europe into media attention due to the extreme flooding events from May and June; 25 fatalities occurred in Czech Republic, Austria,

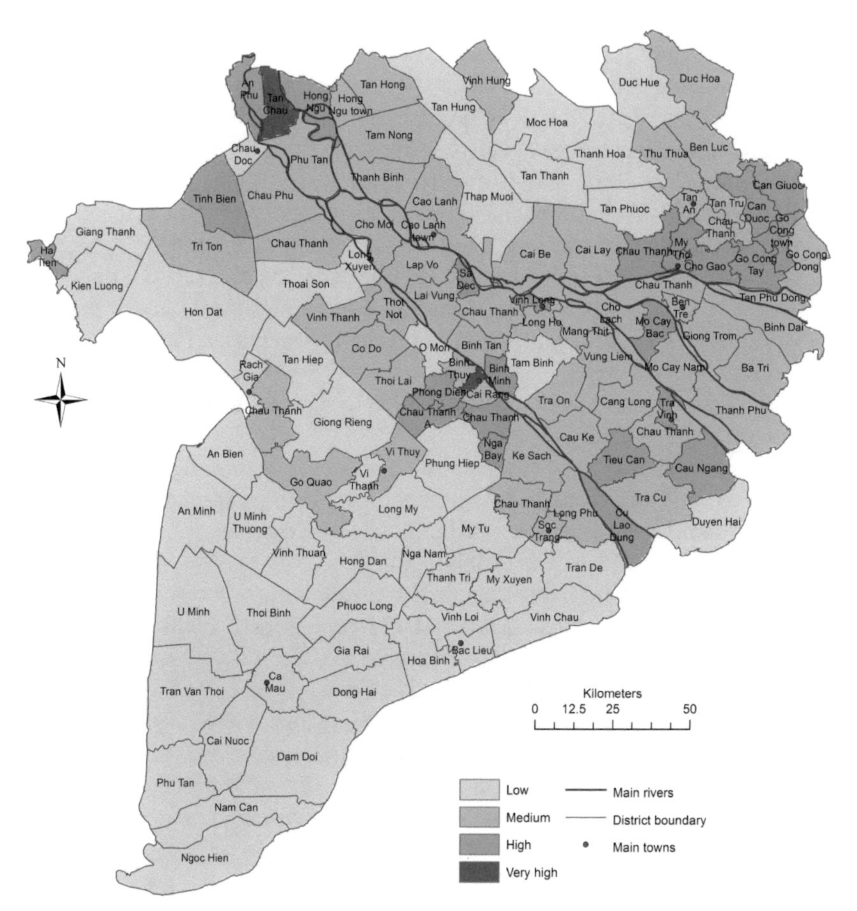

FIGURE 5.3 The Vietnam Mekong Delta (VMD) flood-exposure map for the year 2000. *After Balica et al. (2013).*

and Germany. Floods in Europe mainly occur as a result of several meteorological circumstances such as heavy rainfall, storms, and snow melting. Other factors can increase the occurrence and hence increase their effect, such as lack of suitable river banks, deforestation, and in case of existence of a reservoir not having the flood protection volume available for use.

Only in the first half of the year 2013, flooding caused about EUR 17 billion in economic losses, out of which there were insurance for losses for approximately EUR 4.0 billion. Most of the losses were in Germany (AON, 2013). Swiss Re (2013) mentioned "as a result, 2013 is already the second most expensive calendar year in terms of insured flood losses on sigma records. This year's flooding in Europe has also been more expensive than the 2002 floods in the same region which cost the industry over \$2 billion (\$3 billion at current prices)."

In USA, Sandy Hurricane caused \$10–\$20 billion in losses, which were insured by private companies, or by the National Flood Insurance Program, as everywhere there were uninsured losses, as well, which were estimated to \$30–\$50 billion (Eqecat, Inc., 2013). Comparing, the two costliest storms on record Katrina, in 2005, and Hurricane Andrew, in 1992, which generated losses of over \$100 billion and \$46 billion in inflation-adjusted respectively, the first one is the costliest in US history. Sandy Hurricane killed 54 people, whereas Katrina 1,200 (Jonkman et al., 2009). Hallegate et al.(2013) suggested that by 2050, under climate change conditions and soil subsidence, the present protection needs to be augmented "to avoid unacceptable losses of US\$1 trillion or more per year." The study suggested to "drastically raise" flood defenses. This amount of raise is mainly due to high economically developed areas that can be affected.

It seems that flood exposure and vulnerability is increasing in developed countries as well; therefore, a need occurs for modern nonstructural measures to be used to mitigate and protect from floods worldwide.

Italy's HDI in 2013 is 0.881 ranking number 25 in the Very High Human Development category. Affected number of people due to floods from 1994 to 2002 in the whole Italy was 70,300 and 128 people lost their lives. Between 1984 and 2002, Italy suffered an estimated economic loss, due to floods, of US\$20.03 billion (PreventWeb, 2013). The most significant floods in Italy occurred on the Po River (1951, 1994, and 2000) (Ministry for the Environment, Land and Sea of Italy (2007)). The Po River is the longest river in Italy with a length of 652 km, springing from the Cottian Alps and discharging on the Adriatic Sea (See Figure 5.4). The Po River's annual regime is considered to have two low-water periods (winter and summer) and two flood periods (late

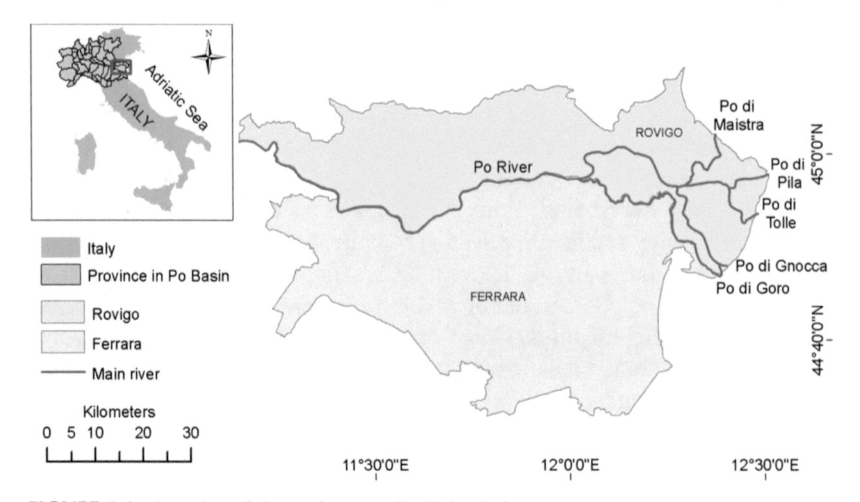

FIGURE 5.4 Location of the study area—Po Delta, Italy.

fall and spring), which are highly influenced by the seasonal precipitations (Zanchettin et al., 2008). The average Po River discharge is about 1,500 m^3/s with a maximum of 13,000 m^3/s (Carraro and Sgobbi, 2008).

The 2000 October flood was an extreme event in Northern Italy. More than 600 mm of precipitation was felt in less than 96 h. With this event the Piedmont area underwent one of the largest floods on record (Carraro and Sgobbi, 2008). Another major flooding event took place in 2003 when 300,000 people were affected and produced an economic damage of more than €2 million. Moreover, on the River Po numerous smaller flooding events occur, which harm agricultural and urban areas, causing significant economic loss. The Po Plain in Northern Italy is a fast subsiding sedimentary basin where about 30 percent of the Italian population lives and 40 percent of Italy's total productive activities are carried out (Carraro and Sgobbi (2008)). The Po River network is divided into four main stretches: the Upper Po (37,000 km^2), the Middle Po (68,000 km^2), the Lower Po (70,000 km^2), and the Po Delta (Zanchettin et al. (2008). Subsidence rates range from 0 to 7 cm/year, the maximum occurring in synclinal areas at the Po Delta (Carraro and Sgobbi, 2008).

The Po Delta is spread over 320 km^2 and the modern Po Delta system, covers five main delta lobes (Correggiari et al., 2005): Po di Volano, Po di Gnocca, Po di Tolle, Po di Pila, and Po di Maistra (See Figure 5.4). The agricultural area of Polesine (Po Delta and the Laguna) is mainly cultivated with corn (30–50 percent), beet sugar (5–10 percent), soya (5–10 percent) barley (0.1–5 percent), and olive trees; the industrial concentration is between 10 and 100 industries/km^2 (Carraro and Sgobbi, 2008). The topography of the coastal line is made out of small hills "rupi" and cliffs.

River floods increased in occurrence and magnitude and caused several episodes of flooding of the Po alluvial plain. Regular avulsions affected the entire Po alluvial and delta plain (Ministry for the Environment, Land and Sea of Italy (2007)). For this study, the area considered are the two provinces, Rovigo and Ferrara, which embraces the Po River for a length of 108 km.

5.6.1 Flood-Exposure Maps

Exposure is an indication of how humans as individuals or/and as social targets, economy, and ecology may be affected by flood hazards. Figure 5.5 presents Polisine's flood-exposure map, for the year 2010, for each "commune."

From the flood-exposure map it can be seen that 20.2 percent of the study area represents very high exposure (998 km^2, 38 percent situated in Ferrara) and 26.2 percent high (1,200 km^2, 54 percent situated in Ferrara) exposure to floods. The highly exposed communes are: Ferrara, Argento, and Comacchio for Ferrara Province and Loreo, Occhiobello, Taglio di Po, and Porto Viro for Rovigo Province. The 30.1 percent of the Rovigo Province is high exposed to floods; the high exposure occurs in the communes of Adria, Bosaro, Carbola, Crespino, Papozze, Posella, Pontecchio Polesine, Porto Tolle, Rasolina,

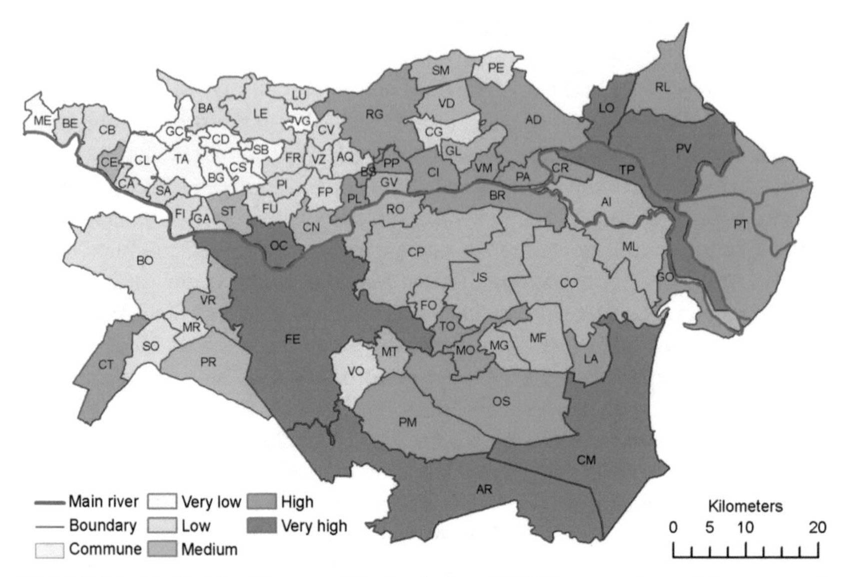

FIGURE 5.5 The Po Flood-Exposure Map for the year 2010 (Please, see Annex 1 for commune's abbreviations).

Rovigo, and Villanova Manchesana. The 25 percent of the high exposure occurs in the Province of Ferrara within the communes Berra, Cento, Goro, Lagosanto, Migliarino, Ostellato, Portomaggiore, and Tresigallo. Mainly the very high and high exposed communes are situated along Po River and the coast. The very high and high exposed province is Ferrara with 1,645km^2, followed by Rovigo with 837 km^2. The 55.6 percent of the whole study area represents high to very high flood exposure. The flood-exposure index is given by the high river discharge, the high number of people living in the area, the high number of cultural heritages exposed and the natural parks, and also to the direct contact with the river or coast.

The 1,192 km^2 (27 percent) of the entire area is under average exposure. Only 17.4 percent, about 1,205 km^2 of the study area are low and very low exposed to floods. The lowest exposed communes from Rovigo are: Villanova del Ghebbo, Trecenta, san Bellino, Melara, Giacciano con Baruchella, Ceneselli, Castelguglielmo, Canda, and Bagnolo di Po. These communes are situated in the Northwestern part of Rovigo's Province. The province of Ferrara has no communes with very low exposure.

5.6.2 Flood Vulnerability Maps

From the flood vulnerability map (Figure 5.6), it can be seen that 2.3 percent (101 km^2) of the whole Po Delta represent very high vulnerability. The very high vulnerable communes (Formignana, Migliaro, Mirabello, Voghiera) are situated

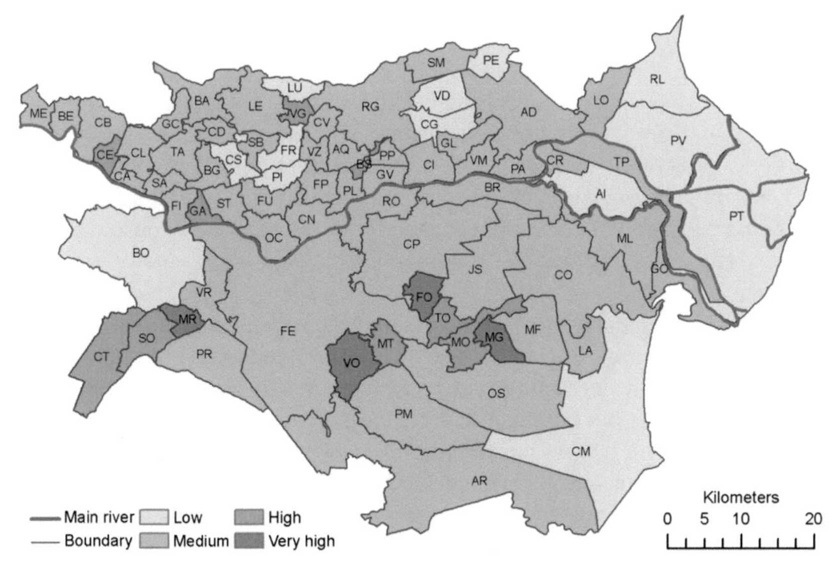

FIGURE 5.6 The Po Flood Vulnerability Map for the year 2010.

only in Ferrara province. High vulnerability to floods, 4.9 percent representing 220 km^2, are mainly the communes (Cento, Masi Torello, Migliatino, Sant'Agostino, Tresigallo) situated in Ferrara Province (176 km^2) along the Po River. High vulnerability to floods also occurs amongst the communes situated in Rovigo, Bosaro, Castemassa, and Villanova del Ghebbo (44 km^2).

The remaining majority of the districts represent average vulnerability to floods (2,967 km^2, 66.4 percent of the total Po Delta). These districts are spread among the two provinces studied, on each bank of Po River. The districts situated on the Laguna have low vulnerability to floods, since almost no population is living in the areas and no industries are developed there, and it is a National Park as well. The 26.4 percent (1,181 km^2) of the whole Po Delta represents low vulnerability to floods. Low vulnerable communes are situated mainly in Rovigo Province (Ariano nel Polesine, Castelguglielmo, Ceregnano, Frata Polesine, Lusia, Pettorazza Grimani, Pincaro, Porto Tolle, Porto Viro, Rosolina, and Viladosa), and some of them are National Parks. Low vulnerability to floods is also a characteristic of the communes of Ferrara, Bondeno, and Comacchio.

5.7 DISCUSSIONS AND PERSPECTIVES

The vulnerability chapter of the book aimed at showing that vulnerability is geographically differentiated and social resilience and exposure do not depend on economy. However both "exposure" and "resilience" are deeply rooted within different lines of vulnerability research.

The comparative analysis of exposure and vulnerability of the two presented case studies, VMD and PRD, can distinguish a set of characteristics that always need to be considered in assessing exposure and vulnerability in two different economic conditions. The analysis focuses on the fact that flood vulnerability happens and affects everyone no matter the type of economy of the society and even more the flood vulnerability is not depending fully on the economy but mainly on the physical exposure of the society and on a population's susceptibility and their socioeconomic resilience.

Measuring and mapping flood exposure and vulnerability could become part of the efforts of the authorities to assess flood-mitigation measures. Knowing the local vulnerability state of a region, resilience can be built. Such kind of assessments can highlight to decision-makers where investments could be made; this could help in prioritizing investments.

Though a lot of technical development can help societies to cope with flood hazard events and to recover quickly, as economies are growing fast and urban areas develop very fast, still a need occurs to better understand water-related vulnerability in both developed and developing countries. Especially in the case of developing countries, where economic development is very fast; population is more exposed to water-related disasters; where recovery processes are relatively slow. The assessment of exposure and vulnerability was analyzed through the critical dimensions of vulnerability in developing and developed delta areas.

Discussing the multiple facets of vulnerability through its hydrogeological, social, and economic components, the case studies offer new perspectives on the use and significance of vulnerability in areas that are having fast and constant economic development. From the results it can be concluded that by establishing the necessary requirements to tackle vulnerability in developing countries, this can lead to strengthening household resilience, building appropriate safeguards against flood risk, and creating and maintaining quality institutions. Vulnerability can be reduced by investing in the future in resilience, by creating resilience plans and by issuing policies that call for flood mitigation and integrating vulnerability into the flood-risk management plans.

Population of VMD experiences fluvial and coastal floods often, mainly due to their hydrogeological exposure they know how to deal with, and how to protect from floods, even if small amounts of investment are given in this way. However, population of the PRD (Italy) has no recent flood experience. PRD is situated in a less exposed area, than VMD; therefore even if high amount of money are invested into building and maintaining dikes or nonstructural measures, it is possible that the population would not be able to deal, cope, or protect themselves from floods, because their memory of floods is very low.

It is important to acknowledge that applying FVI methodology is not revealing important differences between the flood vulnerability and exposure of deltas of developed and developing countries. Additionally, more in-depth assessment is needed to evaluate the various linkages between flood risk and vulnerability. In particular, flood vulnerability assessment applied to different social groups and/or economic sectors requires more in-depth assessment approaches, such as flood modeling.

ANNEX 1. PO'S DELTA COMMUNES ABBREVIATIONS

N°	Commune	Abbreviation	Prov.
1	Adria	AD	RO
2	Argenta	AR	FE
3	Ariano nel Polesine	AI	RO
4	Arqua Polesine	AQ	RO
5	Badia Polesine	BA	RO
6	Bagnolo di Po	BG	RO
7	Bergantino	BE	RO
8	Berra	BR	FE
9	Bondeno	BO	FE
10	Bosaro	BS	RO
11	Calto	CA	RO
12	Canaro	CN	RO
13	Canda	CD	RO
14	Castelguglielmo	CS	RO
15	Castelmassa	CE	RO
16	Castelnovo Bariano	CB	RO
17	Ceneselli	CL	RO
18	Cento	CT	FE
19	Ceregnano	CG	RO
20	Codigoro	CO	FE
21	Comacchio	CM	FE

(Continued)

—cont'd

N°	Commune	Abbreviation	Prov.
22	Copparo	CP	FE
23	Corbola	CR	RO
24	Costa di Rovigo	CV	RO
25	Crespino	CI	RO
26	Ferrara	FE	FE
27	Ficarolo	FI	RO
28	Fiesso Umbertiano	FU	RO
29	Formignana	FO	FE
30	Frassinelle Polesine	FP	RO
31	Fratta Polesine	FR	RO
32	Gaiba	GA	RO
33	Gavello	GL	RO
34	Giacciano	GC	RO
35	Goro	GO	FE
36	Guarda Veneta	GV	RO
37	Jolanda di Savoia	JS	FE
38	Lagosanto	LA	FE
39	Lendinara	LE	RO
40	Loreo	LO	RO
41	Lusia	LU	RO
42	Masi Torello	MT	FE
43	Massa Fiscaglia	MF	FE
44	Melara	ME	RO
45	Mesola	ML	FE
46	Migliarino	MO	FE
47	Migliaro	MG	FE
48	Mirabello	MR	FE

—cont'd

N°	Commune	Abbreviation	Prov.
49	Occhiobello	OC	RO
50	Ostellato	OS	FE
51	Papozze	PA	RO
52	Pettorazza Grimani	PE	RO
53	Pincara	PI	RO
54	Poggio Renatico	PR	FE
55	Polesella	PL	RO
56	Pontecchio Polesine	PP	RO
57	Porto Tolle	PT	RO
58	Porto Viro	PV	RO
59	Portomaggiore	PM	FE
60	Ro	RO	FE
61	Rosolina	RL	RO
62	Rovigo	RG	RO
63	Salara	SA	RO
64	San Bellino	SB	RO
65	San Martino di Venezze	SM	RO
66	Sant'agostino	SO	FE
67	Stienta	ST	RO
68	Taglio di Po	TP	RO
69	Trecenta	TA	RO
70	Tresigallo	TO	FE
71	Vigarano Mainarda	VR	FE
72	Villadose	VD	RO
73	Villamarzana	VZ	RO
74	Villanova Ghebbo	VG	RO
75	Villanova Marchesana	VM	RO
76	Voghiera	VO	FE

ANNEX 2. AWARENESS AND PREPAREDNESS INDICATOR SCALED

Scale	Urban/Sub-Catchment/District/Commune
Score	**Indicating**
1	The population has no concern with floods
2	The population did not experienced flood in recent years
3	The population has little experience with floods; they did not create institutions for flood mitigation. Population does not realize the effects of their actions toward flood protection and are not prepared for emergency situations
4	The population has little experience with floods; institutions have neglected their responsibilities. Population does not realize the effects of their actions toward flood and are not prepared for emergency situations
5	The population has experienced floods a long time ago, so that institutions still exists, population is not aware of these institutions; budget is enough, there is no flood insurance
6	The population has experienced floods; they have recently created institutions to mitigate the harm of floods, budget is scarce, awareness and preparedness is in process of being raised
7	The population has experienced floods long time; they have created and have little trust in institutions to mitigate the harm of floods; population has limited concern over their actions toward flood protection and are not quite prepared for emergency situations
8	The population has experienced floods for long time; they have created and have some trust in institutions to mitigate the harms of floods, there is no flood insurance, population understand the consequences and restrictions of their actions, they are prepared for certain emergency situations
9	The population has experienced floods for long time; they have created and have trust in institutions to mitigate the harms of floods, limited flood insurance, population understand the consequences and restrictions of their actions, they are prepared for certain emergency situations
10	The population has experienced floods for long time (know the potential for floods in the area); they have created and they have high trust in institutions to mitigate the harms of floods, they have flood insurance, they understand the consequences and restrictions of their actions toward flood protection, they are prepared for emergency situations

Source: After Balica, 2007

REFERENCES

Adger, N.W., 2006. Vulnerability, Global Environmental Change, vol. 16, pp. 268–281.

Adger, N.W., 1999. Social vulnerability to climate change and extremes in coastal Vietnam. World Dev. 27, 249–269.

ARPAV, 2010. Misure di Portata Eseguite da ARPAV Nell'anno 2010, Responsabile del Progetto Italo Saccardo. Autori: Antonio Andrich, Silvia Cremonese, Dipartimento Regionale per la Sicurezza del Territorio.

AON, 2013. Aon Benfield, as seen on http://thoughtleadership.aonbenfield.com/Documents/20130904_if_august_global_recap.pdf on August 2013.

ARPAV, 2011. Marco Zasso, Tommaso Settin, SULLA RIPARTIZIONE DELLE PORTATE DEL PO TRA I VARI RAMI E LE BOCCHE A MARE DEL DELTA: ESPERIENZE STORICHE E NUOVE INDAGINI ALL'ANNO 2011 Dipartimento Regionale per la Sicurezza del Territorio. ARPA Emilia Romagna, Servizio Idro-Meteo-Clima: Silvano Pecora, Alberto Agnetti; Progetto e realizzazione, Italo Saccardo.

ASEAN REGIONAL CENTER FOR BIODIVERSITY CONSERVATION, 2012. Mekong Delta. http://www.arcbc.org.ph/wetlands/vietnam/vnm_mekdel.htm (accessed on 11.10.12).

Balica, S.F., 2007. Development and application of flood vulnerability index methodology for various spatial scale, MSc thesis (WSE-HERBD 07–01), UNESCO-IHE, Institute for Water Education. April 2007.

Balica, S.F., Wright, N.G., Van Der Meulen, F., 2012. A Flood Vulnerability Index for Coastal Cities and its Use in Assessing Climate Change Impacts research paper published online.

Brochier, F., Ramieri, E., 2001. Climate change impacts on the Mediterranean coastal zones. NOTA DI LAVORO 27.2001 APRIL 2001 CLIM – Climate Change, Modelling and Policy. Fondazione Eni Enrico Mattei. Thetis, Venice, Italy.

Balica, S.F., Douben, N., Wright, N.G., 2009. Flood vulnerability indices at varying spatial scales. Water Sci. Technol. J. 60 (10), 2571–2580. ISSN 0273–1223.

Balica, S.F., Dinh, Quang, Popescu, I., Vo, Thanh Q., Dieu, Pham Q., 2013. Flood Impact in the Mekong Delta, Vietnam submitted to Journal of Maps. http://dx.doi.org/10.1080/17445647.2013.859636.

Briguglio, L., 2004. Economic Vulnerability and Resilience: Concepts and Measurements home.um.edu.mt/islands/brigugliopaper_version5.doc.

Carraro, C., Sgobbi, A., 2008. Modelling negotiated decision making in environmental and natural resource management. Automatica 44 (6), 1488–1503.

Connor, R.F., Hiroki, K., 2005. Development of a method for assessing flood vulnerability. Water Sci. Technol. 51 (5), 61–67.

Correggiari, A., Cattaneo, A., Trincardi, F., 2005. The modern Po Delta system: lobe switching and asymmetric prodelta growth. Mar. Geol. 222–223, 49–74.

CRED (Centre for Research on the Epidemiology of Disasters), 2013. Floods : the most frequent natural disasters 1993-2012. Disaster Data: A Balanced Perspective, Issue no. 32. August 2013, The report can be downloaded at. http://www.emdat.be/Documents/Publications/ADSR_2012.pdf. THIS IS ALSO the reference for EM-DAT, 2013.

CRED-CRUNCH32, August 2013. Disaster Data: A Balanced Perspective issue #32.

Cutter, S.L., 1996. Vulnerability to environmental hazards. Prog. Hum. Geogr. 20 (4), 529–539.

Cutter, S.L., Boruff, B.J., Shirley, W.L., 2003. Social vulnerability to environmental hazards. Social Sciences Quarterly 84 (2), 242–261.

Cutter, S.L., Boruff, B.J., Shirley, W.L., 2005. Social vulnerability to environmental hazards. Soc. Sci. Q. 84 (2), 242–261.

Davidson, C.M., Robinson, S., Neufeld, V., 2004. Mapping the Canadian Contribution to Strengthening National Health Research Systems in Low and Middle Income Countries. A concept paper, as seen on http://www.ccghr.ca/docs/ConceptPaperPostWrkshp.doc%20on%20September%202010 http://www.ccghr.ca/docs/ConceptPaperPostWrkshp.doc on September 2010.

De Sherbinin, A., Schiller, A., Pulsipher, A., 2007. The vulnerability of global cities to climate hazards. Environ. Urban 19, 39–64.

Di Baldassarre, G., Montanari, A., Lins, H., Koutsoyiannis, D., Brandimarte, L., Bloeschl, G., 2010. Flood fatalities in Africa: from diagnosis to mitigation. Geophys. Res. Lett. 37, L22402. http://dx.doi.org/10.1029/2010GL045467.

Di Mauro, C., 2006. Regional vulnerability map for supporting policy definitions and implementations. In: ARMONIA Conference "*Multi-hazards: Challenges for Risk Assessment, Mapping and Management*", Barcelona.

Dinh, Q., Balica, S., Popescu, I., Jonoski, A., 2012. Climate change impact on flood hazard, vulnerability and risk of the Long Xuyen Quadrangle in the Mekong Delta. Int. J. River Basin Manage. 10 (1), 103–120.

Eqecat Inc, 2013. EQECAT Catastrophe Risk Models. As seen on http://www.eqecat.com/catwatch/hurricane-sandy-insured-losses-initial-estimate-5-billion-more-2012-10-29/ (on 18.09.13).

Fekete, A., 2009. Validation of a social vulnerability index in context to river-floods in Germany. Nat. Hazards Earth Syst. Sci. 9, 393–403. www.nat-hazards-earth-systsci.net/9/393/2009/.

Gabor, T., Griffith, T.K., 1980. The assessment of community vulnerability to acute hazardous materials incidents. J. Hazard. Mater. 8, 323–22.

Gambolati, G., Teatini, P., 1998. Numerical analysis of land subsidence due to natural compaction of the Upper Adriatic Sea basin. ibidem, chapter 5, 103–131.

General Statistics Office of Vietnam, 2010. http://www.gso.gov.vn/default_en.aspx?tabid=491, (accessed 11.08.12).

GSO., 2009. The 1999 Vietnam Population and Housing Census and the 2009 Vietnam Population and Housing. Census, Hanoi. Feb 2011.

GSO., 2010. Statistical Yearbook of Vietnam 2009. General Statistics Office, Hanoi.

GSO., 2011. Prediction of population in Vietnam 2009–2049 and the 2009 Vietnam Population and Housing. Census, Hanoi. Feb 2011.

Gornitz, V., 1990. Vulnerability of the East Coast, USA to future sea level rise. J. Coastal Res. 1 (9), 201–237.

Guillaumont, P., 2008. An Economic Vulnerability Index: Its Design and Use for International Development Policy. WIDER Research Paper, UNUWIDER, 99.

Hallegatte, S., Green, C., Nicholls, R.J., Corfee-Morlot, J., 2013. Future flood losses in major coastal cities. Nature Clim. Change 3, 802–803. http://dx.doi.org/10.1038/NCLIMATE1979. Published online: 18 August 2013.

Holling, C.S., 1973. Resilience and stability of ecological systems. Ann. Rev. Ecol. Syst. 4, 1–25.

Huy, N.S., Toan, T.Q., 2010. Scientific Base for Adaptation Measures for Climate Change and Sea Water Level Rise to the Mekong Delta. Tech. rep., Ministry level project. Hanoi Water Resources University.

IFRC, 2013. International Federation of Red Cross and Red Crescent Societies. 2013, An IFRC Plan of Action for Climate Change 2013–2016, Geneva as seen on. http://www.climatecentre.org/downloads/File/IFRCGeneva/IFRCPlanOfActionForClimateChange.pdf. on May, 2012.

IPCC, 2001. Intergovernmental Panel on Climate Change, Climate Change 2001: The Scientific Basis. Cambridge University Press, Cambridge.

Jonkman, S.N., Maaskant, B., Boyd, E., Levitan, M.L., 2009. Loss of life caused by the flooding of New Orleans after Hurricane katrina: analysis of the relationship between flood characteristics and mortality. Risk Anal. 29 (5), 615–781. As seen on Wiley. Web. On 18th of September, 2015.

Kaufmann, D., Kraay, A., Zoido-Lobatn, P., 1999. Aggregating Governance Indicators. World Bank Policy Research Working Paper 2195.

Lewis, J., 1999. Development in Disaster-prone Places. Intermediate Technology, London.

Lionello, P., et al., 2012. Program focuses on climate of the Mediterranean region. Eos Trans. AGU 93 (10), 105.

McEntire, D., Crocker, C., Peters, E., 2010. An addressing vulnerability through an integrative approach. Int. J. Disaster Resilience Built Environ. 1 (1), 50–64.

McLaughlin, S., Cooper, J.A.G., 2010. A multi-scale coastal vulnerability index: a tool for coastal managers? Environ. Hazard.: Hum. Policy Dimens. 9 (3). Publisher: Earthscan pp. 233–248(16).

Mekong River Commission, 2009. Assessment of Basin-wide Development Scenarios, Technical report.

Mekong River Commission, 2010. Assessment of Basin-wide Development Scenarios (Technical report).

Ministry for the Environment, Land and Sea of Italy, 2007. Fourth National Communication under the UN Framework Convention on Climate Change Italy November, 2007 as seen on. http://unfccc.int/resource/docs/natc/itanc4.pdf. on September 2013.

Mirza, M. Monirul Qader, 2003. Climate change and extreme weather events: can developing countries adapt? Clim. Policy 3, 233–248.

NATURA, 2000. WWF, 2006: Natura 2000 in Europe. An NGO assessment, Budapest, 92 pages.

Neukum, C., Hötzl, H., Himmelsbach, T., 2008. Validation of vulnerability mapping methods by field investigations and numerical modelling. Hydrogeol. J. http://www.springerlink.com/earth-and-environmental-science/. http://www.springerlink.com/earth-and-environmental-science/. Earth and Environmental Science, ISSN1431–2174, April 2007.

Peduzzi, P., Dao, H., Herold, C., Rochette, D., Sanahuja, H., 2001. Feasibility Study Report – on Global Risk and Vulnerability Index – Trends per Year (GRAVITY) (United Nations Development Programme Emergency Response Division UNDP/ERD, Geneva).

Penning-Rowsell, E.C., Chatterton, J.B., 1977. The Benefits of Flood Alleviation: A Manual of Assessment Techniques (The Blue Manual). Gower Technical Press, Aldershot, UK.

Pérez España, H., Arreguín Sánchez, F., 1999. Complexity related to behaviour of stability in modelled coastal zone ecosystems. Aqua. Ecosyst. Health Manage. 2, 129–135.

Pethick, J., Crooks, S., 2000. Development of a coastal vulnerability index: a geomorphological perspective. Environ. Conserv. 27, 359–367.

Pratt, C., Kaly, U., Mitchell, J., 2004. How to Use the Environmental Vulnerability Index. UNEP/SOPAC South Pacific Applied Geo-science Commission. Technical Report 385. www.vulnerabilityindex.net/Files/. EVI%20Manual.pdf (accessed 15.09.13).

PreventWeb, 2013. As seen on http://www.preventionweb.net/english/countries/statistics/?cid=85 on 18th of September 2013.

Quinn, P., Hewett, C., Popescu, I., Muste, M., 2010. Towards new types of water-centric Collaboration: Instigating the upper Mississippi river basin observatory process. Water Manage. 163, 39–51. http://dx.doi.org/10.1680/wama.2010.165.1.39.

Ramade, F., 1989. Ecological catastrophes. Futuribles 1, 63–78.

Sarewitz, D., Pielke, R., Keykhah, M., 2003. Vulnerability and Risk: Some Thoughts from a Political and Policy Perspective submitted to Risk Analysis.

Sebald, C., 2010. Towards an Integrated Flood Vulnerability Index– A Flood Vulnerability Assessment Geo-Information Science and Earth Observation for Environmental Modelling and Management Level: Master of Science 2010-25. University of Southampton, UK.

Sullivan, C.A., 2002. Calculating a water poverty index, Wallingford. World Develop. 7, 1195–1210.

Swiss Re, 2013. News release as seen on http://www.swissre.com/media/news_releases/nr_20130708_floods_europe.html on 1st September 2013.

Syvitski, James P.M., Kettner, Albert J., Overeem, Irina, Hutton, Eric W.H., Hannon, Mark T., Brakenridge, G. Robert, Day, John, Vörösmarty, Charles, Saito, Yoshiki, Giosan, Liviu, Nicholls, Robert J., 2009. Sinking deltas due to human activities. Nature Geoscience 2, 681–686 (2009). Published online: 20 September 2009 | doi:10.1038/ngeo629.

Timmerman, P., 1981. Vulnerability, Resilience and the Collapse of Society: A Review of Models and Possible Climatic Applications. Institute for Environmental Studies, University of Toronto, Canada.

Umberto, Simeoni, Corinne, Corbau, 2009. A review of the Delta Po evolution (Italy) related to climatic changes and human impacts. Geomorphology 107 (2009), 64–71. Elsevier.

UN, 2013. United Nations Statistics Division-Standard Country and Area Codes Classifications. Unstats.un.org. Retrieved on 2013-07-12.

UNDP/BCPR, 2004. United Nations Development Programme and Bureau for Crisis Prevention and Recovery: A Global Report: Reducing Disaster Risk, A Challenge for Development. New York, 146 pp., ISBN 92-1-126160-0. Printed by John S. Swift Co, USA.

UNDP, 2010. Human development report 2010, Palgrave Macmillan United Nations, 1982. In: Proceedings of the Seminars on Flood Vulnerability Analysis and on the Principles of Floodplain Management for Flood Loss Prevention, September, Bangkok.

UNEP, 2004. Manual: How to Use the Environmental Vulnerability Index (EVI) as see on http://www.vulnerabilityindex.net/EVI_Calculator.htm. September 2013.

Walker, B., Holling, C.S., Carpenter, S.R., Kinzig, A., 2004. Resilience, adaptability and transformability in social–ecological systems. Ecol. Soc. 9 (2), 5 http://www.ecologyandsociety.org/vol9/iss2/art5/.

Wilhelmi, O.V., Wilhite, D.A., 2002. Assessing Vulnerability to Agricultural Drought: A Nebraska Case Study Natural Hazards, vol. 25. Kluwer Academic Publishers, 37–58. (Printed in the Netherlands).

World Bank, 2012. Developing Countries seen on http://occupationalhealth2014.conferenceseries.net/Developing-Countries-list.pdf on 1st of October.

Yohe, G., Malone, E., Brenkert, A., Schlesinger, M., Meij, Henk, Xing, Xiaoshi, 2006. Global distributions of vulnerability to climate change. Integr. Assess. 6 (3). ISSN 1389–5176. http://journals.sfu.ca/int_assess/index.php/iaj/article/view/239/210.

Zanchettin, D., Rubino, A., Traverso, P., Tomasino, M., 2008. Impact of variations in solar activity on hydrological decadal patterns in northern Italy. J. Geophys. Res. 113, D12102. http://dx.doi.org/10.1029/2007JD009157.

Chapter 6

Integrated Risk Assessment of Water-Related Disasters

Carlo Giupponi [1], Vahid Mojtahed [1], Animesh K. Gain [1], Claudio Biscaro [1,2] and Stefano Balbi [1,3]

[1] *Dipartimento di Economia and Venice Centre for Climate Studies, Università Ca' Foscari di Venezia, Venezia, Italy,* [2] *Institut für Organisation und Globale Managementstudien, Johannes Kepler Universität, Linz, Austria,* [3] *Basque Centre for Climate Change (BC3), Bilbao, Spain*

ABSTRACT

This chapter presents a conceptual framework (KULTURisk Framework or KR-FWK) and its implementation methods (SERRA or Socio-Economic Regional Risk Assessment) for integrated (physical and economical) risk assessment and evaluation of risk prevention benefits in the field of water-related processes. The KR-FWK (i.e. from the name of the European project within which it originated) and the SERRA approach were developed upon preexisting proposals, with three main innovation aims: (1) to include the social capacities of reducing vulnerability and risk, (2) to provide an operational solution to assess risks, impacts, and the benefits of plausible risk reduction measures, by including a monetary estimation of costs and benefits, and (3) to go beyond the estimation of direct tangible costs. Vulnerability is considered as a result of the interactions between physical (territorial) characteristics and the susceptibility and the capacities of the socioeconomic system to adapt and cope with a specific hazard, expressed as a nondimensional index ranging between 0 and 1. Exposure, is instead assessed in monetary terms, and thus the multiplicative combination of two indices ranging between 0 and 1 (hazard and vulnerability) with a third one (exposure) expressed in monetary terms produces a monetary quantification of risk, which can be used for supporting decisions via cost-benefit analysis. Regarding the third aim of going beyond the estimation of direct tangible damages, operational solutions are proposed to evaluate four possible socioeconomic costs possibly deriving from the adverse consequences of flood, namely direct/indirect and tangible/intangible costs. The proposed methodology aims to be comprehensive with respect to the set of receptors usually considered in the literature of regional risk assessment. The sets of receptors considered are people, economic activities, categorized as (1) buildings; (2) infrastructures; and (3) agriculture and cultural heritage and ecosystems. We show how to apply SERRA and

Hydro-Meteorological Hazards, Risks, and Disasters. http://dx.doi.org/10.1016/B978-0-12-394846-5.00006-0

the KR-FWK in the case of Dhaka/Lower Brahmaputra/Bangladesh, by reusing elaborations already done or in progress and by developing some minimal new work; e.g. to demonstrate indirect/intangible costs.

6.1 INTRODUCTION AND STATE-OF-THE-ART OF RISK ASSESSMENT METHODS OF WATER-RELATED PROCESSES

Several legal documents, including the European Flood Directive (2007/60/EC), call for the development of "flood risk maps showing the potential adverse consequence" of floods with different return times. Those maps, together with the results of other analyses and, in particular, economic valuations, should then be used as planning instruments to support decisions. The need thus emerges for methods and tools to assess the adverse consequences of the flood risk in an integrated, comprehensive, and coherent manner to allow competent authorities to make decisions by identifying the preferable solution(s), by trading off among contrasting technical, economic and environmental performances, social conflicts, uncertainty, etc.

Such efforts are multidisciplinary in essence, requiring contributions ranging from hydrology, and environmental sciences to economic and social sciences. However, the various disciplines, when dealing with natural hazards, have provided different and often contrasting interpretations of the crucial concepts of vulnerability and risk. Therefore, a straightforward solution for disciplinary integration does not exist and conceptual discrepancies and terminological inconsistencies emerging from the various research communities have to be solved preliminarily (Mercer, 2010; Renaud and Perez, 2010; Thomalla et al., 2006).

The review of scientific literature on the risks deriving from natural hazards (floods, in particular) is conducted in support to the development of the methodological framework as presented below and demonstrates that:

1. substantial discrepancies are evident in the risk literature, which is fragmented into many disciplinary streams;
2. at least two distinct research streams are of greater impact in the research arena: disaster risk reduction (DRR) and climate change adaptation (CCA);
3. the ambition of trying to unify the terminologies in use is unrealistic, but significant contributions can be provided in terms of communication interfaces and operational solutions;
4. definitions are evolving within each community;
5. risk assessment is usually focused on damages, i.e. direct tangible costs, but other direct, indirect, and intangible costs should be considered to have a more comprehensive understanding of the costs at stake (Balbi et al., 2013);

6. in general, social, and economic aspects find limited consideration, whereas they are crucial for an accurate and comprehensive assessment of the risk;

The most widely adopted approach for the calculation of risk, in particular within the DRR research community, refers to risk as the expected damages—more precisely direct tangible—which are calculated as a function of hazard, physical, and environmental vulnerability, and exposure (Crichton, 1999):

$$R = f(H, V, E) \quad (6.1)$$

The first element is characterized by probability distributions or specific return times, and together with the second it is usually expressed as a dimensionless index, whereas the latter, exposure, provides the unit of measurement of risk, which can be expressed in physical or monetary terms.

This framework is straightforward and widely adopted, but it finds its limitations mainly in the narrow consideration of the complexity of the various dimensions of risk and in particular of the social ones. To fit within the formula reported above, all the risk dimensions have to be extremely simplified and aggregated to produce two dimensionless indices of hazard and vulnerability. This can be quite challenging when the attention is driven to the social dimensions of vulnerability (Cutter, 1996), where it is distinct from biophysical vulnerability, but later aggregated into a single notion of "place vulnerability".

Although the DRR community drives more emphasis on the concept of risk, other research streams, and in particular the one focused on CCA, were more focused on the assessment of vulnerability, under the auspices of the Intergovernmental Panel on Climate Change (IPCC). In the DRR studies, vulnerability is indeed considered, but it is mainly regarded as an input for the quantification of risk. Instead, CCA research considers vulnerability as an output deriving from social conditions and processes, and in particular by the combination of the status of the adaptive capacity of the social–ecological system and the potential impacts deriving from the combination of local sensitivity and the exposure to a specific hazard (Klein, 2004; Parry, 2007).

As a consequence, although DRR traditionally focuses on the knowledge of hazard by means of risk analysis, CCA is more focused on the importance of understanding the behavior of, and the consequences for the local communities involved by means of vulnerability assessment.

The two main research streams have increasingly interacted in recent times, because the climate change dimension has gained ever-greater attention in the consideration of natural hazards, whereas climate variability and extreme events were brought to the core of both climate change science and CCA in particular. Effects of such interactions are clearly visible in the IPCC publication "Managing the Risks of Extreme Events and Disasters to Advance Climate Change Adaptation" IPCC-SREX (Field et al., 2012). Although in the CCA framework adopted by the IPCC-AR4 (Parry, 2007) the concept of risk

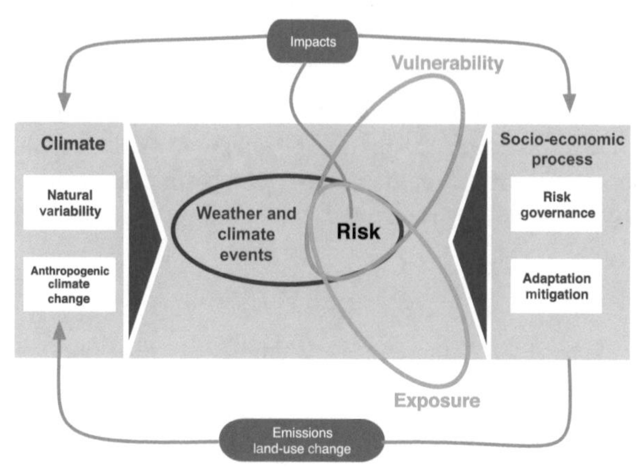

FIGURE 6.1 Managing the risk of extreme events and disasters to advance climate change adaptation (CCA). *Adapted and redrawn from Field et al. (2012).*

was missing and vulnerability resulted from the combination of potential impacts, exposure, and sensitivity, in the IPCC-SREX, risk is at the core (see Figure 6.1). However, the causal chains among the various concepts are not clearly defined, thus limiting the possibility of deriving uncontroversial operational assessment methods.

One widely cited approach for probabilistic risk assessment is proposed by the CAPRA Platform[1] (Cardona et al., 2010) as an operational combination of multiple disciplinary models and cost-benefit analysis (CBA) for decision support. The Methods for the improvement of Vulnerability Assessment in Europe (MOVE, 2010) developed an alternative and rather complex conceptual framework in which risk is determined by the interactions between the environmental dimension in terms of hazards and the society with its specific features in terms of susceptibility, fragility, resilience, etc. A more mechanistic approach based on system dynamics and the notion of socioecosystem modeling can be found in (Turner et al., 2003), with a focus on the analysis of vulnerability in relation again to the notions of resilience, exposure, sensitivity, etc., but without explicit consideration of risk.

In general, it is clear from the analysis of the literature that no practical solution exists for integrating and synthesizing the main references without facing the need to decide among contrasting definitions. Moreover, it is evident that often the proposed frameworks, besides providing a conceptual and pictorial representation of relationships among the various elements, do not identify the causal and functional relationships, needed for any attempt to develop operational algorithms for risk assessment. In some cases, some sort of index is proposed (e.g. a vulnerability index), but the functional structure is

1. See http://www.ecapra.org/.

very simplistic, typically an additive combination of dimensionless indicators, without proper consideration of fundamental issues, such as normalization effects, internal compensation, weighting, independence of variables, etc.

Furthermore, risk assessment is usually focused on the valuation of the potential consequences, but very often these are limited to the expected damages in terms of direct and tangible expected costs, even if indirect and intangible costs are proven to be a quite relevant component of the potential consequences of a natural hazard (Cochrane, 2004; Okuyama and Sahin, 2009).

Given the current state of the art, briefly described above, the KULTURisk project[2], an EU-funded research aimed at developing a culture of risk prevention by evaluating the benefits of different risk-prevention initiatives, has approached the development of a novel methodology for integrated risk assessment, focused on different types of water-related catastrophes, such as inundations, urban flash floods, and rainfall-triggered debris flows[3].

The KULTURisk methodological framework and its operational approach SERRA (Socio-Economic Regional Risk Assessment), are developed upon the well-established Regional Risk Assessment literature (Landis, 2004), with specific focus on: (1) the integration of physical/environmental (P/E) dimensions and the socioeconomic ones; (2) the consideration of social capacities of reducing risk; (3) the economic valuation of risk that goes beyond the direct tangible costs for decision support on risk mitigation measures; and (4) the integration of CCA in DRR[4].

6.2 METHODOLOGICAL FRAMEWORK FOR INTEGRATED RISK ASSESSMENT

A long process of collaboration and recursive exchange of intermediate drafts within the KULTURisk consortium brought to a shared glossary (see Table 6.1), developed mainly upon the IPCC-SREX (Field et al., 2012) and United Nations Office for Disaster Risk Reduction (UNISDR) Hyogo Framework (UNISDR, 2005).

Therefore, the KULTURisk Framework (KR-FWK) implemented in SERRA has the ambition to offer an effective interface and a common ground for experts working across diverse disciplines, with the common aim of supporting decisions for risk mitigation actions.

To have an impact on decision making the proposed approach should become part of the policy implementation process, and more specifically to flood mitigation and CCA measures. Willows et al. (2003) proposed a cyclic

2. KULTURisk: Knowledge-based approach to develop a cULTUre of Risk prevention. FP7-ENV-2010 Project 265,280 (http://www.kulturisk.eu/).
3. The present paper is the result of author's elaborations on the contents of project deliverable 1.6.
4. For details about the KULTURisk approach to RRA, see project deliverables 1.2 and 1.7 at http://www.kulturisk.eu/results/wp1.

TABLE 6.1 The Comparison between KULTURisk Framework (KR-FWK) and Other Frameworks Terminologies

	KR-FWK	Field et al., 2012 UNISDR (2009)
Adaptive capacity	Field et al. (2012)	*IPCC*: The combination of the strengths, attributes, and resources available to an individual, community, society, or organization (ex-ante hazard) that can be used to prepare for, and undertake actions to reduce adverse impacts, moderate harm, or exploit beneficial opportunities. *UNISDR*: N/A
Attenuation	Considers structural and explicit, manufactured barriers to the hazard, which may affect exposure.	N/A
Coping capacity	Field et al. (2012)	*IPCC*: The ability of people, organizations, and systems, using available skills, resources, and opportunities, to address, manage, and overcome (ex-post hazard) adverse conditions. *UNISDR*: The ability of people, organizations, and systems, using available skills and resources, to face and manage adverse conditions, emergencies, or disasters.
Direct costs	The costs due to the damages provoked by the hazard and which occur during the physical event (Merz et al., 2010).	N/A

Exposure	Field et al. (2012)	*IPCC*: The presence of people; livelihoods; environmental services and resources; infrastructure; or economic, social, or cultural assets in places that could be adversely affected. *UNISDR*: People, property, systems, or other elements present in hazard zones that are thereby subject to potential losses.
Hazard	Field et al. (2012)	*IPCC*: The potential occurrence of a natural or human-induced physical event that may cause loss of life, injury, or other health impacts, as well as damage and loss to property, infrastructure, livelihoods, service provision, and environmental resources. *UNISDR*: A dangerous phenomenon, substance, human activity or condition that may cause loss of life, injury or other health impacts, property damage, loss of livelihoods and services, social and economic disruption, or environmental damage.
Indirect costs	Those induced by the hazard but occurring, in space or time, outside the physical event (Merz et al., 2010)	N/A
Intangible costs	Values lost due to a disaster, which cannot, or are difficult/controversial to be monetized because they are non-market values (Merz et al., 2010).	N/A
Pathway	The geomorphological characteristics of the region under assessment, which affect the way hazards propagate and therefore exposure (e.g. digital elevation model). It includes natural barriers to the hazard.	N/A

Continued

TABLE 6.1 The Comparison between KULTURisk Framework (KR-FWK) and Other Frameworks Terminologies—cont'd

	KR-FWK	Field et al., 2012 UNISDR (2009)
Potential consequences	Are expressed in the form of the total cost matrix.	N/A
Receptor	A Physical entity, with a specified geographical extent, which is characterized by particular features (e.g. human beings, protected areas, cities, etc.).	N/A
Resilience	Not applied in our framework but can be interpreted as opposite to the definition of vulnerability.	*IPCC*: The ability of a system and its component parts to anticipate, absorb, accommodate, or recover from the effects of a hazardous event in a timely and efficient manner, including through ensuring the preservation, restoration, or improvement of its essential basic structures and functions. *UNISDR*: The ability of a system, community or society exposed to hazards to resist, absorb, accommodate to and recover from the effects of a hazard in a timely and efficient manner, including through the preservation and restoration of its essential basic structures and functions
Risk	The combination of the probability of a certain hazard to occur and of its consequences.	*IPCC*: The likelihood over a specified time period of severe alterations in the normal functioning of a community or a society due to hazardous physical events interacting with vulnerable social conditions, leading to widespread adverse human, material, economic, or environmental effects that require immediate emergency response to satisfy critical human needs and that may require external support for recovery. *UNISDR*: The combination of the probability of an event and its negative consequences.

Risk perception/ awareness	The overall view of risk as perceived by a person or group including feeling, judgment, and culture (ARMONIA project, 2007).	*IPCC*: N/A *UNISDR*: The extent of common knowledge about disaster risks, the factors that lead to disasters and the actions that can be taken individually and collectively to reduce exposure and susceptibility to hazards, while increasing the adaptive capacity
Susceptibility	Susceptibility brings in a physical/environmental assessment of the receptors, i.e. the likelihood that receptors could potentially be harmed by any hazard given their structural factors, typology of terrain and characteristics (in physical and non-monetary terms)	N/A
Tangible costs	The costs, which can be easily specified in monetary terms because they refer to assets, which are traded in a market (Merz et al., 2010).	N/A
Value factor	The social, economical, and environmental value of the exposed receptors.	N/A
Vulnerability	The combination of susceptibility and social capacities determining the propensity or predisposition of a community, system, or asset to be adversely affected by a specific hazard.	IPCC: The propensity or predisposition to be adversely affected. *UNISDR*: The characteristics and circumstances of a community, system, or asset that make it susceptible to the damaging effects of a hazard.

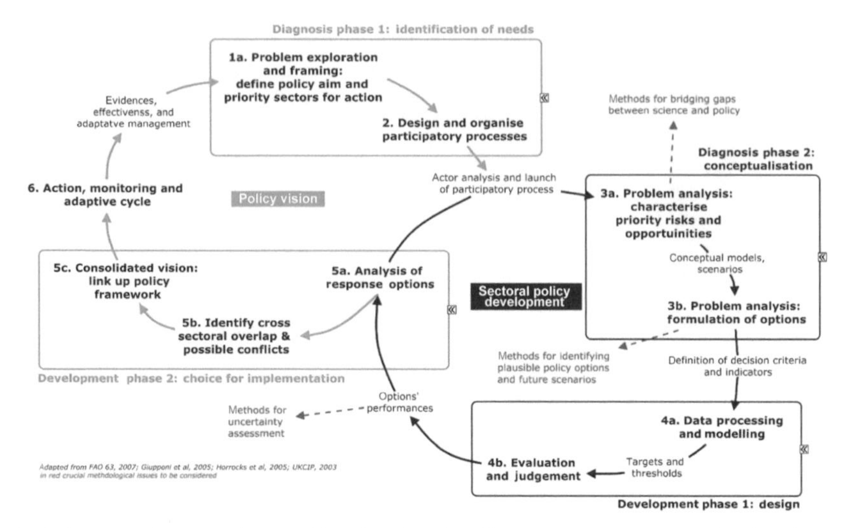

FIGURE 6.2 Cyclic decision-making flowchart for climate change adaptation (CCA) and risk management.

approach for dealing with climate change risks, others proposed frameworks for CCA (Harley et al., 2009), or for the management of participatory decisions (Giupponi et al., 2008). Figure 6.2 shows how the assessment of measures for flood risk management could be framed within two coupled cyclic processes to mainstream sectoral measures in the policy implementation.

The implementation of the methodological framework into an operational procedure is developed upon the formalization of risk being a function of *hazard, exposure, and vulnerability.* Overall formula (6.1) holds in the various processes proposed in SERRA (e.g. risk being necessarily null, when hazard is zero), even if not necessarily the algorithm is forced to produce two independent and dimensionless indexes (H and V) to be used in a multiplicative combination with one monetary index of exposure.

Figure 6.3 depicts how the variables of formula (6.1) are assessed to produce a quantification of risk. In the case of a flood event, the *hazard* outcomes are typically one or more maps of intensity (expressed in terms of depth, persistence, and/or velocity) of the flood, provided by the hydrological analysis and modeling, with reference to different return times and measures of intensity. Additionally, multiple receptors have to be considered. The European Flood Directive identified four categories of receptors: people, economic activities, cultural goods, and the environment component (EC, 2007).

Exposure identifies the presence of people and assets and as much as possible the social, environmental, and economical value of them. It may represent the term of eqn (6.1) expressed in monetary terms to be multiplied with normalized indices of hazard and vulnerability in case of full

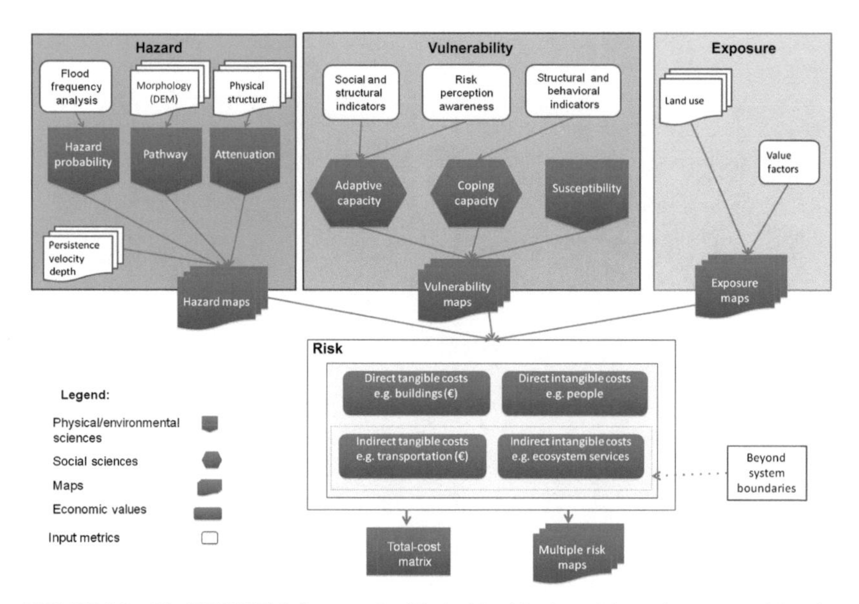

FIGURE 6.3 The KULTURisk framework with the identification of the main sources of data for the quantification of nodes.

monetization of damages and costs. Vulnerability maps result from the combination of P/E and social components. The P/E component is captured by the likelihood that receptors located in the area considered could potentially be harmed (susceptibility of receptors). The social components can be described through the assessment of the ex-ante preparedness of society given their risk perception and awareness, to combat hazard and reduce its adverse impact (Adaptive Capacity) and the ex-post skills to cope with and overcome the impacts of the hazard considered (Coping Capacity). A list of social indicators that can proxy adaptive and coping capacities is proposed in the following section.

The above-described elements allow calculating the expected damages related to the risks associated to different hazardous scenarios. Risk is composed of four components constituting the total cost matrix (TCM) and deriving from the combinations of indirect/direct tangible/intangible costs (Figure 6.4). Direct costs include all the tangible/intangible costs in the geographical location during the hazardous event. All the costs generated outside the time frame or the geographical location of the hazardous event are represented by indirect costs.

Hazard, vulnerability, and exposure are usually reported as maps. Therefore, they are spatially explicit, and typically managed in a GIS context (Geographical Information System), such as, for example, raster-based maps of inundated areas produced by hydrologic models, or census-based maps

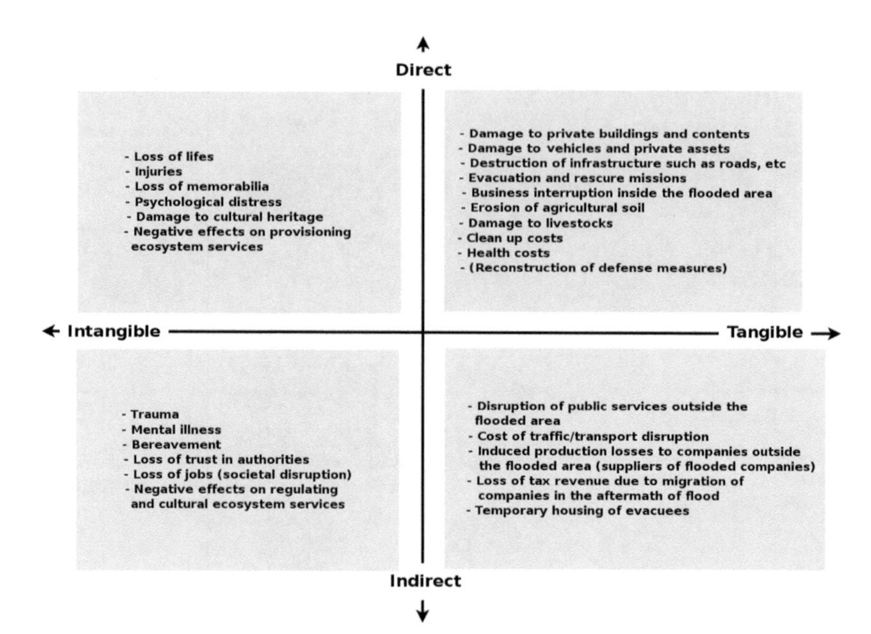

FIGURE 6.4 The total cost matrix (TCM). *Adapted from: Penning-Rowsell et al. (2005), Jonkman et al. (2008), and Merz et al. (2010).*

representing the distribution of buildings or people, and other indicators used to calculate adaptive and coping capacities.

The foreseen usage of the proposed approach is mainly in supporting decision making and in particular in the process of identifying risk-mitigation measures, which is usually a comparative exercise in which plausible solutions to be implemented are compared with a "Baseline scenario". The effects expected from the measures are thus expressed either in terms of monetary benefits (avoided costs), or by means of effectiveness indicators and they are compared together with their expected costs, by means of CBA or cost-effectiveness analysis (CEA), respectively, as described below.

6.3 THE EVALUATION OF BENEFITS OF RISK REDUCTION

In case of full monetization of risk, the first step consists of identification of potential costs of a given disaster (e.g. a flood with a specific return time), without considering any of the preventive measures to be assessed (baseline scenario). A second step is to estimate the expected costs of the same disaster with the risk-reduction measures in place (alternative scenarios) and their implementation costs. The preferred measure is the one that provides the best benefit/cost ratio. In case of CEA, instead, the preferred measure is typically the one meeting the risk prevention objectives at the minimum cost.

As stated above, traditional risk assessments have been primarily dealing with direct tangible costs in a very detailed fashion, however, a whole set of neglected costs exist that should be considered in view of providing a comprehensive quantification of risk (Balbi et al., 2013). For example, an early warning system might only partially reduce the amount of direct tangible costs (e.g. you can move your car but not your house), but it can:

1. save many people's lives (direct intangible costs);
2. change behavior of people by avoiding long-lasting traumas (indirect intangibles costs);
3. prevent evacuation costs following the disaster (indirect tangible costs).

In order to go beyond the direct tangible costs, the following methodological and operational requirements emerge:

1. a more comprehensive and less reductionist notion of risk, but with a disaggregated structure as shown by the TCM (Figure 6.4);
2. a functional description of the expected consequences of the hazard considered according to the quadrants of the TCM;
3. the consideration of other types of impacts, for which the issue of expression in monetary terms emerges;
4. the implementation of methods for economic valuation of nonmarket goods, in order to provide monetary values to intangibles, whenever possible and desired for supporting the decisions based upon the implementation of CBA methods;
5. the consideration of alternative evaluation methods in those cases in which CBA is not possible or desired, such as CEA.

6.4 THE SOCIAL DIMENSION: ADAPTIVE AND COPING CAPACITIES FOR RISK PREVENTION

One of the main innovations of the proposed framework is the detailed consideration of social capacities (adaptive and coping capacities) in the process of measuring risk by means of the TCM, thus providing an attempt to make the assessment of social vulnerability operational and as far as possible quantifiable.

The main challenges for the analysis of social capacities in a risk assessment context can be identified with respect to:

1. tailoring the set of indicators to the context;
2. defining empirical functions for the estimation of indicators and for their aggregation in a single vulnerability index.

Social scientists usually investigate these capacities at the case study level by means of questionnaires and interacting with local stakeholders, with a semiquantitative research approach (Steinfuhrer et al., 2009). Indeed, the variables measuring those capacities should be chosen according to the context of application. However, as shown in (Cutter et al., 2003), a minimal set of indicators based on secondary data can be selected to approximate the magnitude of social vulnerability.

We propose a list of variables and indicators, as shown in Table 6.2, which may compose a minimal set of data to approximate those capacities. Furthermore, we declare our assumptions about their contributions to the TCM. Some of the indicators may affect both adaptive and coping capacities, such as income level. A society with higher income level could have had a higher adaptive capacity by incorporating early warning systems at the community level, or at the individual level by taking precautionary actions such as fortifying their residential building. Equally, higher income can affect coping capacity when the communities' or individuals' ability in coping with flood is increased. Therefore, a careful scrutiny is necessary for empirically testing the significance of each indicator on adaptive or coping capacities, while avoiding double counting and internal correlations. Although most of the indicators can be derived from secondary data or from the census and regional accounts, some variables might be difficult to derive without ad hoc activities. This is particularly evident for trust or risk perception, which is an important component of the project and of the framework. Depending on the geographical scale, level of detail, available time, and financial resources, proxies could always be considered as substitutes to the proposed variables.

6.5 THE IMPLEMENTATION OF THE KR-FWK

To deal with heterogeneous issues and application contexts, the proposed framework is necessarily generic, simplified in its overall conceptualization, but still rather complex in its practical implementation. To make it operational, the KR-FWK's various nodes reported in Figure 6.3 must be quantified and contribute to the calculation of the vulnerability index, as prescribed in the SERRA approach.

As it has been demonstrated by the application of SERRA in the various case studies of the project, not a single procedure exists for quantification of the nodes, instead, each should be defined according to the specific objectives and conditions (e.g. data availability, study boundary, etc.) of each implementation. For instance, simpler solutions can consider aggregate costs and/or indicators of social capacities, instead of spatial ones.

The process of tailoring has various degrees of freedom summarized as follows (Giovannini et al., 2008):

- identification of the application context in terms of scenarios and measures;
- selection of indicators and data mining;
- choice of normalization procedures;
- choice of weighting methods;
- selection of aggregation algorithms.

TABLE 6.2 Adaptive and Coping-Capacity Indicators

				Assumed Relationship with Costs:			
Variable	**Adaptive Capacity**	**Coping Capacity**	**Indicators/*Proxies***	**Direct Tangible**	**Direct Intangible**	**Indirect Tangible**	**Indirect Intangible**
Age		X	Percent of population under five years old; Percent of population over 65 years; median age;		+		+
Gender		X	Percent of females; percent of female headed households;		+		+
Family structure		X	Percent of single parents households, percent of households with more than four individuals,		+		+
Disabled	X	X	Percent residents in nursing homes, percent ill or disabled residents,		+		+
Income level	X	X	Per capita income; median monetary value of owner-occupied housing; median rent for renter-occupied housing units; credit rating of inhabitants	+		+	
Social disparity	X	X	Gini index of income; percent of households earning more than X; percent of households earning les than Y; dependents on social services	+	+		

Continued

TABLE 6.2 Adaptive and Coping-Capacity Indicators—cont'd

				Assumed Relationship with Costs:			
Variable	**Adaptive Capacity**	**Coping Capacity**	**Indicators/*Proxies***	**Direct Tangible**	**Direct Intangible**	**Indirect Tangible**	**Indirect Intangible**
Education	X		Negative of percent of population 25 years or older with no high school diploma; percent of population with higher education;			–	+
Employment	X		Percentage of labor force unemployed; type of employment (full time, part time, self employed, etc.)			+	+
Safety network		X	Negative of quality of relationships within the community; *percent of isolated population; percent population change; negative of percent with 1st to 2nd level connections to civil protection;*		+		+
Trust		X	Experts elicitation/measure of trust	+			–
Risk perception	X	X	Experts elicitation/measure of perception				
Risk governance	X	X	*Per capita number of community hospitals; per capita number of physicians;*	–	–	–	–

			local gov. debt to revenue ratio; access to places (number) of safety during the event; number of red cross volunteers; hours spent on training and maneuver				
Early warning capacity	X	X	*Number of early warning systems in place for typology of hazard;*	–	–	–	–
Risk spreading	X	X	% of hazard insured households; % of hazard insured economic activities;	–			
Economic diversification	X	X	Normalized Herfindahl index of sectorial (i.e. coarse: primary, secondary, tertiary) contribution to GDP and or to employment;	+		+	
Interconnectivity of economy	X	X	Net trade in goods and services (exports +); percent of resident that travel to work outside the modeled area;	–		+	+
New comers		X	Percent renter-occupied housing units; percent of recent residents/immigrants; percent of people living in informal houses;	–	+		+

The first step towards the implementation of the framework is the *identification of the application context*. This is a strategic choice, which depends not only on the considered socioecosystem, but also on the application purpose thus affecting the detail level of analysis and the evaluation method putting KR-FWK into operation. Fundamental elements of such introductory step are the definition of the normative frame (the flood directive in Europe, EC/60/2007, in its national implementations), the identification of information sources and management systems, the ambitions and preferences of stakeholders in terms of economic-valuation methods.

Indicators' selection and data mining: the traditional assessment of risk grounded in tangible costs focuses on historical records of river flows, precipitation, digital elevation models, land use maps, maps of infrastructures, etc., as part of data-mining activity. These data are usually available from regional or national authorities, or river-basin districts. They are used to calculate indicators such as return time of extreme events. More challenging is the identification of indicators for social capacities. The list of indicators summarized in Table 6.2 is considered as a reference, pointing to information usually available from the national census.

Normalization is the procedure of transforming indicators' values expressed in different units of measurement into dimensionless figures, for allowing comparative valuation and aggregation (Giupponi et al., 2013a; Nardo et al., 2005). Normalization issues emerge for two components of the risk-assessment formula (6.1) that are expressed as dimensionless indices (H and V), but they may be also needed in E (exposure), whenever full monetization is not performed, as in many cases in which multicriteria analysis decision rules are adopted. Some of the different available normalization functions are mentioned below:

1. Ranking;
2. Standardization (z-score);
3. Min−max normalization;
4. Value functions;
5. Distance to a reference measure;
6. Categorical scales.

The type of normalization function depends on the indicators under consideration and on the preferences of the experts and decision makers involved in the evaluation process. The simplest normalization method consists of *ranking* each indicator value. The main advantages of ranking approach are its simplicity and the independence from outliers. Disadvantages are the loss of information on absolute levels and the impossibility to draw any conclusion about differences in performance. One of the most commonly used normalization procedures is *Standardization* (z-score) in which all indicators can be converted into a common scale with an average of zero and standard deviation of one. The *min–max* normalization is achieved through determining desirable and least acceptable (best and worst) values and linear scaling of the measured value between the two thresholds. *Value function* is one of the most widely used normalization procedures, using mathematical representations of human judgments, which offers the possibility of treating people's values and judgments explicitly, logically, and systematically (Beinat, 1997). *Distance to a reference measure* takes the ratio of the indicator for a generic value with respect to the reference value. The reference could be a target to reach, for example, in terms of required effectiveness of the measure. In determining *categorical scale*, first, we select the categories. They can be numerical, such as one, two, or three stars, or qualitative, such as "fully achieved", "partly achieved", or "not achieved". Scores are then assigned to each category, to express valuation judgments.

Weighting is the procedure to express the relative importance of individual indicators to calculate composite indices, such as a vulnerability index. Weights are essentially value judgments, thus essentially subjective, and have the property to make the objectives underlying the construction of a composite index explicit. Depending on the subjective judgment, different weights may be assigned to different indicators and no uniformly agreed methodology exists to weight individual indicators before aggregating them into a composite index. Therefore, weights are always a matter of debate and this is why weighting models need to be made explicit and transparent through involving the relevant stakeholders. In many cases multiple weight vectors provided by different stakeholders are applied to assess the sensitivity of final results to this source of subjectivity. Commonly used weighting procedures are the following:

1. Statistical weighting methods:
   a. Equal weights
   b. Principal component analysis
   c. Factor analysis
   d. Multiple regression models
2. Participatory weighting methods
   a. Expert judgment
   b. Public opinion
   c. Pair-wise comparison
   d. Conjoint analysis

In the indicator-based assessment, the outcome (i.e. the index) is the result of *aggregation*, i.e. a—often hierarchical—combination of several indicators in one or more converging nodes, or even into a single overall index of risk. Aggregation of indicators is a crucial task because the chosen methodology has substantial impacts on the computation of the final index; furthermore, the choice of the aggregation method typically involves trade-offs between loss of information, computational complexity, adherence to decision makers' preferences, transparency of procedure, etc. Among the available aggregation operators are the following:

1. Averaging operators
   a. Quasiarithmetic means
   b. Order weighted average (OWA)
2. The "AND" and "OR" operators
3. Nonadditive measures (NAM).

Averaging operators are still the most commonly used operator in practice, given the simplicity of their computation, immediacy, and transparency of the aggregation process. Nevertheless, averaging operators are typically compensatory (i.e. a bad score in one criterion can be offset by a good score in another one) and more importantly they are not able to consider any interaction among the criteria. *Quasi-arithmetic means* includes not only the simple arithmetic mean, but also geometric and harmonic means. OWA is still based on weighted sums, but the criteria are ordered by magnitude, and weights can be modeled to express vague quantifiers. "*AND*" *operators* are a family of operators that express logical conjunction (pessimistic behavior assigning the lowest value of the criteria to the aggregation), whereas "*OR*" *operators* consider logical disjunction (optimistic behavior). *NAM* approaches such as *Choquet Integral* have been introduced to overcome the main drawbacks of the averaging operators (Giupponi et al., 2013a).

6.6 A DEMONSTRATION OF SERRA APPLIED TO FLOOD RISK IN THE CITY OF DHAKA

The KR-FWK described above has been implemented for demonstration purposes to flood-risk assessment of Dhaka city, the capital of Bangladesh, where the combination of seasonal monsoon rainfall under the effects of climate change and transboundary upstream flow (Gain et al., 2011, 2013) make the region particularly vulnerable to severe flooding. Because Dhaka city is an internationally recognized case for flood risk, it has been selected for implementing the KR-FWK and the SERRA method, for the sake of concreteness in presenting our proposed approaches. We focus in particular on representing microlevel flood risk of the eastern part of Dhaka city that is surrounded by a network of rivers and, contrary to the western part, is unprotected from flooding with structural measures (Gain and Hoque, 2013). The spatial extent of the study area is 124 km^2. The map of Dhaka city is reported in Figure 6.5.

6.6.1 The Social Dimension of Risk: Vulnerability, Adaptive, and Coping Capacities for Risk Prevention

Vulnerability is defined as the propensity or predisposition of exposed receptors to be negatively affected by hazard events (Field et al., 2012).

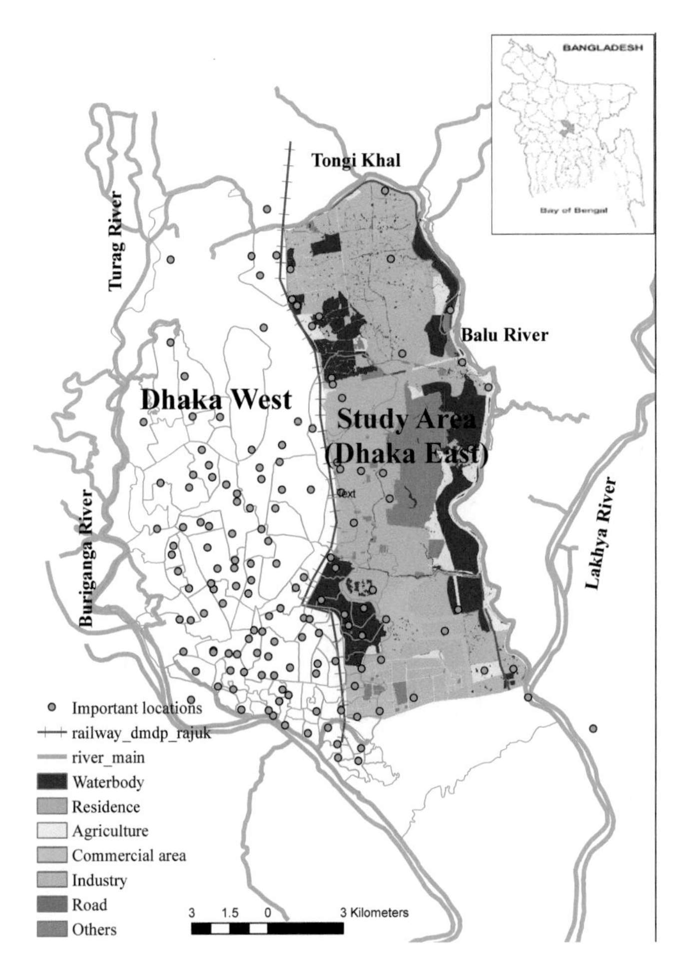

FIGURE 6.5 Dhaka city: the study area selected for demonstration of implementation of the KULTURisk framework and the SERRA is the eastern part of the city which is not protected against flood by structural measures.

Vulnerability considers both human and physical dimensions. According to the KR-FWK, the physical dimension of vulnerability encompasses the susceptibility of the man-made structures and infrastructure, that is the likelihood of these receptors to be negatively affected by hazard events, whereas the human dimension of vulnerability consists of both adaptive capacity and coping capacity as defined before. In order to provide a spatial quantification of vulnerability to be used for the calculation of risk, the dimensions of adaptive capacity, coping capacity, and susceptibility need to be assessed and aggregated. Assessing and aggregating these dimensions into a vulnerability index have already been discussed in the previous sections and they will be demonstrated in the following paragraphs (Table 6.3).

TABLE 6.3 Indicators for Adaptive Capacity, Coping Capacity, and Susceptibility

Components	Variables	Baseline	Improved Scenario
Adaptive capacity (AC)	Early warning system (EWS)	Lead time	EWS with high lead time, detail information content and high reliability leads to increase AC, and consequently decrease vulnerability
		Information content (no to detailed)	
		Reliability (low to high)	
	Risk spread	Flood insurance (presence or absent)	AC is high is if people have flood insurance and vice versa
	Income	Income dependent on agriculture (% of total people)	Values with lower number leads to increase AC, as income from agriculture is more vulnerable to flood
Coping capacity	Education	Literacy ratio, LR (%)	Population with higher LR leads to increase CC and hence decrease vulnerability
	Demography	Dependency ratio, DR (%)	Population with higher DR leads to decrease CC and hence increase vulnerability
Susceptibility	Building properties	Age (% of total buildings more than 30 years old)	Values with higher percentage are highly susceptible to floods
		Types (% of buildings with lower structural type)	
		Materials (% of buildings of earth and wood)	

TABLE 6.4 Definition of Normalized Scores

Normalized Value	Vulnerability Level
0	Not vulnerable
0.25	Slightly vulnerable
0.50	Highly vulnerable
0.75	Extremely vulnerable
1	Fully vulnerable

6.6.1.1 Selection of Indicators

As mentioned before, the choice of indicators for vulnerability components depends on the application context. These indicators mainly vary with hazard types, the spatial scale considered for the study and data availability (Gain et al., 2012). For example, the indicators for assessing flood risk are obviously different from the indicators of seismic risk, even if they may be used to calculate the same notion of vulnerability. In the case of water-related hazards, a generalized list of indicators representing adaptive capacity, coping capacity, and susceptibility is proposed in Table 6.2. Considering the data availability and microlevel spatial extent, the indicators proposed in Table 6.4 are selected.

6.6.1.2 Normalization

For normalizing selected indicators, we use a generalized approach of value function, i.e. a mathematical representation of human judgments through determining upper and lower thresholds and a series of values representing different significant levels of performance with reference to defined goals. Normalized values can be further categorized as reported in Table 6.5, in which 0 represents not vulnerable cases and 1 represents fully vulnerable ones, thus producing the discretized normalization of the selected indicators represented in Figure 6.6.

6.6.1.3 Weighting and Aggregation

The vulnerability index is the result of the combination of several indicators. To combine subsets of indicators in each node where they converge, the hierarchical aggregation follows the framework shown in Figure 6.3. Among the different aggregation methods described in (Giupponi et al., 2013a,b), in this demonstration study we adopt for simplicity the weighted averages method.

In the SERRA method, weights should be provided by stakeholders for each node in which indicators converge. As an example, a vulnerability index

TABLE 6.5 Normalized Values for Baseline and Improved Scenario of Adaptive Capacity (Values Ranging from 0 = Worst Case to 1 = Optimum)

Components	Variables	Indicators	Baseline (Normalized)	Improved Scenario (Normalized)
Adaptive capacity (AC)	Early warning system (EWS)	Lead time (h)	0.75	1
		Information content (no to detailed)	0.25	0.75
		Reliability (low to high)	0.25	0.75
	Risk spread	Flood insurance (presence or absent)	0	1

v_i can thus be calculated as a function of the normalized indicators x_1, x_2, x_3, which can be aggregated by Eqn (6.2).

$$v_i(x_1, x_2, x_3) = w_1 \cdot x_1 + w_2 \cdot x_2 + w_3 \cdot x_3 \tag{6.2}$$

The weights w_1, w_2, and w_3 represent the relative weights of x_1, x_2, and x_3 respectively. The hierarchical combination of the indicators adopted for the demonstration case of Dhaka is provided in Figure 6.7, allowing us to eventually calculate the aggregate vulnerability.

By applying weights to normalized indicators and aggregating them by means of the preferred aggregation rule, one vulnerability index is calculated per each receptor with a score ranging between 0 and 1, where 0 represents no vulnerability, whereas 1 indicates the highest vulnerability. The SERRA method is designed to assess vulnerability for supporting decisions aimed at implementing measures for risk reduction and then typically a baseline scenario is compared to one or more improved ones deriving from the impletion of plausible alternative measures.

In the baseline conditions for Dhaka city, the current early warning system and flood insurance scheme of study area are considered. The current warning system can be described as having 24–48 h lead time, low reliability and little information content. In addition, currently a flood insurance scheme does not exist. The spatial results of vulnerability for the baseline condition of an early warning system and insurance is provided in Figure 6.8(a). In the alternative scenario, an improved early warning system is simulated together with a new flood insurance scheme. Figure 6.8(b) depicts the results for improved condition of early warning system and existence of flood insurance scheme.

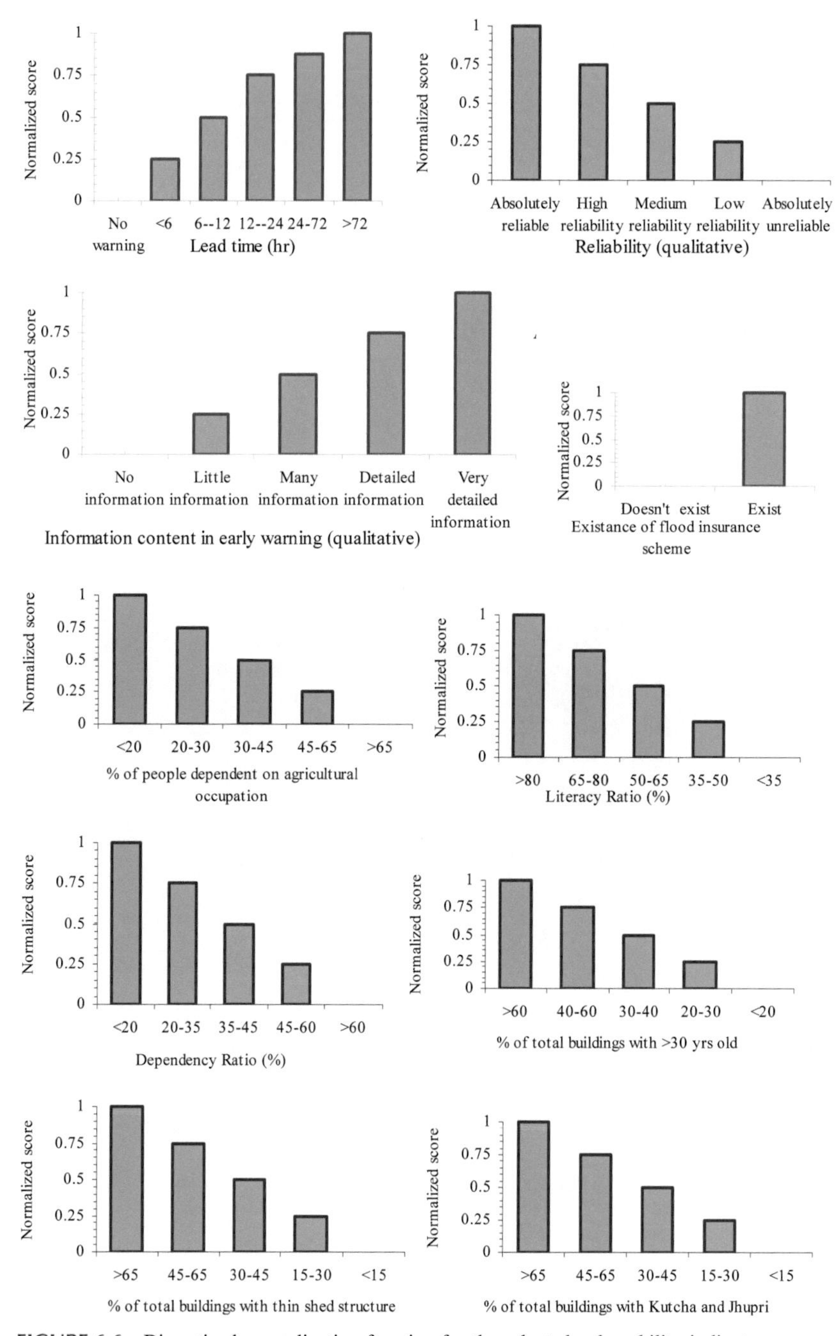

FIGURE 6.6 Discretized normalization function for the selected vulnerability indicators.

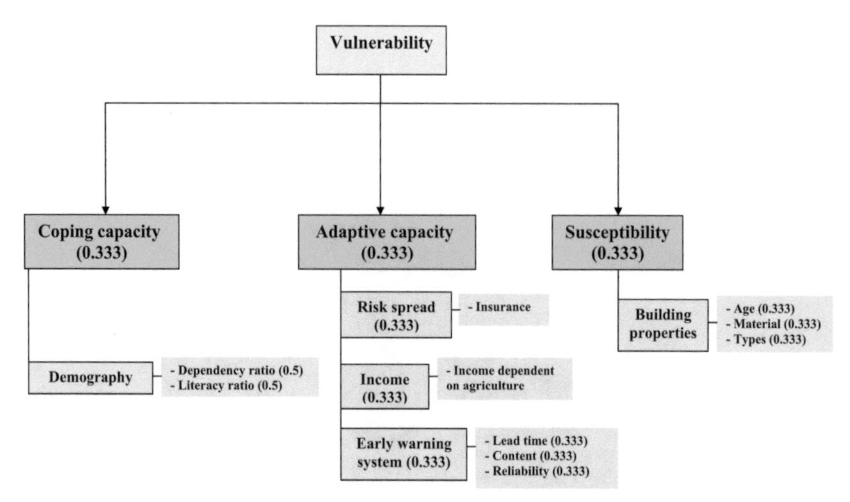

FIGURE 6.7 Hierarchical combination of indicators with relative weights.

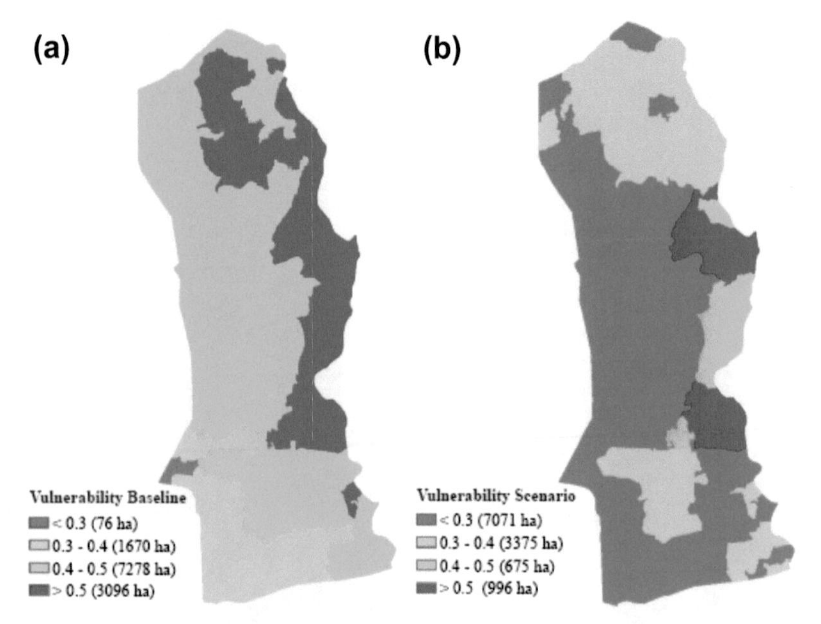

FIGURE 6.8 The results of aggregated vulnerability for (a) baseline condition and (b) improved scenario of early warning system and flood insurance. In the bracket, affected area (in hectare) is indicated for each class of vulnerability.

Normalized values of early warning system and insurance for both baseline and improved scenario are given in Table 6.6. In the baseline condition, vulnerability range is 0.28–0.70, whereas in the improved scenario the vulnerability is reduced to the range 0.12–0.54. These values, as reported in

TABLE 6.6 Summary Results of Risk to People for 100 Year Return Flood

Types	Baseline Vulnerability	Improved Scenario of Vulnerability
Number of people exposed to risk	18,740	12,394
Number of injured people	8686	3805
Number of deaths	500	217

the respective maps, will represent the spatial multipliers of the monetary values calculated in the following section.

6.6.2 The Economic Dimensions of Risk: Direct Damage Costs on People, Economic Activities and the Environment

6.6.2.1 Risk to People

We start our economic assessment by considering the damages to people, which constitute loss of lives and injuries as direct and intangible costs.

Following the results of experiments presented in the guidance document proposed by DEFRA and EA (2006), we can define *flood severeness* as:

$$FS_i = \frac{(d_i \times (v_i + 0.5) + DF_i)}{10} \tag{6.3}$$

where d_i is the depth of water measured in meters, v_i is the velocity of the flood (m/s) and DF_i is a debris factor (1 = urban, 0.5 = Woodland) all considered for any unitary cell of land *i*. The above formula is estimated for human receptors based on field experiments and the coefficients are subject to variations based on characterization of body masses of the samples.

Following the characterization of flood severeness, we identify the number of people exposed to risk, $n.p.r._i$ as:

$$n.p.r._i = N_i \times FS_i \times VI_i \tag{6.4}$$

where N_i is the number of people, and VI_i is the vulnerability index in cell *i*.

The number of injuries, $n.inj._i$, is similarly calculated by:

$$n.inj._i = \lceil n.p.r_i \times \alpha \times VI_i \rceil \tag{6.5}$$

where α is calibrated by means of the historical data for a given hazard with certain return time, and then rounded to the closet integer.

The number of deaths is calculated as follows

$$n.dth._i = \left\lceil \frac{n.inj._i \times \beta \times FS_i}{10} \right\rceil \tag{6.6}$$

where β is also calibrated with the historical data and then rounded up. For this assessment, we assumed $\alpha = 1$ and $\beta = 1.5$.

For estimating the cost of injuries, we need to identify the benefits of injury prevention. Currently, there are two main methods in use: (1) the human capital or lost output (Rice and Cooper, 1967), and (2) the willingness to pay (WTP) (Landefeld and Seskin, 1982).

The first approach consists of valuing injury in accordance with the economic impact. This ex-post evaluation is based on discounted present value of the injurer's future output forgone due to injury. To this value we add the market costs such as cost of medication. This method focuses only on the economic cost of injuries, neglecting the cost of traumas; thereby it underestimates the true value of prevention of injuries from environmental disaster. The differences between the two approaches is further discussed in Rice and Hodgson (1982).

The WTP is based on how individuals trade-off risk and economic resources and is hence measured by the marginal rate of substitution of wealth for risk of death or injury. The estimation of WTP uses *contingent valuations* methods to evaluate the *revealed preferences* derived from actual purchase of risk reduction devices or the *stated preferences* derived from hypothetical choices using questionnaires (see David et al., 2006).

Applying the above methods in the eastern part of Dhaka city, we have estimated the number of exposed, injured and dead people for a several-years return period. The summary of results for both baseline and an alternative scenario of a 100-year return-period flood are presented in Table 6.7.

As a consequence of improving the adaptive capacity of the Dhaka city meaning early warning system and risk spread, the number of exposed, injured and dead people will reduce by 33 percent, 55 percent, and 56 percent, respectively. The number of the saved people is hence interpreted as the benefit of the investment in risk mitigation measures. In this study the cost of implementing these measures is not considered, however, if these data were

TABLE 6.7 Costs to People: Income Loss and Rent

Cost Category	Baseline Vulnerability	Improved Scenario of Vulnerability	Reduced
1. Number of households exposed to risk	4358	2882	33%
2. Cost of rent in BDT	130,740,000 ৳	86,460,000 ৳	33%
3. Cost of foregone income in BDT	196,110,000 ৳	129,690,000 ৳	33%
Total cost (3 + 2) in BDT	326,850,000 ৳	216,150,000 ৳	33%

available we could have arrived at the net benefit of improving risk mitigation measures for receptor people.

A key hazard metric in further assessing the flood damage is the flood duration, which enables us to estimate the cost of accommodation of households and their forgone income. These two damages are examples of direct and tangible costs to receptor people.

We calculate the number of people who need to evacuate and be accommodated in another place after the event, and based on the number of exposed people as calculated before and average household size in Bangladesh, which is 4.3 (BBS, 2001), we estimate the number of exposed households.

Flood duration in Bangladesh varies from weeks to months (del Ninno et al., 2001). For the sake of demonstration, we consider the average between two flood events of 1988 and 1998 floods, which is 6 weeks. The cost of accommodation is similarly estimated based on the number of exposed households multiplied by the average rent per month (i.e. 20,000 Bangladesh Taka (BDT) based on Internet Survey) for the given duration of the flood.

Furthermore, we calculate the forgone income (i.e. the income households will lose, as they will not be able to work during and after the event. This is derived from the number of households affected (assuming only one person working in each household) and the average monthly income of 30,000 BDT provided by Bangladesh Bureau of Statistics (BBS, 2001). Table 6.8 summarizes the direct tangible to receptor people as explained above.

By investing in risk mitigation measures, the cost due to rent and income loss is reduced by 33 percent, which is a gross benefit of 110,700,000 BDT for receptor people.

TABLE 6.8 Data Summary

	West	East	
Population Density (People/km^2)	25,144	9861	
	Cluster M	**Cluster N**	
Employees number of firms	94,350	34,498	
	225	75	
	Surface (km^2)	**Distance to M (km)**	**Distance to N (km)**
North Khilket and Kamalapur (K)	0.6	1.9	3.3
Center (K)	14.3	4.2	7.2
South (K)	14.3	4.4	7.4

6.6.2.2 Risk to Economic Activities

For calculating the monetary damage to different land-use categories such as residential, commercial, industrial, agricultural, transport, etc., we use the depth-damage functions as a function of depth and duration (Dutta et al., 2003; Mojtahed et al., 2013). The depth-damage functions are a way to define the susceptibility of the man-made capital thus it integrates the social dimension of vulnerability as per KR-FWK.

The expected total damage in each cell of the flooded land i is hence:

$$D_{ki} = \sum_i \sum_j \sum_k \sum_s P_j \left[Suc_{sk} \cdot Asm_{ski} \cdot Vul_{si} \right] \tag{6.7}$$

where P_j is the probability of flood with a return-time j, Suc_{sk} is the susceptibility of type k of receptor s, $Asms_{ki}$ is the average square meter area of type k of receptor s in grid i, and Vul_{si} is the vulnerability of receptor s in grid i.

The susceptibility function for the Dhaka city case study is presented by the Japan International Cooperation Agency as reported in Gain and Hoque (2013). The expected damages of different land-use categories for the 100-year, return- period flood are shown spatially in Figure 6.9. Figure 6.9(a) represents risks for the baseline scenario, whereas Figure 6.9(b) refers to risks for the improved scenario. The expected risk near the river, water body and regions with economic zones, as shown in both Figure 6.9(a) and (b), is higher due to inundation and high economic value. The results demonstrate that expected damages are significantly reduced in the improved scenario of early warning and flood insurance compared to the baseline.

In brief, the total costs of damages to land-use categories in baseline and alternative scenarios are, according to this first approximation study carried

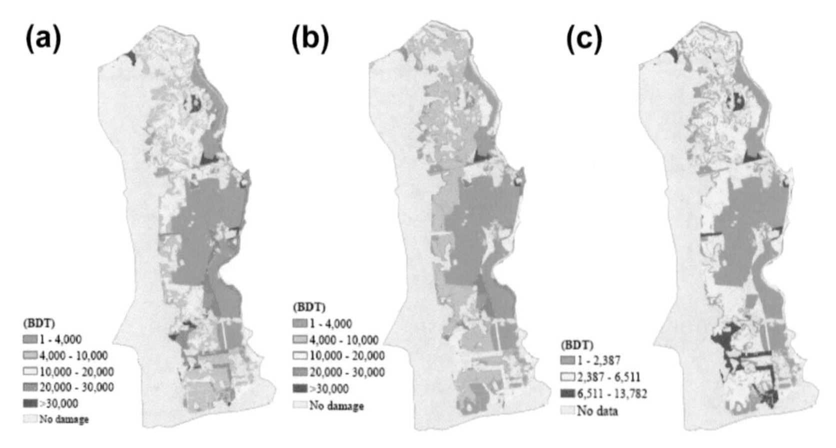

FIGURE 6.9 Risk to different categories of land uses for baseline (a), improved scenario (b) and benefits of risk-reduction measures (c) for 100-year return flood.

out for demonstration purposes, 1.33 billion BDT and 900 million BDT, respectively. Therefore, the gross benefit of investing in adaptive capacities is a reduction of 32 percent in damages to receptor economic activities such as residential and commercial buildings.

In a concrete application of the proposed method, those values should be compared with estimated costs for the implementation of the nonstructural measures considered here in order to derive the figures to be utilized in a CBA, to be applied with consideration of all the other relevant receptors. For example, in this costing exercise, we also considered the damage to vehicles, which is a category exemplifying economic assets not reported in land-use maps. To this end, first, we calculated the number of vehicles per household for the considered area. Second, we apply the generic depth-damage functions reported by US Army Corps of Engineers (USACE, 2009). Although the depth-damage functions are varying for each type of vehicle, we focused only on sedans, which are the dominant type of vehicle also in Bangladesh. Based on the survey carried by USACE (2009), households usually are able to move their vehicles to safe places with higher elevation if they are warned in advance. We have taken into account this information in estimating the damages to vehicles based on how much time the households have in advance before flood happens when the early warning system (EWS) is active. For any warning sent at least 12 h in advance by the EWS, it appears that 88.1 percent of household will be able to move their vehicles to the higher ground. Similarly to what was presented above for assets related to land use classes, maps of vehicles exposure were produced for baseline and scenario conditions with a 100-year return time.

In operational applications of SERRA, direct tangible and intangible costs as reported in Tables 6.7 and 6.8 and Figure 6.9 should be extended to all the relevant receptors, thus producing a series of maps, which can be utilized within a GIS context to produce a comprehensive picture of risks within the study area. But, as stated above, costs are quite often not limited to the area directly affected by risk and they can instead propagate outside as indirect risks. An example of assessment of such component of the TCM is provided in the following paragraph.

6.6.3 The Economic Dimensions of Risk: Indirect Damage Costs

In this section, we demonstrate how to compute the indirect costs taking the example of economic activities, which are not directly hit by a natural hazard, but part of its labor force is impeded in reaching the workplace because of a flood event. The indirect effects can be computed by estimating the daily lack of production in a labor–intensive activity such as the one occurring in the Textile and Clothing (TC) industry in Bangladesh in a determinate area. A strong point of this method is that it uses publicly available data to estimate indirect-tangible costs.

We can infer workers' commuting flows and daily loss of factories using demographic data and knowledge about the industry production function, firms' location and the number of their employees.

We select the TC industry, because it is the most important economic sector in Bangladesh for export value and because, after agriculture, it is the economic activity that employs the most people (Islam et al., 2013; Keane and te Velde, 2008). Only in Dhaka city, the number of employees in the TC industry exceeds 800,000 (data of 2005, Hoque et al., 2006). In this study we focus on the area of Motijheel (hereafter M) and another area closed by to the north (hereafter N), located in the southwest of the Dhaka city and flood-secured by an embankment. The economic activities consist in a cluster of 225 garment factories (Hoque et al., 2006): 146 factories in Motijheel C/A, 43 nearby Purana Paltan, and 36 in the Mojitheel Circular Road. They employ more than 94,000 workers. Workers walk to work from both the protected part of the city and from the nonprotected part in the east, as shown by a preliminary study on workers' travel pattern (Hoque et al., 2006). In Figure 6.10, we display the location of the two clusters of firms in the flood-secured part of the city together with the area in the eastern flood prone area from which workers commute.

In absence of detailed data on traffic flows of the urban population in Dhaka, we estimated people's mobility using two different models that embed the preference for individuals to shorten their commuting distance (Clark et al., 2003). The results from a gravity model used to estimate relative preference of workers to commute to two different workplaces on the secure side of the embankment are validated by the radiation model (Simini et al., 2012), that is less biased towards short commuting distances[5]. Grounded in physics, gravity models are used to predict flows of goods and people (Eaton and Tamura, 1995; Vanderkamp, 1971). In our case, we use a gravity model to show the force of attraction on workers of particular neighborhoods of the two clusters of TC firms (N and M). The force of attraction is proportional to the size of firms, i.e., number of employees, and decays strongly with the distance. We compute that the attraction of the two clusters of firms M and N to estimate the flow of people moving from the flood-impacted eastern part (hereafter K, as denoted also in Figure 6.10) that is situated on the eastern and non-secured part of Dhaka to the workplaces located in M and N.

We divide the zone around K in three parts, north, center, and south to accurately estimate the attraction force, see Table 6.8.

5. The gravity model (GM) is biased towards closer points of attraction, whereas the radiation model (RM) seems to better estimate population flows and commuting habits overcoming biases of the GM (Simini et al., 2012). We can use RM only to infer the attraction of the two clusters of firms on the people living in the unprotected part. Using population densities as suggested by Simoni and colleagues, RM's flows around 12 times stronger towards M with respect to N. Thereby in our case, GM seems to be more conservative.

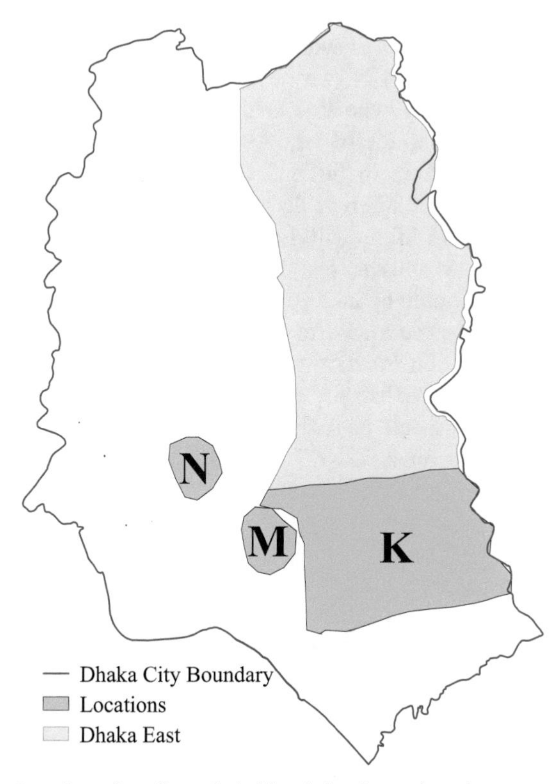

FIGURE 6.10 Location of textile and clothing industries and workers commuting from the eastern part of the city. The exercise considers two clusters of industries located in N and M (west Dhaka), and examine the case of workers commuting from the area K situated in the eastern part of the city.

The force of attraction of the cluster i is given by

$$aa_i = \frac{size_i}{f(r)} \tag{6.8}$$

where r is the distance to a cluster of firms, and size is the number of employees within the cluster.

We computed this with r^2 that seems to be conservative with respect to walking as means for commuting—the higher the exponent of r, the higher is the preference for a closer location (see Simini et al., 2012 for a brief discussion on $f(r)$ in western societies). Due to the proximity to the unsecured area, the attraction for the M cluster with respect to the N cluster is between 7.5 and 8.5 times higher, with higher values in the north and center, due to the proximity with M and the exponential decay with distance. The force of attraction can be seen as a proportion of workers moving from the unsecured area towards M rather than in N.

To understand the amount of workforce living in K and moving west towards the clusters of M and N, we need to assume that the proportion of workers in the TC industry who live in the eastern part of Dhaka does not differ from that of the whole city, i.e., 8.9 percent of the population. In this case nearly 25,670 workers in the TC industry live in K: around 22,800 commute to the M district, whereas the remainder go to N.

A shrinking of the workforce will negatively affect the output of the firms in the cluster. In the M cluster, a reduction of nearly 22,800 workers will decrease the total workforce by 24.2 percent, whereas in N the reduction is 8.4 percent. We use these proportions to compute the reduction of output on a Cobb–Douglas production function that has been recently estimated for the industries in Bangladesh (Hossain et al., 2012). In the cluster of M, the reduction in production would be 6.4 percent for each day of flood, whereas in N it would be only 2 percent.

In the alternative scenario, a more efficient early warning system would enable the population to get a different accommodation in the western and flood-protected part of the city by family, friends or in structures that could be temporarily set up and allocated to provide shelter for people. These actions would certainly reduce the number of people who cannot reach the workplace, and hence reduce the negative effect of the flood on the economic activity of the TC industry.

6.7 FINAL REMARKS

As mentioned in the introduction, a need exists for a holistic and scientifically sound approach towards risk assessment of water-related hazards. In this work we focused on developing a framework based on the integration of different components of risk from a multidisciplinary perspective, with innovative solutions in particular for the social and economic dimensions of risk. We propose that the estimation of risk should not only be based on direct tangible costs, whereas it should also go beyond to contain indirect and intangible costs. In particular we propose to consider social indicators, which have been often neglected in the literature of risk assessment, to take into consideration the capacities of local communities to cope with risks and adapt to them.

An effective (more successful with a lower cost of implementation) risk-reduction policy that is mainly based on developing a culture for risk abatement requires more emphasis on social capacities of individuals or society (whether it be coping or adaptive) to abate all damages as summarized in the TCMof the KR-FWK.

Another distinction from the consolidated approaches for risk assessment consists in enhancing cost estimation well beyond the tangibles and direct damages. For this purpose, we proposed solutions for estimating total costs that reckon with indirect, beyond time and geographical limit of hazard, and intangible costs. The TCM can thus provide a solid basis for CBA of alternative strategies and measures to deal with water-related risks. However, the proposed method can be adopted also in those cases in which the full monetization of costs and benefits (including in particular intangibles) is not envisaged or desired. In particular, in cases in which, for example, the use of the statistical value of life is not considered, the KR-FWK can be adopted in different contexts and in particular for supporting the implementation of CEA.

The experience developed during the tests of the methods in the various case studies of the KULTURisk Project suggests that the proposed approach can be applied at various scales and in different situations, for example, in terms of data availability. Indeed it requires tailoring to the various cases, for example, in order to meet the specific institutional and legislative contexts. The Dhaka case presented above, though being just a first-approximation application for demonstration purposes and its further developments not reported here showed in particular that the SERRA method can be applied in contexts of limited data availability for the preliminary screening of risk mitigation measures to be further developed according to local legislation and with the involvement of local stakeholders and decision makers.

ACKNOWLEDGMENTS

The authors gratefully acknowledge the financial support of the European Commission, 7th Framework Programme, KULTURisk Project (Knowledge-based approach to develop a culture of risk prevention), coordinator Giuliano Di Baldassarre.

REFERENCES

Balbi, S., Giupponi, C., Olschewski, R., Mojtahed, V., 2013. The Economics of Hydro-Meteorological Disasters: Approaching the Estimation of the Total Costs. BC3 Working Paper Series 2013–12. Basque Centre for Climate Change (BC3), Bilbao, Spain.

BBS, B.B.o.S., 2001. Household, Population, Sex Ratio and Literacy Rate Census Report. Bangladesh Bureau of Statistics.

Beinat, E., 1997. Value Functions for Environmental Management. Kluwer Academic Publishers.

Cardona, O.D., Ordaz Schroder, M.G., Reinoso, E., Yamín, L., Barbat Barbat, H.A., 2010. Comprehensive Approach for Probabilistic Risk Assessment (CAPRA): International Initiative for Disaster Risk Management Effectiveness. Macedonian Association for Earthquake Engineering (MAEE).

Clark, W.A., Huang, Y., Withers, S., 2003. Does commuting distance matter?: commuting tolerance and residential change. Reg. Sci. Urban Econ. 33 (2), 199–221.

Cochrane, H., 2004. Economic loss: myth and measurement. Disaster Prev. Manag. 13 (4), 290–296.

Crichton, D., 1999. Ingleton, Jon, ed. Natural Disaster Management: A Presentation to Commemorate the International Decade for Natural Disaster Reduction (IDNDR) 1990–2000. Leicester: Tudor Rose, 1999.

Cutter, S.L., 1996. Vulnerability to environmental hazards. Prog. Human Geogr. 20, 529–539.

Cutter, S.L., Boruff, B.J., Shirley, W.L., 2003. Social vulnerability to environmental hazards*. Soc. Sci. Q. 84 (2), 242–261.

David, P., Atkinson, G., Mourato, S., 2006. Cost-benefit analysis and the environment: Recent Developments. Organisation for Economic Co-operation and Development. http://www.oecd.org/environment/tools-evaluation/36190261.pdf.

DEFRA, Environment_Agency, 2006. Flood and coastal defence R&D program: flood risk to people. Phase 2. FD2321/TR2 Guidance Document.

Dutta, D., Herath, S., Musiake, K., 2003. A mathematical model for flood loss estimation. J. Hydrol. 277 (1), 24–49.

Eaton, J., Tamura, A., 1995. Bilateralism and Regionalism in Japanese and US Trade and Direct Foreign Investment Patterns. National Bureau of Economic Research.

EC, 2007. Directive 2007/60/EC of the European Parliament and of the Council of 23 October 2007 on the assessment and management of flood risks. Off. J. Eur. Union L 288/27. The European Parliament and the Council of the European Union. http://eur-lex.europa.eu/legal-content/EN/TXT/?uri=CELEX:32007L0060.

Field, C.B., Barros, V., Stocker, T.F., Dahe, Q., 2012. Managing the Risks of Extreme Events and Disasters to Advance Climate Change Adaptation: Special Report of the Intergovernmental Panel on Climate Change. Cambridge University Press.

Gain, A., Hoque, M., 2013. Flood risk assessment and its application in the eastern part of Dhaka city, Bangladesh. J. Flood Risk Manag. 6 (3), 219–228.

Gain, A., Immerzeel, W., Sperna Weiland, F., Bierkens, M., 2011. Impact of climate change on the stream flow of lower Brahmaputra: trends in high and low flows based on discharge-weighted ensemble modelling. Hydrol. Earth Syst. Sci. 15 (5), 1537–1545.

Gain, A.K., Apel, H., Renaud, F.G., Giupponi, C., 2013. Thresholds of hydrologic flow regime of a river and investigation of climate change impact—the case of the lower Brahmaputra river basin. Clim. Change 120 (1–2), 463–475.

Gain, A.K., Giupponi, C., Renaud, F.G., 2012. Climate change adaptation and vulnerability assessment of water resources systems in developing countries: a generalized framework and a feasibility study in Bangladesh. Water 4 (2), 345–366.

Giovannini, E., Nardo, M., Saisana, M., Saltelli, A., Tarantola, A., Hoffman, A., 2008. Handbook on Constructing Composite Indicators: Methodology and User Guide. Organisation for Economic Cooperation and Development (OECD). http://www.oecd.org/std/42495745.pdf.

Giupponi, C., Giove, S., Giannini, V., 2013a. A dynamic assessment tool for exploring and communicating vulnerability to floods and climate change. Environ. Model. Softw. 44, 136–147.

Giupponi, C., Mojtahed, V., Gain, A.K., Balbi, S., 2013b. Integrated Assessment of Natural Hazards and Climate Change Adaptation: I. The KULTURisk Methodological Framework. University Ca' Foscari of Venice, Dept. of Economics Research Paper Series No. 06/WP/2013. http://dx.doi.org/10.2139/ssrn.2233310.

Giupponi, C., Sgobbi, A., Mysiak, J., Camera, R., Fassio, A., 2008. NETSYMOD-an Integrated Approach for Water Resources Management, Integrated Water Management. Springer, pp. 69–93.

Harley, M., Horrocks, L., Hodgson, N., Van Minnen, J., 2008. Climate change vulnerability and adaptation indicators. European Topic Centre on Air and Climate Change (ETC/ACC) Technical Paper 2008, 9.

Hoque, M.S., Debnath, A.K., Mahmud, S., 2006. Impact of garment industries on road safety in metropolitan Dhaka. In: Proceedings of International Conference on Traffic Safety in Developing Countries. Accident Research Center, Bangladesh University of Engineering and Technology.

Hossain, M.M., Majumder, A.K., Basak, T., 2012. An application of non-linear Cobb-Douglas production function to selected manufacturing industries in Bangladesh. Open J. Stat. 2 (4), 460–468.

Islam, M.M., Khan, A.M., Islam, M.M., 2013. Textile industries in Bangladesh and challenges of growth. Res. J. Eng. Sci. 2 (2), 31–37.

Jonkman, S.N., Vrijling, J.K., Vrouwenvelder, A.C.W.M., 2008. Methods for the estimation of loss of life due to floods: a literature review and a proposal for a new method. Nat. Hazards 46, 353–389.

Keane, J., te Velde, D.W., 2008. The Role of Textile and Clothing Industries in Growth and Development Strategies. Overseas Development Institute Investment and Growth Programme, Mimeo. London.

Klein, R.J., 2004. Vulnerability indices—an academic perspective. In: Proc. of the Expert Meeting "Developing a Method for Addressing Vulnerability to Climate Change and Climate Change Impact Management: To Index or Not to Index. Germanwatch. http://germanwatch.org/download/klak/ws04vuln/klein.pdf.

Landefeld, J.S., Seskin, E.P., 1982. The economic value of life: linking theory to practice. Am. J. Pub. Health 72 (6), 555–566.

Landis, W.G., 2004. Regional Scale Ecological Risk Assessment: Using the Relative Risk Model. CRC Press.

Mercer, J., 2010. Disaster risk reduction or climate change adaptation: are we reinventing the wheel? J. Int. Develop. 22 (2), 247–264.

Merz, B., Kreibich, H., Schwarze, R., Thieken, A., 2010. Assessment of economic flood damage. Nat. Hazards Earth. Syst. Sci. 10, 1697–1724.

Mojtahed, V., Giupponi, C., Biscaro, C., Gain, A.K., Balbi, S., 2013. Integrated Assessment of Natural Hazards and Climate-change Adaptation: II. The SERRA Methodology. University Ca' Foscari of Venice, Dept. of Economics Research Paper Series No. 07/WP/2013. http://dx.doi.org/10.2139/ssrn.2233312.

MOVE, 2010. Assessing vulnerability to natural hazards in Europe: from principles to practice. A Manual on Concept Methodology and Tools. Seventh Framework Programme of European Commission. http://www.move-fp7.eu/documents/MOVE_Manual.pdf.

Nardo, M., Saisana, M., Saltelli, A., Tarantola, S., Hoffman, A., Giovannini, E., 2005. Handbook on Constructing Composite Indicators: Methodology and User Guide (No. 2005/3). OECD publishing.

del Ninno, C., Dorosh, P.A., Smith, L.C., Roy, D.K., 2001. The 1998 Floods in Bangladesh: Disaster impacts, household coping strategies and response. Research Report 122. International Food Policy Research Institute, Washington D.C.

Okuyama, Y., Sahin, S., 2009. Impact estimation of disasters : a global aggregate for 1960 to 2007. Policy Research working paper; no. WPS 4963. World Bank, Washington, DC. http://documents.worldbank.org/curated/en/2009/06/10690399/impact-estimation-disasters-global-aggregate-1960-2007.

Parry, M.L., 2007. Climate Change 2007: Impacts, Adaptation and Vulnerability: Working Group II Contribution to the Fourth Assessment Report of the IPCC Intergovernmental Panel on Climate Change. Cambridge University Press.

Penning-Rowsell, E., Johnson, C., Tunstall, S., Tapsell, S., Morris, J., Chatterton, J., Green, C., Wilson, T., Koussela, K., Fernandez-Bilbao, A., 2005. The benefits of flood and coastal risk management: a manual of assessment techniques. Middlesex University Press, London. ISBN 1-904750-52-4. XII, 238 pp.

Renaud, F., Perez, R., 2010. Climate change vulnerability and adaptation assessments. Sustainability Sci. 5 (2), 155–157.

Rice, D.P., Cooper, B.S., 1967. The economic value of human life. Am. J. Pub. Health Nations Health 57 (11), 1954–1966.

Rice, D.P., Hodgson, T.A., 1982. The value of human life revisited. Am. J. Pub. Health 72 (6), 536–538.

Simini, F., González, M.C., Maritan, A., Barabási, A.-L., 2012. A universal model for mobility and migration patterns. Nature 484 (7392), 96–100.

Steinfuhrer, A., Kuhlicke, C., DeMarchi, B., Scolobig, A., Tapsell, S., Tunstall, S., 2009. Towards Flood Risk Management with the People at Risk: From Scientific Analysis to Practice Recommendations (and Back). CRC Press, Taylor and Francis Group.

Thomalla, F., Downing, T., Spanger-Siegfried, E., Han, G., Rockström, J., 2006. Reducing hazard vulnerability: towards a common approach between disaster risk reduction and climate adaptation. Disasters 30 (1), 39–48.

Turner, B.L., Kasperson, R.E., Matson, P.A., McCarthy, J.J., Corell, R.W., Christensen, L., Eckley, N., Kasperson, J.X., Luers, A., Martello, M.L., 2003. A framework for vulnerability analysis in sustainability science. Proc. Natl. Acad. Sci. 100 (14), 8074–8079.

UNISDR, January 2005. Hyogo Framework for Action 2005–2015: Building the Resilience of Nations and Communities to Disasters. World Conference on Disaster Reduction. pp. 18–22. United Nations Office for Disaster Risk Reduction. http://www.unisdr.org/2005/wcdr/intergover/official-doc/L-docs/Hyogo-framework-for-action-english.pdf.

UNISDR, 2009. 2009 UNISDR Terminology on Disaster Risk Reduction. United Nations Office for Disaster Risk Reduction. http://www.unisdr.org/files/7817_UNISDRTerminologyEnglish.pdf.

USACE, U.S.A.C.o.E., 2009. Economic guidance memorandum, 09–04. Gen. Depth–Damage Relat. Veh. Department of the Army, U.S. Army Corps of Engineers. http://planning.usace.army.mil/toolbox/library/EGMs/egm09-04.pdf.

Vanderkamp, J., 1971. Migration flows, their determinants and the effects of return migration. J. Political Econ. 1012–1031.

Willows, R., Reynard, N., Meadowcroft, I., Connell, R., 2003. Climate Adaptation: Risk, Uncertainty and Decision-making (UKCIP Technical Report. UK Climate Impacts Programme). UK Climate Impact Programme (UKCIP), Union House, 12–16 St., Michael's Street, Oxford. ISBN: 0-9544830-0-6.

Chapter 7

KULTURisk Methodology Application: Ubaye Valley (Barcelonnette, France)

Micah Mukolwe[1,2], Giuliano Di Baldassarre[1,3] and Thom Bogaard[1,2]
[1] *UNESCO-IHE Institute for Water Education, Delft, The Netherlands,* [2] *Delft University of Technology, Civil Engineering and Geosciences Water Management, Water Resources, Delft, the Netherlands,* [3] *Department of Earth Sciences, Uppsala University, Uppsala, Sweden*

ABSTRACT
The growing coincidence of occurrence of natural hazards and vulnerable societies, leading to economic damages and fatalities, has triggered more studies on benefits of prevention measures. This chapter describes a study that aims to demonstrate benefits of risk-prevention measures by applying the KULTURisk methodology (see Chapter 6). The demonstration was implemented in the Ubaye Valley (Barcelonnette town), France. Our findings show that the methodology is an adaptable decision-making tool that may be used to support the analysis of alternative scenarios for flood-risk reduction.

7.1 INTRODUCTION

7.1.1 Motivation and Objective

Floodplain habitation has occurred throughout history, notably due to associated economic benefits for a number of activities, such as trade and agriculture. In many countries, the occupation of floodplains has grown after construction of flood protection structures (such as levees). This process has led to a reduction in the frequency of flooding; on the other hand, there is a corresponding increase of potential adverse consequences ("levee effect"; Di Baldassarre et al., 2013; Klijn et al., 2004). As a matter of fact, many megacities are located in floodplains or deltas.

Population growth in high-risk areas has often resulted in remarkable damages and fatalities. To assess risk, it is normative to evaluate the potential natural hazards, calculate the susceptibility of the potentially affected population (in relation to the hazard) and combine this output with the

Hydro-Meteorological Hazards, Risks, and Disasters. http://dx.doi.org/10.1016/B978-0-12-394846-5.00007-2

consequences (e.g., Apel et al., 2004; Jongman et al., 2012; Moel, 2012; Winsemius et al., 2012; Genovese, 2006; Genovese et al., 2007).

This chapter shows a demonstration of the benefits of risk-prevention measures by applying the KULTURisk methodology (see Chapter 6) to potential flood risk in Barcelonnette (France).

7.1.2 The Case Study: Ubaye Valley, France

The Ubaye Valley, located in the French Alps (Figure 7.1), is a popular tourist destination for ski-sports and draws upon a rich historical heritage. Protection of the city of Barcelonnette from flooding has led to levee construction along the river. Also, in the upper catchment there have been efforts to improve slope stability and reduce debris flows from steep unstable alpine slopes, achieved by construction of check dams and planting of trees (Flageollet et al., 1996; Weber, 1994).

In June 1957, an approximately 1-in-100-year flood wave, resulting from a combination of snowmelt and spring rainfall, caused exceptionally high discharges in the Ubaye River. This resulted in levee overtopping and flooding of surrounding areas of Barcelonnette town in the Ubaye Valley (Flageollet et al., 1996).

Consequently, different flood-risk-reduction measures were being planned in the Ubaye Valley because of the potentially disruptive nature of flood hazard. Thus, the KULTURisk methodology was applied to determine benefits of these preventive options.

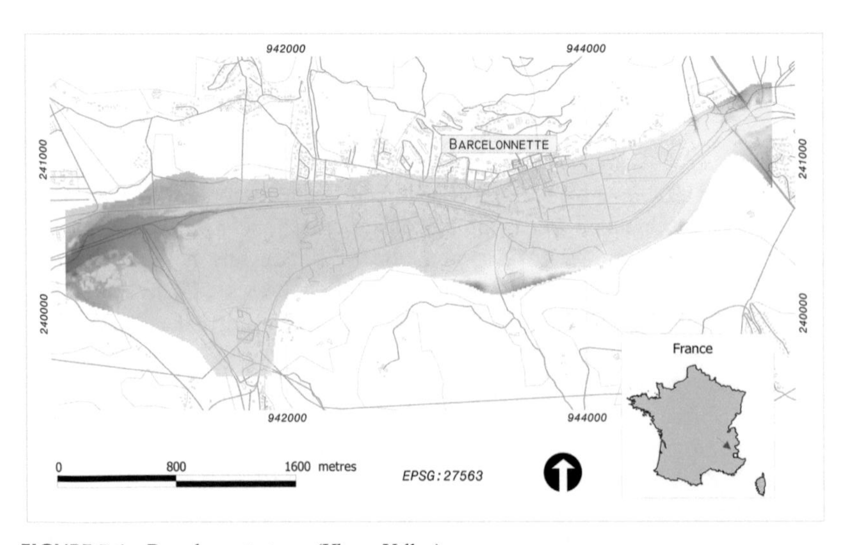

FIGURE 7.1 Barcelonnette town (Ubaye Valley).

7.2 METHODOLOGY

7.2.1 Hydraulic Modeling

At the core of any flood-risk assessment is the determination of potential flood-event characteristics and resulting intersection with different receptors in the floodplain to assess damages (based on susceptibility and vulnerability). Increasing availability of computing power, along with the development of several hydraulic-modeling algorithms has led to growth in the use of flood-inundation models in flood-risk management (Dottori and Todini, 2011; Hunter et al., 2007). This growth has been also facilitated by new data sources of topographic information (Di Baldassarre and Uhlenbrook, 2012; Bates, 2012). For instance, technology such as LiDAR (Light Detection and Ranging) facilitates relatively accurate representation of floodplain topography (Cracknell, 2007) that can be used as input for flood-inundation modeling at a high resolution.

In this study, a two dimensional flood inundation model, LISFLOOD-FP (Neal et al., 2012; Bates et al., 2010) was built for a reach of the Ubaye Valley River covering the Barcelonnette town area. Topography information was derived from a LiDAR survey carried out in November 2011, which was supplemented by additional river-bathymetry, cross-sectional data obtained by RTM (*Restauration des Terrains en Montagne*). Model calibration was based on observed water levels during a high flow event in May 2008. This resulted in a main channel Manning's roughness value of 0.040 $m^{1/3}/s$, for which we found the lowest root mean square error (RMSE) between observed and simulated water levels.

This model was then used to simulate the June 1957 flood event, whose characteristics were derived from a postevent analysis based on an analysis of sediment deposition and expert-knowledge (Lecarpentier, 1963). In particular, the peak river discharge of the June 1957 flood event was estimated between 420 and 480 m^3/s.

7.2.2 Scenario Selection

Implementation of the KULTURisk methodology was based on different receptors: people and economic activities (consisting of buildings, road network, and agricultural assets). Potential flood-risk mitigation measures that may be adopted by civil protection authorities were considered as alternative scenarios. We considered two scenarios: structural measures and nonstructural measures, namely, improvement of conveyance capacities of the bridge cross-sections, and implementation of an early warning system (EWS). A third scenario considers combination of the two. Benefits of these risk-reduction measures were evaluated by comparing potential flood losses with losses resulting from the baseline scenario, which was representative of the situation then.

7.2.2.1 Regional Risk Assessment

Potential risk to people was defined by the nature of cultural practices and social behavior that characterize the exposure of people to flooding. In this respect, it is important to consider a large number of tourists and occupancy of secondary homes during high season. To this end, exposure of people was calculated by taking the average house occupancy and average floor area. Furthermore, the use of spatial exposure data sets was beneficial in characterizing areas with high damages.

Damage to the road network was calculated as a percentage area of the road network that was affected by flooding. Important aspects of connectivity and inaccessibility of vital location of the valley were taken into account mainly in the social assessment (see below).

Building damage was taken as a function of flood metrics intensity (flood velocity and water depth) and the capacity of exposed buildings to resist destructive forces defined by a depth damage curve. Lastly, for calculation of damage to agricultural fields, crop damage thresholds were used to determine potential damage.

7.2.2.2 Social—Regional Risk Assessment

French Statutory considerations govern and give opportunity for public participation in the risk-mitigation process, which raises risk awareness of the population (Schwarze et al., 2011). Weight factors representing the state of receptor vulnerability were determined from stakeholder discussions, field visits, and a literature review. However, it was vital to discuss final indicator weights with stakeholders, given that expert judgment was used to normalize indicator values.

Indicators were categorized as those based on adaptive capacity, coping capacity, and susceptibility of receptors. Following the determination of normalized indicator values, importance factors weighing the human dimension versus the physical dimension, and those weighing the adaptive capacity, coping capacity, and the susceptibility were determined. To avoid bias, these factors were equally distributed prior to initial discussion with stakeholders.

A major consideration during the analysis was that the population remained relatively stable, approximated at a growth of −0.6% and 1.1% from 1990 to 1999 and 1982 to 1990 respectively (INSEE, 2013). In addition to this, substantial public participation and contribution to risk awareness (e.g., Angignard, 2011) contributes to resilience of the population.

7.2.2.3 Economic—Regional Risk Assessment

Value factors and *willingness to pay* were the main considerations in determining damage to receptors. Average house rent and contraction costs were readily available, however, for building content value, an estimated *content to value ratio* equal to 50% was used (USACE, 1996). Similar to advice by Messner et al. (2007) agricultural losses were calculated based on characteristics of wheat (Brisson et al., 2010). With regards to damage to roads, losses were calculated arising from debris deposition and road-surface damage, requiring minor road maintenance (Doll and van Essen, 2008).

Physical damage and repair costs of receptors were relatively straightforward to determine, on the other hand, service-disruption costs were difficult to determine due to lack of data. Thus, rough estimate of ratios of structural damage to service-disruption costs were used in the analysis.

Finally, to demonstrate benefits of preventive measures, a relative benefit was calculated that was defined as the relative reduction of flood losses and expressed as a percentage. Thus, the relative benefit in this case can be interpreted as the benefit of *taking preventive action* versus *inaction*.

7.3 RESULTS AND DISCUSSION

7.3.1 Risk Assessment

7.3.1.1 Flood Hazard Metrics

Scenario 1, which included channel-conveyance capacity improvement (by reshaping the geometry of the bridge), resulted in a significantly reduced floodplain inundation (Werner et al., 2005). The flood hazard to the receptors was based on the two flood-inundation extents shown in Figure 7.2.

7.3.1.2 Regional Risk Assessment

Results of the Regional Risk Assessment (RRA) show a significant reduction of the risk to receptors as a result of reducing flood water volume in the floodplain (Figure 7.3). This reduction results in lower (less threatening) flood velocities and water depth, hence lower physical damage (Tables 7.1 and 7.2).

Application of nonstructural measures (EWS) does not have an effect on the flood-wave propagation dynamics, but rather on inherent receptor characteristics, such as alert and evacuation, mitigate effects of flooding. EWS mainly facilitates evasive action to lessen exposure in terms of magnitude and extent (Figure 7.4).

7.3.1.3 Social—RRA

Changes in receptor vulnerability were based on the implementation of a reliable EWS (Scenario 2). Being an alpine river recharged by creeks characterized by steep slopes, the Ubaye River basin is a fast responding catchment. Hence the ability to adequately warn inhabitants of changing river conditions would improve the resilience of floodplain inhabitants. Given that

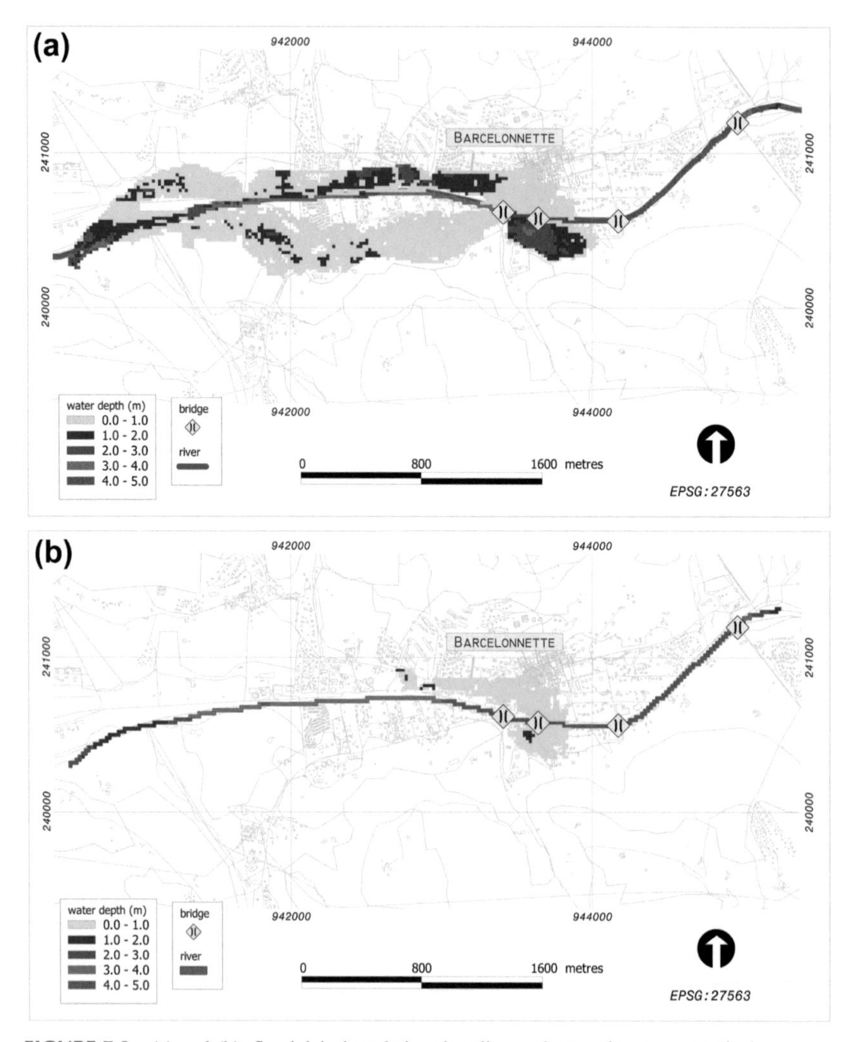

FIGURE 7.2 (a) and (b), floodplain inundation; baseline and scenario one respectively.

there is no formal EWS implemented, receptors were categorized as being *highly vulnerable* (Indicator value 0.5). Following the application of Scenario 2, including a functioning EWS, the indicator value is expected to improve to a lower vulnerability, *slightly vulnerable* (depending on characteristics of the installed system), i.e., indicator value equal to 0.25, according to nomenclature in the methodology. These factors were then combined in a hierarchical structure as specified in the methodology.

Figure 7.4 shows the resulting decrease in vulnerability of receptors as a result of EWS implementation. Adjusting of factors and weights in the SRRA

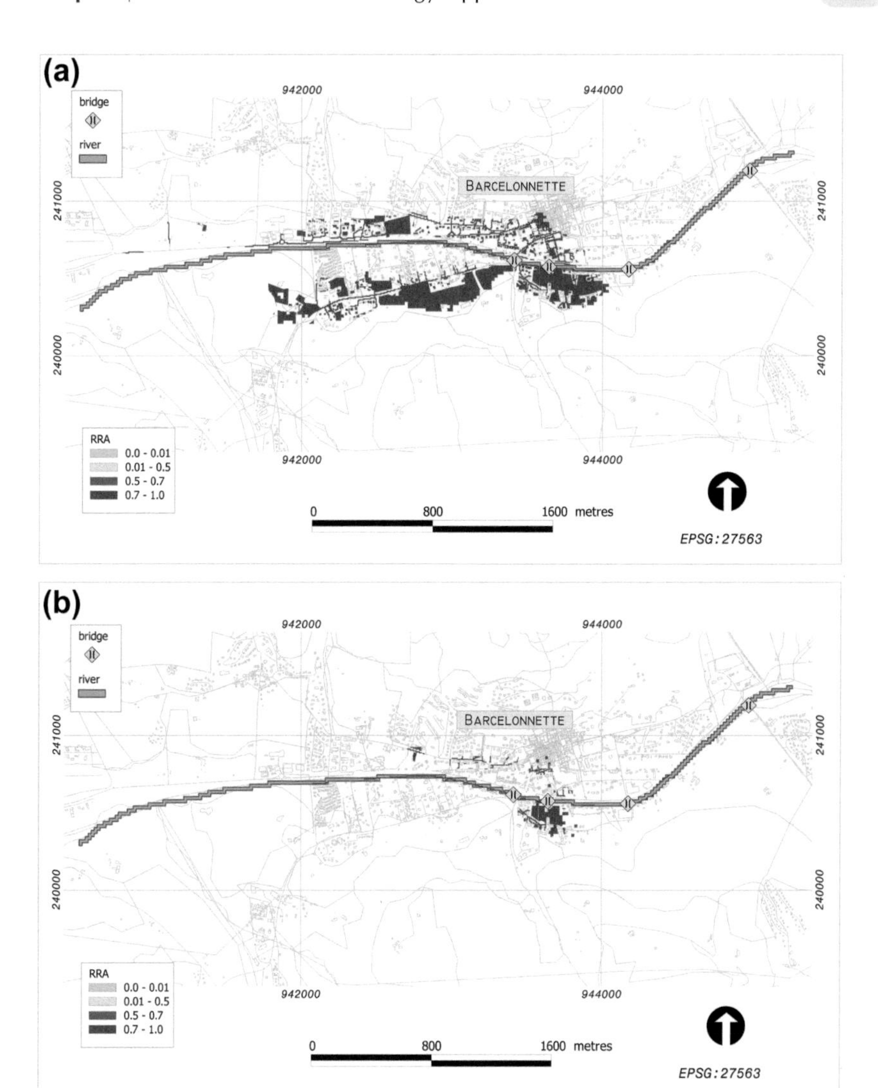

FIGURE 7.3 RRA: (a) baseline and scenario two, (b) scenario one and scenario three.

allowed for inherent study-area characteristics and social sensitivities to be accounted for, while focusing on areas of importance.

7.3.2 Social Economic RRA

Application of the social economic RRA resulted in aggregated values and damage maps. Clearly, a reduction of flood volumes, that cause floodplain inundation, results in the least damage in monetary terms.

TABLE 7.1 Percentage of Physical Damage to Receptors

Receptor	Damage Level	Baseline and Scenario 2 (%)	Scenario 1 and Scenario 3 (%)
Buildings	Inundation	31.83	6.04
	Partial damage	0.00	0.00
	Destruction	0.00	0.00
Roads	Inundated	20.11	6.45
Agriculture	Inundation	10.32	1.08
	Destruction	7.40	0.73

TABLE 7.2 Potential Fatalities and Injuries (Total Exposure = 3,380 People)

	Injuries		Potential Fatalities	
Scenario	Number (–)	Percentage (%)	Number (–)	Percentage (%)
Baseline and Scenario 2	60	1.74	3	0.075
Scenario 1 and Scenario 3	11	0.32	1	0.010

(The values displayed are as a result of the application of the model in a worst-case scenario application considering maximum possible exposure).

Table 7.3 shows that relative benefits of structural measures are significantly higher than nonstructural measures for this case study. However, it should be noted that costs of implementing the two scenarios are different. Thus, although a reduction of potential flood losses corresponding to the introduction of a reliable EWS is less than that corresponding to structural measures, it may be less expensive and, therefore, appropriate in a cost-benefit consideration.

7.4 CONCLUSION

This chapter presented an application of the KULTURisk methodology to an alpine catchment located in the Ubaye Valley. The methodology was found to be adaptable to needs expressed by stakeholders, and the type and resolution of available data. Most notable was the need for continuous interactions with local stakeholders (e.g., Refsgaard et al., 2007) at each stage of application,

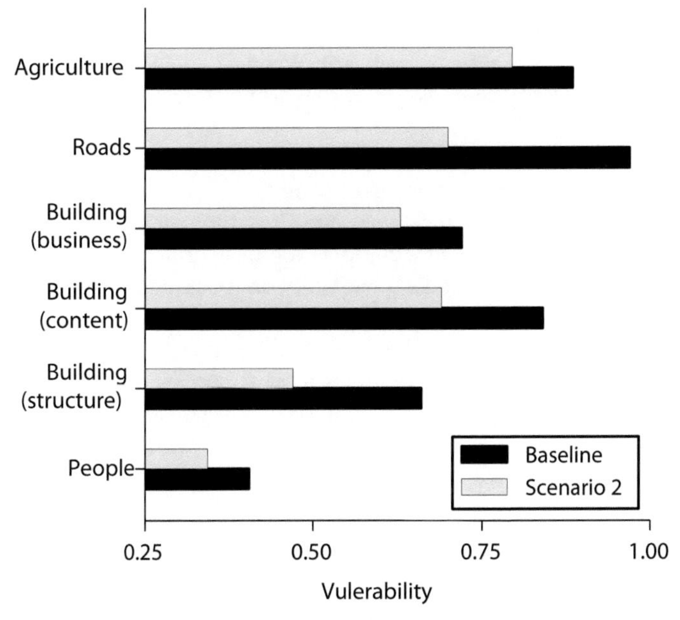

FIGURE 7.4 Vulnerability comparison; baseline and scenario two.

TABLE 7.3 Relative Benefit of the Different Scenarios (%)

Receptor	Scenario 1	Scenario 2	Scenario 3
People	64.3	15.4	69.8
Buildings	80.6	18.5	84.3
Infrastructure (roads)	67.9	27.8	76.8
Agriculture	89.6	10.2	90.6

which helped build-up stakeholder confidence in the output and understanding of model assumptions and procedures applied. The KULTURisk methodology was found to be a useful decision-making tool, which gives insight into (physical, social, and economic) benefits of proposed measures for flood risk reduction. Stakeholder participation in the determination of weights resulted in a transparent process and yielded a better understanding of inherent uncertainties in the analysis.

In terms of damage cost, the value of statistical life (VOSL) was found to be the highest value factor. Sensitivity of the VOSL, as compared to damage costs of the other receptors, was attributed to a high value estimated at €3.1 million (OECD, 2012), especially when multiplied by the affected population. It should be mentioned that even without use of VOSL, which is questionable for ethical reasons (e.g., van Wee and Rietveld, 2013), potential losses of human lives is the most significant factor triggering efforts for flood-risk reduction.

Results show that risk-prevention measures that bring more benefits would be a reduction of the flood-hazard metrics in the floodplain, coupled with the implementation of nonstructural measures to improve receptor resilience, however, further alternatives and combination of measures may be tested.

The effect of hydraulic-model uncertainty on the flood-risk analysis output is not well understood. As an extension to this work, accounting for hydraulic-model uncertainty within the KULTURisk methodology application, would yield insight into suitable methodologies to treat uncertainty.

The methodology may be further improved by accounting for the projected cost of implementation of the proposed measures, which would enable an analysis of the cost-benefit ratios.

ACKNOWLEDGMENTS

This work was funded by the European Union; FP7 KULTURisk project (no. 265280). The authors are deeply grateful to Dr J-P. Malet and the RTM staff based in the Ubaye Valley for provision of data and critical discussions regarding this work.

REFERENCES

Angignard, M., 2011. Applying Risk Governance Principles to Natural Hazards and Risks in Mountains. Dortmund University of Technology.

Apel, H., Thieken, A.H., Merz, B., Blöschl, G., 2004. Flood risk assessment and associated uncertainty. Nat. Hazards Earth Syst. Sci. 4 (2), 295–308.

Bates, P.D., 2012. Integrating remote sensing data with flood inundation models: how far have we got? Hydrol. Processes 26 (16), 2515–2521.

Bates, P.D., Horritt, M.S., Fewtrell, T.J., 2010. A simple inertial formulation of the shallow water equations for efficient two-dimensional flood inundation modelling. J. Hydrol. 387 (1–2), 33–45.

Brisson, N., Gate, P., Gouache, D., Charmet, G., Oury, F.-X., Huard, F., 2010. Why are wheat yields stagnating in Europe? A comprehensive data analysis for France. Field Crops Res. 119 (1), 201–212.

Cracknell, A.P.H.L., 2007. Introduction to Remote Sensing. CRC Press, Boca Raton, FL [u.a.].

Di Baldassarre, G., Uhlenbrook, S., 2012. Is the current flood of data enough? A treatise on research needs for the improvement of flood modelling. Hydrol. Processes 26 (1), 153–158.

Di Baldassarre, G., Viglione, A., Carr, G., Kuil, L., Salinas, J., Blöschl, G., 2013. Socio-hydrology: conceptualising human-flood interactions. Hydrol. Earth Syst. Sci. Discuss. 10 (4), 4515–4536.

Doll, C., van Essen, H., 2008. Road Infrastructure Cost and Revenue in Europe. Produced within the Study Internalisation Measures and Policies for All External Cost of Transport (IMPACT) Deliverable, 2. CE Delft Report Publication number 08.4288.17, Delft, The Netherlands.

Dottori, F., Todini, E., 2011. Developments of a flood inundation model based on the cellular automata approach: testing different methods to improve model performance. Phys. Chem. Earth, Parts A/B/C 36 (7–8), 266–280.

Final National Report Flageollet, J., Maquaire, O., Weber, D., 1996. The Temporal Stability and Activity of Landslides in Europe with Respect to Climatic Change (TESLEC). Final Report: Part II. National Reports. European Commission-Environment Programme (Ct. EV5V-CT94-0454).

Genovese, E., 2006. A Methodological Approach to Land Use-based Flood Damage Assessment in Urban Areas: Prague Case Study. Technical EUR Reports, EUR, 22497.

Genovese, E., Lugeri, N., Lavalle, C., Barredo, J., Bindi, M., Moriondo, M., 2007. An Assessment of Weather-related Risks in Europe DA2. 1. Preliminary month 18 report on maps of flood and drought risks. ADAM.

Hunter, N.M., Bates, P.D., Horritt, M.S., Wilson, M.D., 2007. Simple spatially-distributed models for predicting flood inundation: a review. Geomorphology 90 (3–4), 208–225.

INSEE, 2013. National Institute of Statistics and Economic Studies (Institut national de la statistique et des études économiques) [Online]. Available: www.insee.fr (accessed 18.06.13.).

Jongman, B., Kreibich, H., Apel, H., Barredo, J.I., Bates, P.D., Feyen, L., Gericke, A., Neal, J., Aerts, J.C.J.H., Ward, P.J., 2012. Comparative flood damage model assessment: towards a European approach. Nat. Hazards Earth Syst. Sci. 12 (12), 3733–3752.

Klijn, F., Van Buuren, M., Van Rooij, S.A., 2004. Flood-risk management strategies for an uncertain future: living with Rhine river floods in the Netherlands? AMBIO: A J. Human Environ. 33 (3), 141–147.

Lecarpentier, C., 1963. La crue de juin 1957 et ses conséquences morphodynamiques (Thèse de Doctorat). Centre de Géographie Appliquée, Faculté des Lettres et des Sciences Humaines, Universite De Strasbourg, Strasbourg, France.

Messner, F., Penning-Rowsell, E., Green, C., Meyer, V., Tunstall, S., Van Der Veen, A., 2007. Evaluating Flood Damages: Guidance and Recommendations on Principles and Methods. FLOODsite-Report T09-06-01.

Moel, D.H., 2012. Uncertainty in Flood Risk (Ph.D. thesis). VU University, Amsterdam.

Neal, J., Schumann, G., Bates, P., 2012. A subgrid channel model for simulating river hydraulics and floodplain inundation over large and data sparse areas. Water Resour. Res. 48 (11), W11506.

OECD, 2012. The Value of Statistical Life: A Meta Analysis. ENV/EPOC/WPNEP(2010)9/FINAL. Organisation for Economic Co-operation and Development, Paris, France.

Refsgaard, J.C., Van Der Sluijs, J.P., Højberg, A.L., Vanrolleghem, P.A., 2007. Uncertainty in the environmental modelling process – a framework and guidance. Environ. Modell. Software 22 (11), 1543–1556.

Schwarze, R., Schwindt, M., Weck-Hannemann, H., Raschky, P., Zahn, F., Wagner, G.G., 2011. Natural hazard insurance in Europe: tailored responses to climate change are needed. Environ. Policy Governance 21 (1), 14–30.

USACE, 1996. Engineering and Design; Risk-based Analysis for Flood Damage Reduction Studies. EM 1110-2-1619. U.S. Army Corps of Engineers, Washington, DC.

van Wee, B., Rietveld, P., 2013. Using value of statistical life for the ex ante evaluation of transport policy options: a discussion based on ethical theory. Transportation 40 (2), 295–314.

Weber, D., 1994. Research into Earth Movements in the Barcelonnette Basin. Temporal Occurrence and Forecasting of Landslides in the European Community. Final Report 1) 321–326.

Werner, M.G.F., Hunter, N.M., Bates, P.D., 2005. Identifiability of distributed floodplain roughness values in flood extent estimation. J. Hydrol. 314 (1–4), 139–157.

Winsemius, H.C., Van Beek, L.P.H., Jongman, B., Ward, P.J., Bouwman, A., 2012. A framework for global river flood risk assessments. Hydrol. Earth Syst. Sci. Discuss. 9 (8), 9611–9659.

Chapter 8

Floods and Storms Practical Exercises

Amy Dabrowa, Jeffrey C. Neal and Paul D. Bates
School of Geographical Sciences, University of Bristol, Bristol, UK

ABSTRACT

These computer-based practical exercises compliment the preceding chapters by allowing users to put into practice some of the ideas they have been reading about. The tasks in this chapter introduce users to numerical flood modeling using both test data and a real-life example. In Task 1 users explore the effects as well as pros and cons of including varying degrees of physical complexity in flood modeling by simulating a simple test case using the four different two-dimensional solvers available in LISFLOOD-FP. In Task 2 users model river and floodplain dynamics using a real-world example reach. The different ways in which LISFLOOD-FP represents river channels in simulations are introduced and results are evaluated quantitatively by comparison with satellite data. Additional exercises are available in electronic form covering a wider range of themes. These guide users through the creation of risk maps and estimation of uncertainty and begin to consider flood prevention measures.

8.1 INTRODUCTION TO FLOOD MODELING

Government organizations, such as the Environment Agency of England and Wales, are increasingly required to produce flood hazard and risk maps for the rivers they manage. For example, the European Floods Directive requires that European Union Member States produce flood risk assessments, hazard and risk maps, and management plans. Due to the lack of sufficient observational data for flood events, their work is often aided by predictive hydraulic models (e.g., DEFRA, 2006).

Hydraulic models vary greatly in their physical complexity and dimensionality, though most commonly they solve either the full or approximated St Venant equations in up to two dimensions. Application of the two-dimensional (2D) St Venant equations assumes that flow velocity can be depth averaged (e.g., vertical velocities are not important), which is a reasonable assumption in many cases of fluvial flooding (Bates et al., 2010). Though most

Hydro-Meteorological Hazards, Risks, and Disasters. http://dx.doi.org/10.1016/B978-0-12-394846-5.00008-4

models share a volume-conserving continuity equation they differ in the physical terms included in their momentum equations. The most appropriate model will depend on the case study and although accuracy is the key objective, other issues all play their part and must be considered. For example, what spatial resolution or physical process representation is needed to accurately capture flood dynamics? Conversely, how much uncertainty is already in the model inputs and boundary conditions (with implications for the accuracy of results)? Are multiple simulations needed? For example, if Monte Carlo analysis is required, simulation time may become the most important factor.

The development of flood inundation models has been driven by a number of factors. For practical purposes, a compromise has traditionally been made between the physical complexity of the model, its accuracy, its spatial resolution, and computing time. Also of importance is the availability of observation data to force, calibrate, and validate models. Hydraulic models that solve the full St Venant equations in one-dimension include MIKE11, ISIS, and HEC-RAS. The one-dimensional (1D) St Venant equations are solved at specific river cross-sections along the reach (perpendicular to flow direction) to predict cross-sectional, averaged water velocity and surface elevation. The use of such models was established relatively early (e.g., Fread, 1985; Samuels, 1990) because the hydrometric values can be calibrated and verified using stage and flow data from gauging stations, whereas river cross-sections could be characterized using traditional field survey techniques. The 1D solutions so obtained are computationally efficient and describe within-bank flow well but at the expense of the representation of floodplain dynamics. Flood extent across the floodplain is commonly obtained by linearly interpolating from the water elevations at river cross-sections onto the floodplain and then overlaying the output water surface elevations onto a digital elevation model (DEM), or by creating a series of polygonal storage areas on the floodplain through which water can flow into and out. The latter method requires modelers to specify the direction of flow between these storage areas a priori, rather than allowing flow patterns to develop during the simulation.

As computing power increases 2D modeling of floodplain inundation has become feasible (e.g., JFLOW, TUFLOW). These models are computationally expensive at high resolution, meaning approximations of the St Venant equations and other simplifications are often used. For example, the river—floodplain system can be modeled as a 1D channel which overflows to a 2D floodplain or the resolution of the model may be varied depending on the complexity of the topography. Along with their increased cost, 2D models require topographic data at suitable resolution and accuracy as input to the model and can more easily take advantage of 2D inundation extent data for model calibration and evaluation. For this reason such models developed later with the advent of remote sensing methods such as airborne LiDAR for reach scale, or of satellite data for larger, basin-wide studies (see Chapter 1.2 for more details). Despite these difficulties, 2D models more accurately represent the dynamics of floodplain inundation, floodplain flow pathways, and flow velocity. The 2D models commonly utilize storage cell or finite difference methods (e.g., LISFLOOD-FP which will be described later), finite volume methods (e.g., INFOWORKS-2D), and finite element methods (e.g., TELEMAC 2D). Storage cell and finite volume methods are particularly popular because they are easy to implement and their raster grids readily integrate with remotely sensed data. For these reasons the exercises in this chapter uses a readily available 2D model: LISFLOOD-FP.

8.1.1 Flood Modeling in LISFLOOD-FP

LISFLOOD-FP is a storage-cell-based 2D flood modeling program first developed in 2000 using a diffusive wave approximation of the St Venant equations. Starting as a 1D–2D coupled model solving the 1D kinematic wave equation for channel flow and the 2D diffusive wave equation for floodplain flow, the latest formulation calculates 2D floodplain and 1D channel flow (represented as a subgrid feature) using an approximation of the St Venant equations in which only advection is assumed negligible. It will form the basis of the practical tasks in this exercise and is described in greater detail in the following section. The controlling equations for each of the solvers described can be found in the appendix and further explanation is presented in Bates and De Roo (2000), Hunter et al. (2005), Trigg et al. (2009), Bates et al. (2010), Neal et al. (2012), and De Almeida and Bates (2013).

8.1.1.1 Modeling Floodplain Flow

In its most simple form the LISFLOOD-FP model takes a raster DEM and water inflow details (location and rate) and simulates inundation dynamics. The model uses hydraulic continuity principles to calculate the water depth in each cell of the raster grid and ensure mass balance. Water movement across floodplains is calculated at each cell face at each model time-step based on the difference in hydraulic head between adjacent cells (Figure 8.1) using a momentum equation.

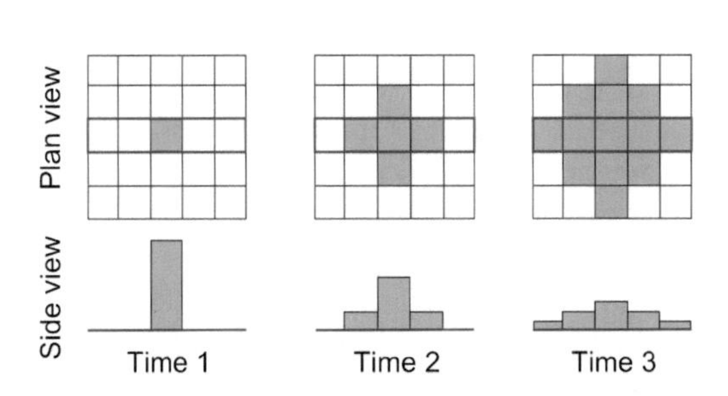

FIGURE 8.1 How water flows across the domain by storing water in cells and then allowing water to flow to neighboring cells in the $x-y$ Cartesian directions over each time-step.

Several solvers are available for calculating how much water will flow between adjacent cells over a time-step. In Task 1 we will look at four different solvers described in chronological order of their development. The first, which will be referred to as the "flow-limited" model, uses an approximation of the diffusion wave equations based on the Manning's equation. It calculates flow between cells during a time-step of fixed duration as a function of the friction slope and water slope (composed of the channel bed gradient and water surface gradient). Because its governing equations do not include inertia, a limit on the maximum flow volume per time-step must be implemented to ensure that upstream cells do not empty too quickly during one time-step (leading to a reversal in flow direction). The second solver (referred to as the "adaptive" model) uses the same momentum equation as the flow-limited solver but overcomes the flow reversal problem by using a time-step which varies in time throughout the simulation according to the water slope and grid resolution, ensuring that cells cannot empty too quickly. The third solver is referred to as the "acceleration" model. It is a simplified form of the shallow water equations, where advection is assumed negligible and calculates flow between cells as a function of the friction slope, water slope, and local acceleration. It uses a time-step which varies according to the well-known Courant−Friedrichs−Lewy condition. Finally, the "Roe" solver includes all the terms in the shallow water equations, using a first-order finite volume method based on the TRENT model presented in Villanueva and Wright (2006). Like the acceleration solver it uses a time-step based on the Courant−Friedrichs−Lewy condition.

8.1.1.2 Modeling Channel Flow

In addition to the 2D floodplain solvers described above, Task 2 will also use the 1D diffusive, kinematic, and subgrid channel solvers.

The simplest channel flow solvers is the 1D kinematic wave approximation of the shallow water equations, which assumes all terms except that the friction and bed slopes are negligible. The bed gradient is a simplification of the water slope term which takes into account the effect of changes in bed height with distance, but not changes in the water-free surface height. In contrast, the "diffusive" solver uses the 1D diffusive wave equation, which includes the friction slope and the free surface slope term and thus is able to simulate backwater effects. Once water in the channel reaches bank-full height, the 1D channel solver's water flows into adjacent floodplain cells as per the chosen floodplain solver.

The subgrid solver is the most recently developed method for representing rivers. Like the 2D acceleration solver, flow between channel segments is calculated based on the friction and free surface water slopes and local water acceleration. Only convective acceleration is assumed negligible. For any cell containing a subgrid channel segment, the solver calculates the combined flow of water within the cell, contained both within the channel located in that cell and across the adjacent floodplain. Water is distributed over the floodplain using the acceleration solver once water has overflowed from subgrid channels.

8.2 EXERCISE 1: NUMERICAL FLOOD MODELING IN LISFLOOD-FP

This is the first in a series of four exercises on flood modeling and risk mapping available with this book. Each exercise provides background information, further reading, and a detailed worked example including all necessary data files. These files and additional exercises are available to download from the Elsevier Website at store.elsevier.com/product.jsp?isbn=9780123948465 under resources tab.

Exercises in this chapter require LISFLOOD-FP which is provided. In other exercises, instructions are included for completing the tasks using Excel, MatLab, and ArcMap; however, calculations are generally simple and other programs could be used instead.

8.2.1 Task 1: Using a Simple Test Case to Explore 2D Flood Modeling

8.2.1.1 Data Provided

All the input files necessary to run simulations of the following test case with high- and low-friction scenarios are provided in the zip file for this exercise. This task was developed using LISFLOOD-FP version 5.8.6 and this version of the program has also been provided in the zip folder for consistency; however, if you would like a more up-to-date copy of LISFLOOD-FP (and for versions compatible for use on Mac or Linux machines) it can be download from the University of Bristol Hydrology Website[1]. In the folders for each test case you will find files providing the data which will be read in to the simulation: a DEM, the boundary conditions for the domain (**.bci*), and details of time-varying boundary conditions (**.bdy*). Importantly, each folder will also contain a file which instructs LISFLOOD-FP what data to read in and which options to use for the simulation (**.par*).

The test case consists of a model domain (the **.dem* file) which is rectangular in plan view, with a time-varying water height specified at one boundary which will cause water to flood across the domain (Figure 8.2). The domain's remaining boundaries are closed (as is default).

8.2.1.2 Task 1.1: Exploring Model Setup: Viewing and Understanding .par *Files*

Take a look at one of the .par files using your favorite text editor.

The model is controlled by the steering files **.par*. These give the names of key files and controlling parameters. You should see that they are set up for a simulation (*sim_time*) of 3,600 s duration and save intervals (*saveint* and *massint*) of 360 and 10 s. This means they will produce 10 grids of water depth

1. http://www.bris.ac.uk/geography/research/hydrology/models/lisflood.

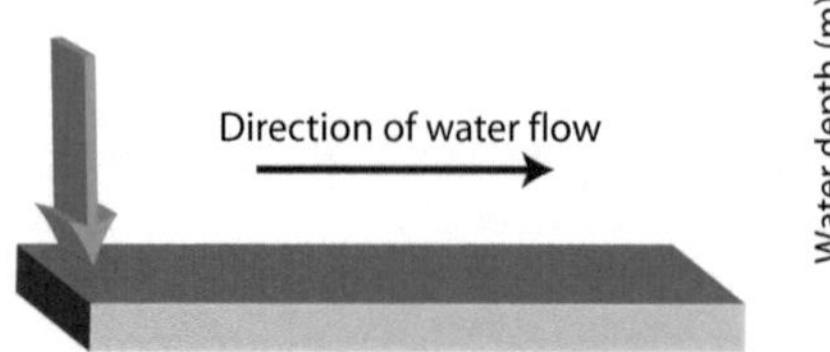

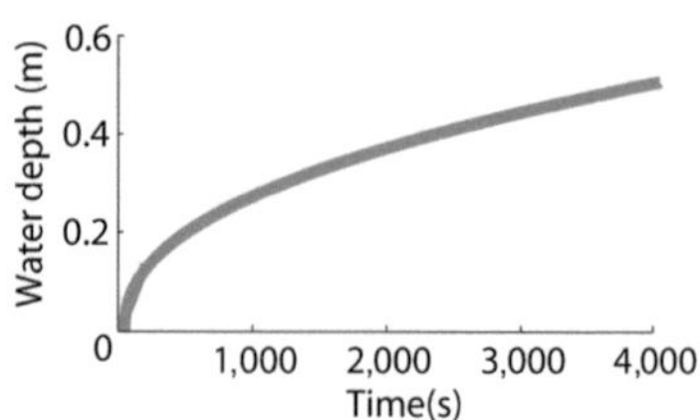

FIGURE 8.2 Left panel shows schematic of the model domain for the test case with the location of the time-varying water height boundary indicated by the blue arrow. Right panel shows the time-varying water depths specified at the boundary, which are the same for the high- and low-friction scenarios.

value as flood extent (.wd file) develops through the simulation and will write additional results to a text file (**.mass* file) every 10 s. These files are saved in the directory (*dirroot*) and with prefix (*resroot*) specified in the *.par* file. A spatially uniform Manning's n for the floodplain has been specified (*fpfric*) and the model is instructed to read in the files which specify the location and rate of water input (*bcifile* and *bdyfile*, respectively).

View and compare the .par files for the different models and scenarios to see how the model setup is specified.

8.2.1.3 *Task 1.2 Running Simulations*

The data for this practical includes *.par* files for running simulations using four different LISFLOOD-FP 2D model solvers under both high- and low-friction scenarios.

Run each simulation; all except the adaptive models should take only a matter of seconds to run.[2]

This is a research code so it runs in COMMAND PROMPT or UNIX to make Monte Carlo analysis easier.

- Copy and paste the *lisflood.exe* and *.dll* file into the folder containing the *.par* file for the model you want to run.
- Open a DOS prompt and move to the directory in which the *.par* file is located.[3]
- To run the model, at the command prompt type: *lisflood -v your_chosen_parfile.par.*

---

2. Due to the shallow water surface gradients and consequently very small time-steps used by the adaptive model, simulations using the adaptive solver may take a *long* time to run. Consider leaving them to run overnight (to stop the model running type "ctrl-c" at the command line). Alternatively, results files have been provided for these simulations in the answers folder.
3. To do this open a Command Prompt window: Start > All Programs > Accessories > command prompt. To move to the right directory type cd (change directory) followed by the full directory name (e.g., cd c:\Documents and SETTINGS\a User\My Documents\Lisflood). If you are not already in the right drive you will first need to enter the drive letter followed by a colon.

"*-v*" stands for verbose mode and is used to increase the amount of information communicated by the code about what it is doing. The code should take a few seconds to run and will have saved the results as specified in the *.par* file.

8.2.1.4 Task 1.3: Viewing Results and Evaluating Models: Water Depth Files

Open and inspect the .wd text files in the results directory using a text editor or excel.[4]

Water depth files consist of a header containing details of the size (1,4000) and location (0,0) of the grid, as well as the "NoData" value, followed by a grid of water depth values for each cell in the domain (in this case the domain is only one cell wide).

Compare the final water depth results with each other and with the analytical solution.

The analytical solution is also provided in an Excel file in the zip folder. You could copy this and the simulated water depths into a single excel file and plot them on the same graph. *Using this model setup, which model compares most favorably with the analytical solution? How long does each simulation take to run?* These will be the major considerations facing computer modelers as they decide which method to use. For more details see the papers suggested at the beginning of this document or the introduction section of the LISFLOOD-FP user manual.

8.2.1.5 Task 1.4: Further Options to Explore...

Explore LISFLOOD-FP by either changing the values of model parameters, outputting additional results files, modifying files such as the DEM or boundary condition files, or even creating additional files such as a *manningsfile* to specify spatially varying manning's values.

Here are some suggestions (you may have to look up some of these items in the instruction manual).

- Move the location of the point sources to the opposite side of the domain.
- Increase *sim_time*—what happens (or doesn't happen) in the horizontal plane test case when the water reaches the opposite side of the domain? Why is this? Can you change this outcome by varying the input files?
- Vary the *fpfric* value (other example values could range from ~0.016 for asphalt, 0.035 for short grass, and up to 0.15 for wooded floodplain).

4. To open in excel use File > Open, and ensure All files (*.*) is selected as the file type. Then choose "Delimited" and Tab for the delimiter.

- Create a manningfile with spatially varying floodplain friction. Try bands of low- and high-friction values perpendicular to the flow direction (you may need to use somewhat extreme values to see the effects).
- For the acceleration model look at the effect of setting the value of theta to 1 for the low-friction scenarios (for more details see De Almeida and Bates (2013)).
- Using the flow-limited solver, try varying the initial time-step (*initial_tstep*) to try and improve results, or vary the *Qlim* value.

Brief explanations of the expected effects of some of these are given in the answers.xls file.

8.2.2 Task 2: Simulating Flooding in a Real-world Example Reach

8.2.2.1 Data Provided

During this exercise we simulate flooding due to river bank overflow in a real-world example reach. The simulated flood event is expected to have a 1 in 100 year recurrence rate. The folder "for simulation" contains all the input files necessary for simulating this event using a number of model setups (Table 8.1). Three *.par* files have been provided to simulate the event using either the 1D diffusive or kinematic solver for channel flow in combination with the flow-limited 2D solver for floodplain flow, or the subgrid solver for channel flow which utilizes the acceleration solver for floodplain flow. From the DEM shown in Figure 8.3 one can see there is highland to the south of

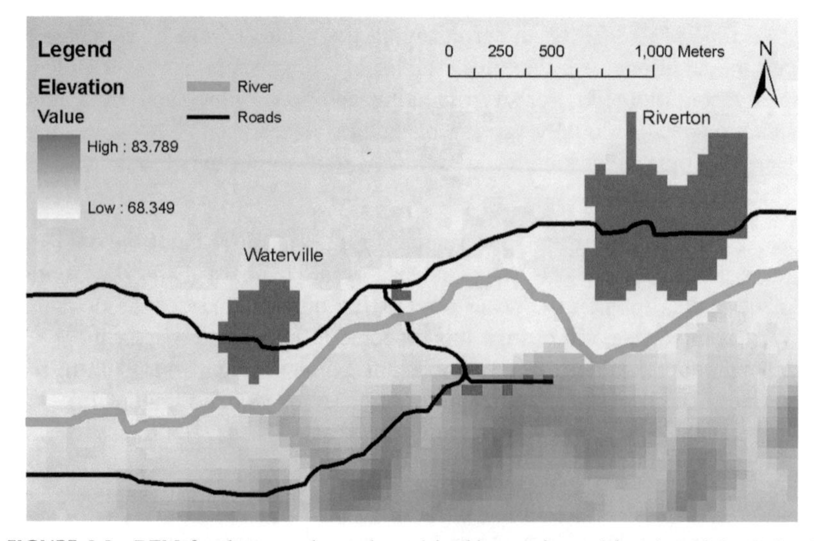

FIGURE 8.3 DEM for the example reach used in this exercise and location of the fictional settlements of Waterville and Riverton used in subsequent exercises available online.

TABLE 8.1 Details of Files Provided with This Task

.dem file	This provides the 2D raster elevation grid which is the model domain
.par files	These tell LISFLOOD-FP which settings to use and what input files to read in; there is one each for the three methods of modeling channel flow
.bci files	These tell LISFLOOD-FP about boundary conditions and point water sources in the domain. There are two files provided, one for use with the diffusive and kinematic solvers and one for use with subgrid
.river files	These tell LISFLOOD-FP where the river channel is located, and what its boundary conditions are. There are two files, one for use with the kinematic solver and one for the diffusive
.width and *.bed* files	These are used by the subgrid solver to input information about the river channel width and bed elevation
.weir file	Used by all solvers to input the characteristics of a weir located on the river toward the eastern edge of the domain

the river and an extensive floodplain to the north. The DEM is provided as a raster grid with a resolution of 50 m and vertical accuracy of approximately 25 cm, whereas the position of the river channel has been digitized from a map of the area and is therefore available at a higher resolution. The model has spatially uniform distributions for floodplain Manning's friction, channel Manning's friction, and channel dimensions. Given its short length, the assumption that the flow is in steady state (inflow = outflow) is reasonable. Also visible in Figure 8.3 are the urban areas of Riverton and Waterville. These are not considered in this exercise but are used in the further exercises available online.

In addition to the DEM, the zip file contains a number of other input files which will be used by LISFLOOD-FP depending on the options chosen for simulation. The letter K in the filename stands for the kinematic solver, D for the diffusive, and SGC for subgrid solver.

8.2.2.2 Task 2.1: Exploring Model Setup: Additional Input Files

Open and inspect the .par files using your favorite text editor and examine the contents.

Following Task 1.1 you should be familiar with the layout and workings of the *.par* files. You will see that the first few lines of each *.par* file are very similar: you should be able to recognize that the file is instructing LISFLOOD-FP to load the DEM and weir files, to save results to a named folder with specific file names, to run the simulation for a set length of time saving results in various formats at various intervals, and to set a Manning's friction value for the floodplain. If you cannot pick out this information then look again at Task 1.2.

After this the *.par* files begin to differ. River channel properties are set in the *.river* files when using the kinematic and diffusive solvers and using the files denoted by *SGCwidth*, *SGCbank*, and *SGCbed* and the channel property *SGCn* while using subgrid.

As default, LISFLOOD-FP uses the 1D kinematic solver and the 2D adaptive solver. You should see that the 1D diffusive channel solver is activated using the keyword *diffusive*. In the *.par* files for both the kinematic and diffusive solver you should see that the flow-limited solver for floodplain flow is activated (keyword *adaptoff*). The subgrid solver is activated whenever a *SGCwidth* file is specified and running the subgrid solver for channel flow automatically activates the acceleration solver for floodplain flow.

With the aid of the user manual explore the other input files:

1. Make sure you understand the details of the *.par* files already noted above and look up any items not mentioned to see what they are doing
2. Examine the *.bci* files and understand what they are doing and why are they different. For example, for each of the solvers make sure you can see how LISFLOOD-FP is informed of the location or rivers in the domain and the water input from them.
3. Examine the *.river* files used by the kinematic and diffusive solvers. What information is provided in these files? What are the standard differences between them?[5]
4. Examine the various SGC files and familiarize yourself with their format and use.

Simulate the flood event using each of the three .par files. If you do not know how to do this then refer to Task 1.2.

8.2.2.3 Task 2.2: Viewing Results—Mass Files and Water Depth Animations

The model outputs water depth files at each *saveint* (**-xxx.wd*) and a file containing a variety of diagnostic parameters saved at each *massint* (**.mass*).

The **.mass* files consist of a header line followed by 12 columns of data (Table 8.2). Take a look at the *.mass* text files. In particular look at *Qout* (the discharge in m^3/s leaving the downstream end of the model). Variation in

5. You will notice there are differences in the Manning's friction properties for all the different model setups. This is because the models have been calibrated to give the most accurate results using n as a free parameter.

TABLE 8.2 Details of Results Saved to *.mass* File

Name	Description
Time	Time in seconds at which data was saved
Tstep	Value of the time-step specified by the user in the par file
MinTstep	Minimum time-step duration used so far in the simulation in seconds
itCount	Number of time-steps since the start of the simulation
Area	Area inundated in m^2
Vol,e-6	Volume of water in the domain in m^3
Qin	Inflow discharge in m^3/s
Hds	Water depth at the downstream exit of the model domain
Qout	Calculated outflow discharge at the downstream exit of the model domain in m^3/s
Qerror	Volume error per second in m^3/s
Verror	Volume error per mass interval (*massint* variable in the parameter file) m^3
Rain-Inf+Evap	Cumulative effect of infiltration, evaporation, and rainfall over the simulation in 10^3 m^3

Qout over the simulation should show the model ramping up from the initial conditions to steady-state inflow discharge of 73 m^3/s over the whole domain.

You can view an animation of the flood dynamics using the *FloodView* windows viewer, located in the "for analysis" folder. To open *FloodView*, simply double click the icon. First, load in the DEM using the *File > Load DEM menu*. Then choose a set of water depth results to view and load in the 10 output files ending with the extension *.wd* using the *File > Open menu* (using the *ctrl* key to select all 10 files at once). As the model has ramped up from initial conditions to steady state, the final *.wd* file should be the most accurate. You should carry out the steps in this order to ensure *FloodView* works correctly.

The *.mass* and *.wd* files are the most commonly used LISFLOOD-FP outputs. Details of a variety of additional outputs from the simulation, from stage heights to water velocities, can be found in the instruction manual.

8.2.2.4 Task 2.3: Evaluating Models—Comparison to Satellite Data

Calculate the fit between the real and simulated flood extent.

Data from a synthetic-aperture radar overpass conducted during a real "1 in 100 year" flood in the area is available and allows quantitative evaluation of how well the models are performing. The file sargrid.asc located in the "for analysis" folder consists of a grid of 0s and 1s where the former represents dry areas and the latter represents inundated areas. To make the comparison easier we have provided a small program called *fstat.exe* (also in the "for analysis" folder) which calculates the fit, *F*, between the model and the data. The program first populates the following contingency table with the number of pixels in each category (Table 8.3).

The fit, *F*, between the model and the data is then calculated using the following formulas:

$$F = \frac{D}{B + C + D} \tag{8.1}$$

This divides the number of pixels correctly predicted as wet by the total number of "floodplain" pixels. It does not account for the pixels correctly predicted as dry as this might bias the measure according to domain size (e.g., it is easy to predict a small flood in a large domain as most pixels will be dry). The value of *F* goes from 0 for a model with no overlap between observed and modeled data, to 1 for a model with perfect overlap.

To use the program, copy the final water depth file from the model simulation you want to calculate *F* for (*res_?-0010.wd*) into the directory with the *fstat.exe* executable in it, open a command prompt window and at the command line type:

fstat sargrid.asc res_*your chosen model-0010.wd*

This prints to the screen the number of pixels in categories A–D and the final *F* value.

For this example, which model is best at predicting the flood inundation event?

This is the end of Exercise 1. Answers to the questions posed in this exercise are available in the Exercise 1 answers document.

TABLE 8.3 Calculation of Model Fit to Observations

	Observed = Dry	Observed = Wet
Model = dry	A = dry/dry	B = predicted dry but observed wet
Model = wet	C = predicted wet but observed dry	D = wet/wet

8.3 FURTHER EXERCISES

Three further exercises in this series covering a wider range of flood hazard problems are also available as online resources to download from the Elsevier Website at store.elsevier.com/product.jsp?isbn=9780123948465 under resources tab.

Exercise 2: Flood hazard risk mapping. Combining results from Exercise 1 with mock socioeconomic data, users are guided through creation of a deterministic risk map. Common concepts in risk analysis are introduced as users calculate the physical risk to people, buildings, and infrastructure and the economic risk to buildings.

Exercise 3: Probabilistic hazard mapping and uncertainty. Introducing probabilistic risk maps as an alternative way of representing flood risk, the exercise guides users through one method of producing a probabilistic flood risk map. The effect of spatial dependence of the flow rates of tributaries and the uncertainty in risk calculations due to small datasets are introduced and calculated.

Exercise 4: Simulating and evaluating flood prevention measures. A number of simple, indicative flood prevention schemes are suggested for the reach. Users modify the original LISFLOOD-FP input files to simulate and evaluate the effects of these schemes on the flood extent. The exercise should familiarize users with LISFLOOD-FP input files, how the model reacts to modifications, and factors that should be considered when planning flood prevention schemes. It is not designed to teach users the technical aspects of flood engineering.

8.4 APPENDIX: GOVERNING EQUATIONS FOR LISFLOOD-FP SOLVERS

8.4.1 Full Shallow Water Equations

All the solvers used in LISFLOOD-FP are approximations to varying degrees of the full shallow water equations. Momentum and continuity equations for the full 1D shallow water equations are given below (Eqns (8.2) and (8.3), respectively).

$$\underbrace{\frac{\partial Q}{\partial t}}_{\text{local acceleration}} + \underbrace{\frac{\partial}{\partial x}\left(\frac{Q^2}{A}\right)}_{\text{convective acceleration}} + \underbrace{gA\frac{\partial(h+z)}{\partial x}}_{\text{water slope}} + \underbrace{\frac{gn^2Q^2}{R^{4/3}A}}_{\text{friction slope}} = 0, \quad (8.2)$$

$$\frac{\partial A}{\partial x} + \frac{\partial Q}{\partial x} = 0, \quad (8.3)$$

where Q is the volumetric flow rate, t is time, x is distance, A is the cross-sectional area of the flow, h is the flow depth, z is the bed elevation, g is gravity, n is the Manning's coefficient of friction, and R is the hydraulic radius (equal to A/P where P is wetted perimeter). Note that the water slope is composed of a channel bed gradient and water depth gradient term.

8.4.2 Kinematic 1D Channel Flow

Channel flow is calculated using a 1D kinematic approach that captures the downstream propagation of a flood wave and the response of flow to the friction and channel bed gradient. It can be described in terms of its momentum equation as:

$$\underbrace{\frac{\partial z}{\partial x}}_{\text{Bed slope term}} - \underbrace{\frac{n^2 P^{4/3} Q_c^2}{A^{10/3}}}_{\text{Friction slope term}} = 0 \tag{8.4}$$

8.4.3 Diffusive 1D Channel Flow

Alternatively, channel flow can be calculated using a 1D diffusive approach that captures the downstream propagation of a flood wave and the response of flow to the friction slope, bed slope, and *water depth slope*. It can be described in terms of its momentum equation as:

$$\underbrace{\frac{\partial z}{\partial x}}_{\text{Bed slope term}} - \underbrace{\frac{n^2 P^{4/3} Q_c^2}{A^{10/3}}}_{\text{Friction slope term}} - \underbrace{\left[\frac{\partial h}{\partial x}\right]}_{\text{Water depth slope term}} = 0 \tag{8.5}$$

The term in brackets is the diffusion term, which in combination with the bed slope makes up the water slope term of the shallow water equations and forces the flow to respond to free surface slope which can vary in time.

Using these first two methods, each channel is discretized as a single vector along its centerline separate from the overlying floodplain raster grid. The channel occupies the floodplain pixels directly above the channel, and interacts with the pixels lying adjacent to the channel during overbank flow. The channel interacts with the floodplain using the 2D solver equations allowing water to flow between channel and floodplain nodes which are adjacent to the channel.

8.4.4 Subgrid Channels

The subgrid channel method captures the propagation of a flood wave and the effect of the friction and water slopes and local water acceleration. It uses an explicit finite difference solution of a simplified shallow water equation to simulate flow in the channel. In the channel the model calculates the flow between cells using:

$$\underbrace{\frac{\partial Q_c}{\partial t}}_{\text{Local acceleration}} + \underbrace{\frac{gA\partial(h+z)}{\partial x}}_{\text{Water slope term}} + \underbrace{\frac{gn^2 Q_c^2}{R^{4/3} A}}_{\text{Friction slope term}} = 0 \tag{8.6}$$

When water depths in the channel exceed the bank height, floodplain pixels over the channel fill with water, depending on how much of the floodplain is

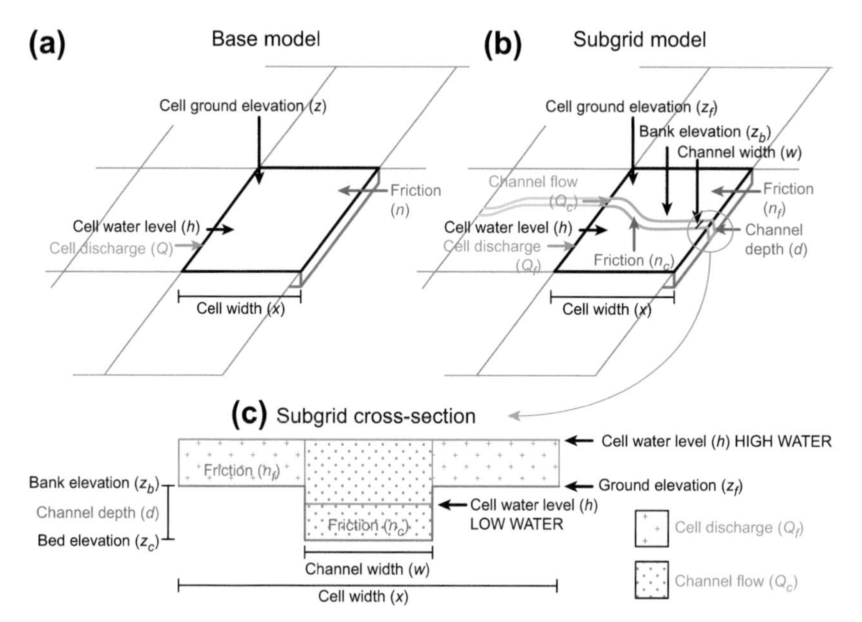

FIGURE 8.4 Model schematics for (a) Floodplain only cell, (b) Floodplain and channel cell, and (c) Interface between cells.

occupied by the channel, which then flows onto the adjacent floodplain cells using the acceleration solver (Figure 8.4). When out of bank, channels and floodplains in the same cell have the same water surface elevation.

8.4.5 Governing Equations for the Flow-Limited Model

The simplest 2D solver is based on the Manning's equation in which flow of water is controlled by only the water slope and the friction slope. Both local and convective acceleration terms are assumed negligible:

$$Q_x^{i,j} = \underbrace{\frac{h_{flow}^{\frac{5}{3}}\Delta y}{n}}_{\text{friction slope}} \underbrace{\left(\frac{(h)^{i-1,j} - (h)^{i,j}}{\Delta x}\right)^{1/2}}_{\text{water slope}} \Delta y \tag{8.7}$$

where symbols are as described earlier except h_{flow}, which is the depth through which water can flow between cells, defined as the difference between the highest water surface elevation and the highest bed elevation, Δx and Δy which are the cell dimensions, and the superscripts i and j that are grid indices. The time-step is user defined and of fixed duration for the whole simulation. However, unless this time-step is very small it may be long enough for all the water to drain from one cell to the next over a single time-step, leading to flow in the opposite direction during the next time-step.

To overcome this problem a "flow limiter" was introduced which sets a limit on the volume of water allowed to flow between cells during a single time-step, as a function of flow depth, grid size, and time-step:

$$Q_x^{i,j} = \min\left(Q_x^{i,j}, \frac{\Delta x \Delta y (h^{i,j} - h^{i-1,j})}{4\Delta t} \right) \tag{8.8}$$

8.4.6 Governing Equations for the Adaptive Model

This solver uses the same equation as the flow-limited solver to calculate flow between cells on the floodplain (Eqn (8.7)). However, it differs from the flow-limited solver by having a time-step which varies in duration throughout the simulation according to the water slope and grid resolution, rather than one with a fixed duration. This overcomes the problem of cells emptying during a time-step without the need of a flow limiter. The equation which governs time-step duration is shown below:

$$\Delta t = \frac{\Delta x^2}{4} \min\left(\frac{2n}{h_{flow}^{\frac{5}{3}}} \left| \frac{\partial h}{\partial x} \right|^{1/2}, \frac{2n}{h_{flow}^{\frac{5}{3}}} \left| \frac{\partial h}{\partial y} \right|^{1/2} \right) \tag{8.9}$$

8.4.7 Governing Equations for the Acceleration Model

Once again, this solver uses an approximation of the shallow water equations to calculate flow between cells. In addition to the water and friction slopes, water flow is also a function of local acceleration and only convective acceleration is assumed to be negligible. Water flow is calculated as below:

$$Q_x^{t+\Delta t} = \frac{\overbrace{q^t}^{\text{local acceleration term}} - \overbrace{gh^t \Delta t \frac{\partial (h_t + z)}{\partial x}}^{\text{water slope term}}}{\underbrace{1 + g\Delta t n^2 \left| q_{flow}^t \right| / [h_{flow}^{t\;7/3}]}_{\text{friction term}}} \Delta x \tag{8.10}$$

where all symbols are as before except q which is the discharge per unit width. As with the adaptive solver, the time-step used with the acceleration solver varies throughout the simulation. In this case it varies according to the cell size and water depth:

$$\Delta t_{\max} = \alpha \frac{\Delta x}{\sqrt{gh_t}} \tag{8.11}$$

8.4.8 Governing Equations for the Roe Model

The "Roe" solver includes all the terms in the full shallow water equations and a full description of its governing equations is beyond the scope of this exercise. The method is based on the Godunov approach and uses an approximate Riemann solver by Roe based on the TRENT model presented in Villanueva and Wright (2006). The explicit discretization is first order in space on a raster grid. It solves the full shallow water equations with a shock capturing scheme. LISFLOOD-Roe uses a pointwise friction based on the Manning's equation, whereas the domain boundary/internal boundary (wall) uses the ghost cell approach. The stability of this approach is approximated by the CFL condition for shallow water models. It is to be noted that this solver has thus far only been tested on a limited number of scenarios and may not be as robust as the other more commonly used solvers.

REFERENCES

Bates, P.D., De Roo, A.P.J., 2000. A simple raster-based model for flood inundation simulation. J. Hydrol. 236, 54–77.

Bates, P.D., Horritt, M.S., Fewtrell, T.J., 2010. A simple inertial formulation of the shallow water equations for efficient two-dimensional flood inundation modelling. J. Hydrol. 387, 33–45.

De Almeida, G.A.M., Bates, P., 2013. Applicability of the local inertial approximation of the shallow water equations to flood modelling. Water Resour. Res. 49 (8), 4833–4844.

DEFRA, 2006. Flood Risk to People Phase 2. Guidance Document FD2321/TR2.

Fread, D.L., 1985. In: Anderson, M.G., Burt, T.P. (Eds.), Hydrological Forecasting. Wiley, Chichester (Chapter 14).

Hunter, N.M., Horritt, M.S., Bates, P.D., Wilson, M.D., Werner, M.G.F., 2005. An adaptive time-step solution for raster-based storage cell modelling of floodplain inundation. Adv. Water Resour. 28, 975–991.

Neal, J., Schumann, J.G., Bates, P., 2012. A sub-grid channel model for simulating river hydraulics and floodplain inundation over large and data sparse areas. Water Resour. Res. 48 (11).

Samuels, P.G., 1990. Cross section location in one-dimensional models. In: White, W.R. (Ed.), International Conference on River Flood Hydraulics. Wiley, Chichester, pp. 339–350.

Trigg, M.A., Wilson, M.D., Bates, P.D., Horritt, M.S., Alsdorf, D.E., Forsberg, B.R., Vega, M.C., 2009. Amazon flood wave hydraulics. J. Hydrol. 374, 92–105.

Villanueva, I., Wright, N.G., 2006. Linking Riemann and storage cell models for flood prediction. Proc. ICE – Water Manage. 159 (1), 27–33.

Section 2

Wind, Heat Waves, and Droughts

Chapter 9

Drought Monitoring and Assessment: Remote Sensing and Modeling Approaches for the Famine Early Warning Systems Network

G.B. Senay [1], N.M. Velpuri [2], S. Bohms [3], M. Budde [1], C. Young [4], J. Rowland [1] and J.P. Verdin [1]

[1] *U.S.Geological Survey (USGS) Earth Resources Observation and Science (EROS) Center, Sioux Falls, SD, USA,* [2] *ASRC InuTeq LLC, Contractor to USGS EROS Center, Sioux Falls, SD, USA (work performed under G13PC00028),* [3] *SGT Inc., Contractor to USGS EROS Center, Sioux Falls, SD, USA (work performed under G10PC00044),* [4] *ERT Inc., Contractor to USGS EROS Center, Sioux Falls, SD, USA (work performed under G10PC00044)*

ABSTRACT

Drought monitoring is an essential component of drought risk management. It is usually carried out using drought indices/indicators that are continuous functions of rainfall and other hydrometeorological variables. This chapter presents a few examples of how remote sensing and hydrologic modeling techniques are being used to generate a suite of drought monitoring indicators at dekadal (10-day), monthly, seasonal, and annual time scales for several selected regions around the world. Satellite-based rainfall estimates are being used to produce drought indicators such as standardized precipitation index, dryness indicators, and start of season analysis. The Normalized Difference Vegetation Index is being used to monitor vegetation condition. Several satellite data products are combined using agrohydrologic models to produce multiple short- and long-term indicators of droughts. All the data sets are being produced and updated in near-real time to provide information about the onset, progression, extent, and intensity of drought conditions. The data and products produced are available for download from the Famine Early Warning Systems Network (FEWS NET) data portal at http://earlywarning.usgs.gov. The availability of timely information and products support the decision-making processes in drought-related hazard assessment, monitoring, and management with the FEWS NET. The drought-hazard monitoring approach perfected by the U.S. Geological Survey for FEWS NET through the integration of satellite data and hydrologic modeling can form the basis for similar decision support systems. Such

Hydro-Meteorological Hazards, Risks, and Disasters. http://dx.doi.org/10.1016/B978-0-12-394846-5.00009-6

systems can operationally produce reliable and useful regional information that is relevant for local, district-level decision making.

9.1 INTRODUCTION

Drought is perhaps the most complex and damaging natural hazard. It can recur frequently and cause considerable damage to agriculture, economy, nature, and property, potentially affecting the lives of a large number of people (Kogan, 1997). Four major types of droughts are broadly defined and agreed upon in the scientific literature (WMO, 1975; Wilhite and Glantz, 1985; White and O'Meagher, 1995; McVicar and Jupp, 1998): (1) meteorological drought is caused by lower than normal precipitation for a prolonged period of time; (2) agricultural drought is caused when plant available water falls below the required limit usually during a critical crop growth stage; (3) hydrologic drought is caused when one or a combination of factors such as stream flow, soil moisture or groundwater is/are not available; and (4) socioeconomic drought is caused when a loss occurs in expected return, usually measured by social and economic indicators. In this chapter, we use a definition provided by Tucker and Chaudhury (1987) that defines—"drought" as a period of reduced plant growth in relation to the historical average caused by reduced precipitation. This definition is similar to the agricultural drought definition, which includes both natural and cultivated vegetation and is what we observe using coarse to medium resolution satellite remote sensing data.

Drought usually involves a deficiency of precipitation that leads to reduced soil moisture and diminished plant growth when prolonged over longer periods of time (Crafts, 1968). The precipitation deficit will have different impacts depending on meteorological conditions, ecosystem type, and socioeconomic circumstances (McVicar and Jupp, 1998). Thus, the onset and impact of droughts is highly variable over space and time and usually occur over large areas. More than half of the terrestrial earth is susceptible to drought each year (Kogan, 1997). Because drought is a recurring phenomenon and common for all climate zones, it is difficult to predict and monitor drought using conventional approaches over large areas.

Drought early warning and monitoring are crucial components of drought preparedness and mitigation plans (Wilhite and Glantz 1985). They are usually carried out using drought indicators that are continuous functions of hydrometeorological variables such as rainfall, vegetation activity, soil moisture availability, etc. The success of drought preparedness and mitigation depends, to a large extent, upon timely information about drought onset, progress and areal extent (Morid et al., 2006). However, several countries (mainly developing nations) have limited institutional and technical capacity to monitor drought and to mitigate its impacts. Moreover, information on drought onset and development is not readily available to agencies responsible for the preparedness and mitigation of droughts. Furthermore, sparse data observation networks result in inadequate spatial coverage, data quality, and time accessibility problems for drought monitoring (Kogan, 1997; Thenkabail et al., 2004). Recent technological advances in satellite remote sensing have improved our ability to address complexities of early warning and efficient monitoring of drought situations (Thenkabail et al., 2004). Satellite remote sensing enables continuous drought monitoring over a variety of spatial and temporal scales that can help to generate timely information on drought onset, progress, and areal extent (Kogan, 1997).

The components of the agro-hydrologic system relevant to drought monitoring that can be estimated or modeled from satellite remote sensing data are: (1) rainfall; (2) vegetation condition; (3) soil moisture; (4) groundwater; and (5) evapotranspiration (ET). Although rainfall and vegetation condition can be directly estimated from satellite remote sensing data, other parameters have to be modeled using agro-hydrologic modeling approaches.

This chapter describes how the operational satellite remote sensing technology is being used by The U.S. Agency for International Development (USAID), Famine Early Warning Systems Network (FEWS NET) to detect drought early enough to assess and produce information for decision support for drought-related hazards. FEWS NET manages an information system designed to identify problems in the food supply system that lead to food insecure conditions in sub-Saharan Africa, Afghanistan, Central America, and Haiti. FEWS NET also provides access to a range of geospatial data, satellite images and derived data and products in support of the monitoring needs throughout the world as a part of the Early Warning and Environmental Monitoring Program. Along with the National Oceanic and Atmospheric Administration (NOAA) and The National Aeronautics and Space Administration (NASA), the U.S. Geological Survey (USGS) is responsible for processing, modeling, and disseminating FEWS NET's operational geospatial products at http://earlywarning.usgs.gov/. Major users of the FEWS NET products are USAID, the World Food Program (WFP), and national governments. Drought products are regularly posted at the Early Warning and Environmental Monitoring Program Web site at (http://earlywarning.usgs.gov).

In accordance with the FEWS NET drought-monitoring initiative, a suite of geospatial products and drought-monitoring indicators are produced. A list of all drought indicators produced by the FEWS NET is presented in Table 9.1. Broadly, based on how they are monitored/estimated using multisource remote sensing data, these indicators can be grouped into the three categories: (1) rainfall-based indicators; (2) vegetation index-based indicators; and (3) model-based indicators. A detailed overview of these drought indicators is presented in this chapter. Furthermore, examples of FEWS NET drought monitoring in several regions and countries around the globe are presented.

TABLE 9.1 List of Drought Indicators Produced Under FEWS NET

				Drought Indicator			
No.	**Drought Index/Indicator**	**Method**	**Resolution/ Scale**	**Short-term**	**Seasonal**	**Annual**	**References**
1	Standardized precipitation index (SPI)	Rainfall based	10 km	Y	Y	Y	McKee et al. (1993)
2	Start of season (SOS)	Rainfall based	10 km	–	Y	–	Agrhymet, 1996
3	Rainfall anomaly	Rainfall based	10 km	Y	–	–	Hiem (2002)
4	Dryness indicators	Rainfall based	10 km	Y	–	–	Hiem (2002)
5	Normalized difference vegetation index (NDVI)	Vegetation based	250 m	Y	Y	Y	Kogan (1997); Anyamba and Tucker (2012)
6	Water requirement satisfaction index (*WRSI*)	Hydrologic model based	10 km	–	Y	–	FAO (1986); Verdin and Klaver (2002); Senay and Verdin (2003)
7	Soil water index	Hydrologic model based	10 km	Y	–	–	Senay and Verdin (2003)
8	Evapotranspiration	Energy balance model	1 km	Y	Y	Y	Senay et al. (2013a) Senay et al. (2011a); Senay et al. (2011b)
9	Water levels	Hydrologic model based	Point scale	Y	Y	Y	Senay et al. (2013b); Velpuri et al. (2014)

9.2 RAINFALL-BASED DROUGHT MONITORING

Since any drought condition is the manifestation of the reduced precipitation or depleted moisture content in the soil, rainfall based, drought indicators offer direct and simple methods to monitor onset, expansion, and intensity of drought situations. Because satellite rainfall products offer precipitation data globally, it is convenient to use rainfall-based indicators for drought monitoring over national and international scales. Although several satellite-based precipitation data sets are available, one of the requirements of FEWS NET is that a rainfall product should be of high spatial resolution and available in near-real time with reasonable accuracy.

Satellite-rainfall estimation (RFE) has been used by the FEWS NET since 1995 with the development of the RFE algorithm (Xie and Arkin, 1997). Although the absolute accuracy of satellite rainfall varies by geographic location and algorithm/sensor used (Huffman et al., 1997), the combination of synoptic coverage of large areas and consistency in estimates (Xie and Arkin, 1997) has allowed the successful implementation of RFE for various agro-hydrologic applications including drought monitoring (Verdin and Klaver, 2002; Senay and Verdin, 2003; Senay et al., 2012).

9.2.1 Standardized Precipitation Index (Multiscale Drought Indicator)

The Standardized Precipitation Index (SPI) developed by McKee et al. (1993) is a valuable tool for the estimation of the intensity and duration of drought events. The SPI presents a rainfall anomaly as a normalized variable that conveys the probabilistic significance of the observed/estimated rainfall. In simple terms, SPI quantifies degree of wetness/dryness by comparing accumulated rainfall over different time periods with the historical rainfall period. For example, a 6 month SPI for February 2014 will be generated by comparing September 2013 to February 2014 rainfall totals with historical totals for the September–February time periods. By expressing anomalies in terms of their likelihood of occurrence, it is easier to evaluate the rarity of the observed event, in the absence of a nuanced understanding of the rainfall regime at a location. To evaluate the likelihood of occurrence, probability distribution functions (PDFs) are fit at each pixel for each accumulation interval. These PDFs are fit to the Collaborative Historical African Rainfall Model (CHARM) (Funk et al., 2003), which provides a 36-year time series with which gamma distribution parameters are estimated. The long-term 36 years of time series of daily precipitation grids with wall to wall coverage is produced by blending national weather prediction model reanalysis fields, interpolated station data, and output from orographic models. The CHARM data establish the shape of the distribution and provide an estimate of the variance. The SPI index presents a better representation of abnormal wetness and dryness than the Palmer indices (Guttman, 1999) (Figure 9.1).

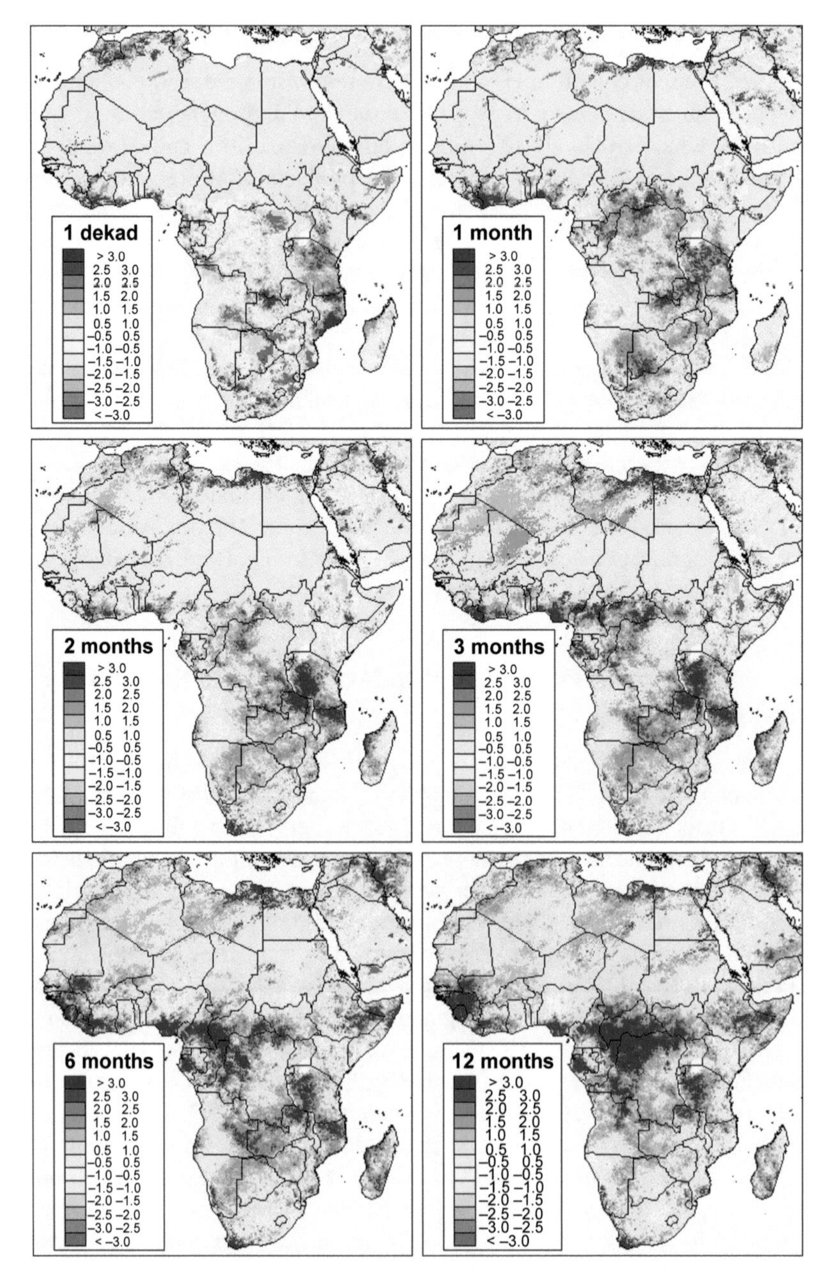

FIGURE 9.1 Africa-wide Standardized Precipitation Index (SPI) for the indicated accumulation periods as of February 20, 2014.

SPI can be produced for different accumulation periods at 10 km resolution. The dekadal (10-day) to 1-month SPI reflects short-term conditions, and its application can be related closely to soil moisture; the 3-month SPI provides a seasonal estimation of precipitation; the 6-month SPI indicates medium-term trends in precipitation patterns; and the 12-month SPI reflects the long-term precipitation patterns, usually tied to stream flows, reservoir levels, and even groundwater levels (NDMC, 2007). SPI data (updated for every dekad) are available for Africa on the FEWS NET portal (http://earlywarning.usgs.gov/fews/africa/index.php).

SPI values that are greater than 0 indicate conditions wetter than the median, whereas negative SPI values indicate drier than median conditions. For drought analysis, an SPI less than −1.0 indicates that the observation is roughly a one-in-six dry event, and is termed "moderate." An SPI less than −1.5 indicates a 1-in-15 dry event, and is termed "severe." Values less than −2.0 are typically referred to as "extreme," indicating the event is in the driest 2 percent of all events. Table 9.2 provides the classification/interpretation of SPI values in terms of drought condition.

On a dekadal scale, it can be seen that much of central and southern Africa are experiencing lower precipitation. However, on the one-monthly to three-monthly time scales, SPI index values show <-2 (severe to extreme drought) over part of East Africa (Horn of Africa). Larger accumulation periods (6−12 months) show positive SPI index values with only a few pockets of east, central and southern Africa showing moderate drought situations (negative SPI values).

9.2.2 Start of Season (Seasonal Drought Indicator)

In addition to producing SPI, one useful application of the RFE is to monitor the "start of season" (SOS). In tropical regions where agricultural practices are strongly tied to the arrival of the seasonal rains, the SOS is defined from field experience in West Africa by researchers in AGRHYMET (AGRHYMET, 1996) as follows: soon after the dry season and beginning of the wet season, the SOS is said to be established when a sum total of at least 25 mm of rainfall has occurred in a given dekad (10-day period) and followed by a total of at least 20 mm of rainfall during the following two consecutive dekads. The information on the dry season and wet season is derived based on the a priori knowledge derived from NDVI phenology information. The SOS anomaly is produced as deviation of SOS from the average of the last 10 years. Although the early arrival of SOS is not considered a hazard, the late arrival of SOS by 3 dekads or more signals a concern, especially in regions where the rainy season is short, i.e., less than 90 days.

An example of SOS and SOS anomaly is shown in Figure 9.2(a) for the Sahel Region of West Africa for the 2013 growing season. The SOS map shows that the SOS generally starts earlier in the southern part of the region by

TABLE 9.2 Classification/Interpretation of SPI Values in Terms of Drought Condition Based on McKee et al. (1993)

SPI Value	Drought Condition
<−2	Extremely dry
−2.0 to −1.5	Severely dry
−1.5 to 1.0	Moderately dry
−0.5 to 0.5	Near normal
1.0 to 1.5	Moderately wet
1.5 to 2.0	Very wet
>2.0	Extremely wet

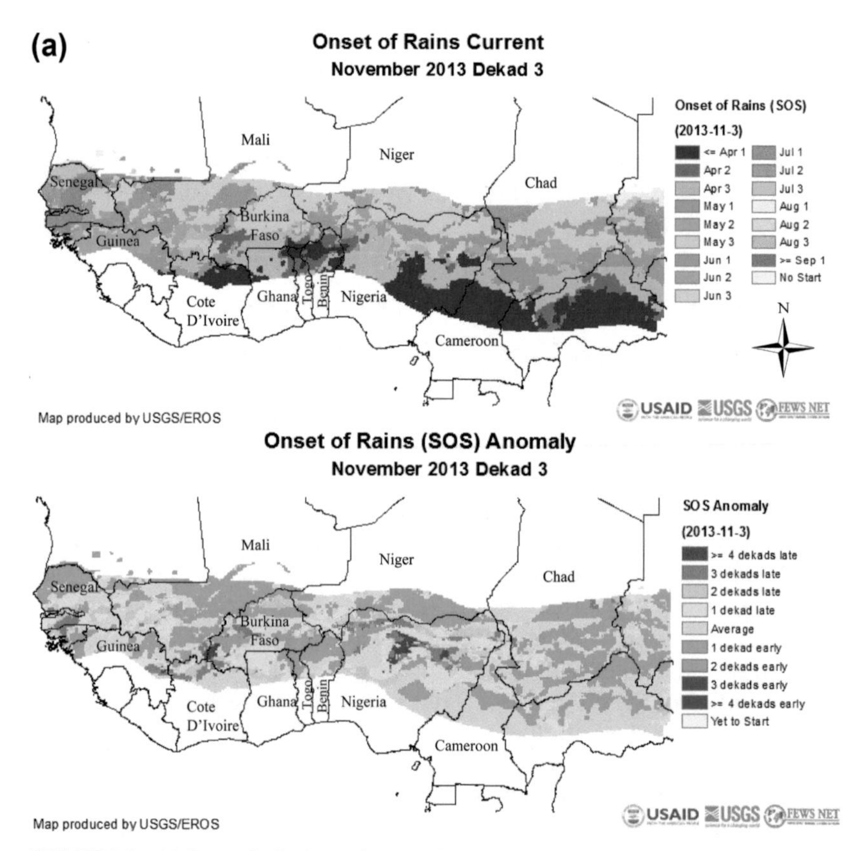

FIGURE 9.2 (a) Onset of rains/start of season (SOS) and SOS anomaly images for 2013 derived using rainfall estimates (RFE) data for West Africa. (b) Onset of rains/SOS and SOS anomaly images for 2013 derived using RFE for Central America.

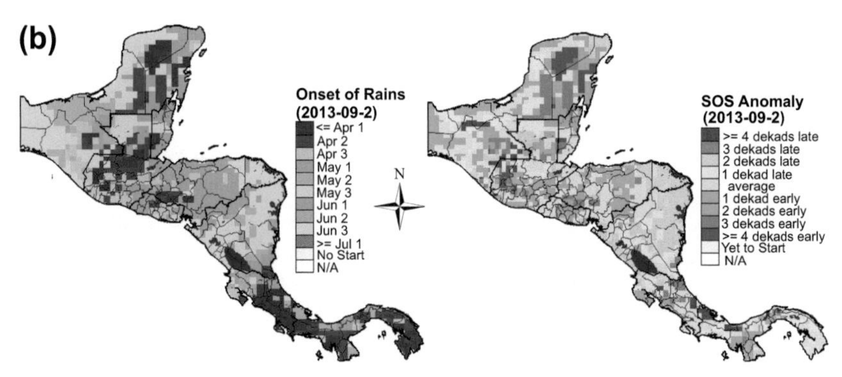

FIGURE 9.2 cont'd

April and progresses north until July. The corresponding SOS anomaly shows that parts of north–central Nigeria and southwestern Burkina Faso experienced a delay of rainfall by more than 3 dekads in the season, signaling a potential drought year leading to reduced crop growth and lower yield. Similar examples of SOS and SOS anomaly are shown in Figure 9.2(b) for Central America for the 2013 growing season. Currently SOS data are being disseminated for Africa, Central America, and the Caribbean regions.

9.2.3 Monthly Dryness Indicators (Short-term Drought Indicator)

Using RFE rainfall data, FEWS NET produces a suite of products to specifically study duration, progression, and intensity of the short-term dryness caused by the deficiency of precipitation. An indication of dryness condition, particularly duration of the dryness can be studied by analyzing the number of rainy or nonrainy days in a region. Hence FEWS NET produces a product that shows number of rainy days (any day with >1 mm of rainfall) by analyzing rainfall occurrence over the previous 30 days (from the current date) (Figure 9.3(a)). This product provides information on the number of rainy days for a region. Generally the higher the number of rainy days, the lower the dryness. However, this product does not show the magnitude of rainfall (in mm) received during a specified time period. The quantification of rainfall is required to understand the condition of wetness/dryness. Figure 9.3(b) shows the total accumulated rainfall over the last 30 days for Central Asia. The accumulated precipitation in mm helps to identify regions that received deficit rainfall over the last 30 days. Such regions are of interest to the FEWS NET as further failure of rains over a prolonged period could lead to drought conditions where humanitarian activities must be concentrated.

Furthermore, it is important to understand the distribution of rainfall. The analysis of distribution of rainfall in terms of consecutive dry days or number of days since a rain event over a specified time period would provide

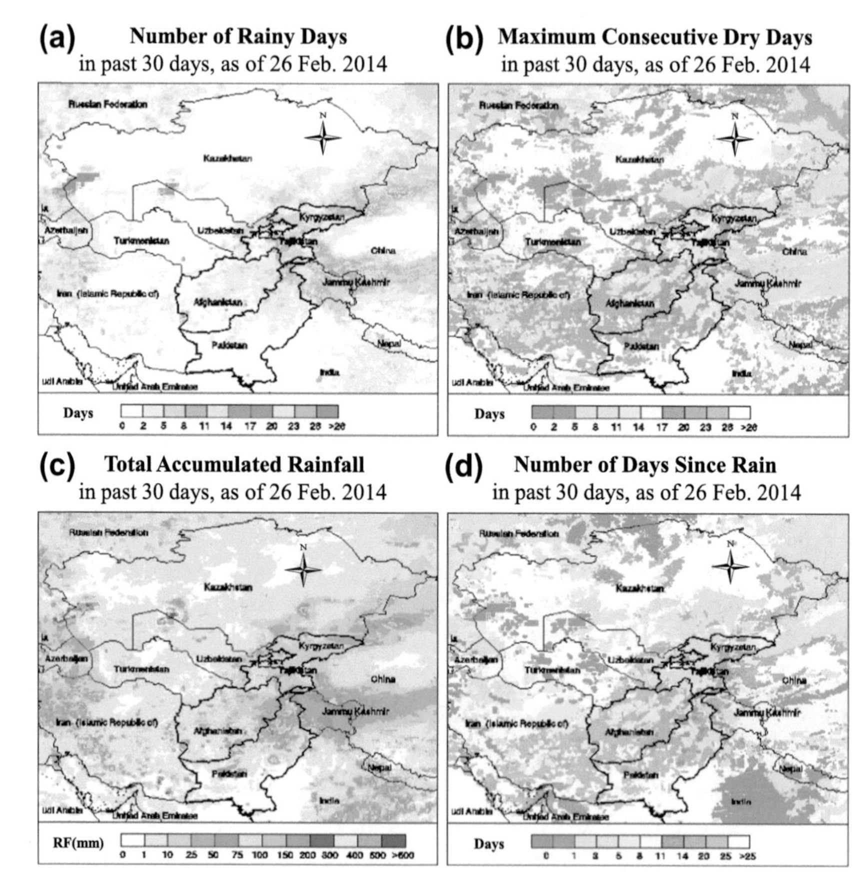

FIGURE 9.3 **Monthly dryness indicators for Central Asia produced by FEWS NET as of the end of February, 2014: (a): number of rainy day; (b) total accumulated rainfall; (c) maximum consecutive dry days; (d) number of days since rain.**

information on the intensity of dryness condition. Generally, the higher the number of consecutive dry days, the higher the dryness intensity. Figure 9.3(c) shows maximum consecutive dry days in the previous 30-day period as of February 26, 2014. Similarly, Figure 9.3(d) shows the number of days since rain in the previous 30-day period. Under various FEWS NET projects, data on these short-term indicators of dryness are available for Africa, Central America, the Caribbean, Mexico, the Middle East, and Central Asia.

9.2.4 Dekadal Dryness Indicator (Short-term Drought Indicators)

Analyzing of dekadal rainfall accumulations and comparing them with the long-term average rainfall for the same period provides useful information on the short-term dryness. Figure 9.4 shows the rainfall accumulations, average

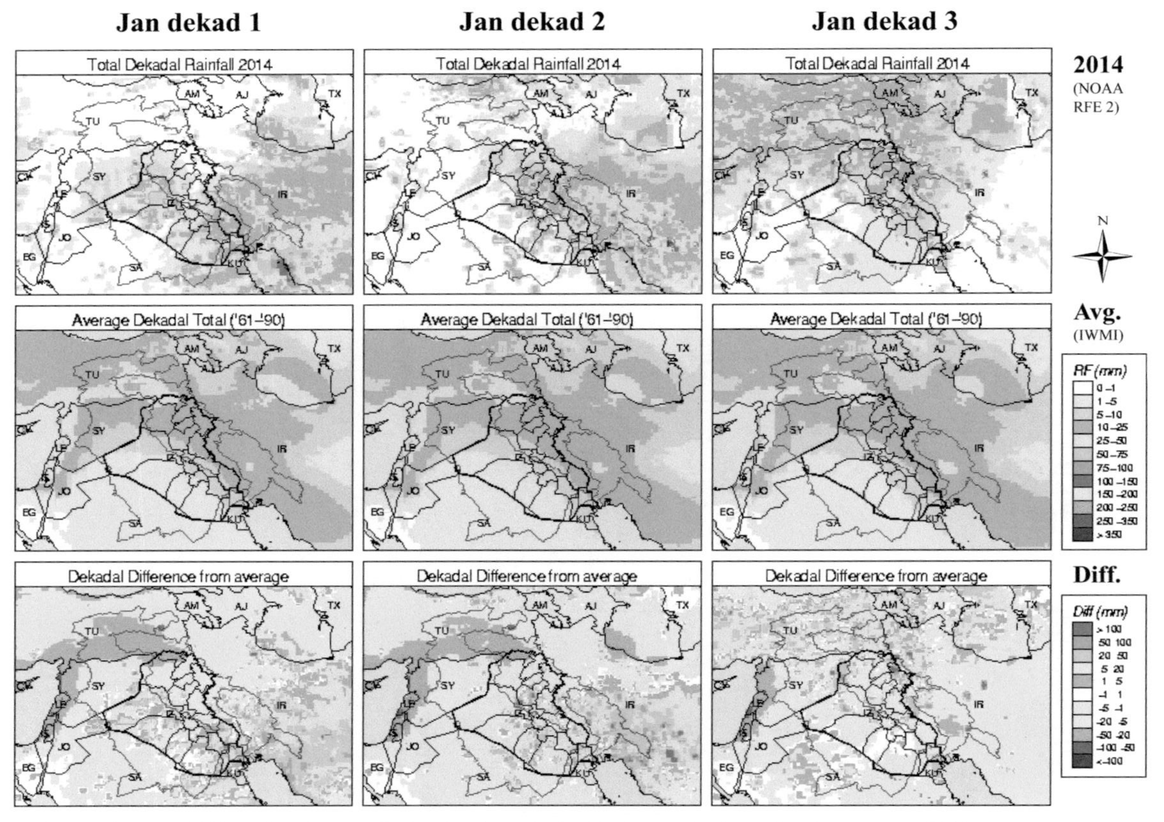

FIGURE 9.4 Dekadal (10-day) RFE total and anomaly for the Iraq/Tigris–Euphrates basin region for January 2014.

dekadal and rainfall anomaly (deficit) for all the three dekads of January 2014. This figure highlights regions with high rainfall deficits (>100 mm/dekad) such as coastal regions of Syria, Lebanon, and Jordan (see SY, LE, and Jo in Figure 9.4). Continuous monitoring of time series information on dekadal rainfall accumulations and anomalies will further provide information on onset and progression of drought/dryness conditions. Furthermore, such information can be combined with other indicators such as NDVI or ET estimates by the FEWS NET managers or stakeholders to assess the potential drought conditions. Currently, dekadal dryness index information is being disseminated over Africa, Central America, the Caribbean, Mexico, the Middle East, and Central Asia.

9.3 VEGETATION INDEX-BASED DROUGHT MONITORING

Generally, vegetation shows a lagged response to drought because of the delayed vegetation response to developing rainfall deficits due to residual moisture stored in the soil (Thenkabail et al., 2004). Theoretically, vegetation response to water stress happens at two separate phases. First, the vegetation responds to the initial short-term/temporary depletion of precipitation/soil moisture through plant water conserving mechanisms that would reduce plant water loss (transpiration) by the closure of stomata (Maselli et al., 2009). However, such mechanism does not show external signs of stress as they do not reduce plant biomass (vegetation condition) as seen by satellites (Running and Nemani, 1988). However, when precipitation deficit is extended over longer time periods, it often results in reduced photosynthetic activity and subsequent reduction in vegetation condition (Maselli et al., 2009). Such reduction in vegetation conditions can be monitored by the analysis of satellite data. Vegetation indices have been extensively used for monitoring the impact of drought on crop growth (Kogan, 1997; Peters et al., 2002; Ji and Peters, 2003; Rhee et al., 2010; Anyamba and Tucker, 2012).

9.3.1 Normalized Difference Vegetation Index (Multiscale Drought Indicator)

Vegetation is constantly monitored for conditions of drought using the Normalized Difference Vegetation Index (NDVI) (Tucker, 1979). This index is computed using near-infrared (NIR) and red channels as.

$$\mathrm{NDVI} = \frac{(\rho_{\mathrm{NIR}} - \rho_{\mathrm{RED}})}{(\rho_{\mathrm{NIR}} + \rho_{\mathrm{RED}})} \tag{9.1}$$

where ρ represents the spectral reflectance at red and NIR bands. Because NDVI can be affected by cloud and cloud contamination, the maximum NDVI in a 10-day period (maximum NDVI composite) is used for a given pixel to minimize the impact of clouds. Current operational drought monitoring for FEWS NET is based on the maximum value composite image of the NDVI.

Real-time and historical NDVI products are generated every 5 days for the previous 10 days on a geographic mapping grid. For operational purposes, FEWS NET uses the NDVI generated from the Moderate Resolution Imaging Spectroradiometer (MODIS). These expedited products are known as "eMODIS" and respond to operational land monitoring applications requiring near-real time NDVI data for comparison against historical records (Jenkerson et al., 2010). The eMODIS NDVI data archive (2001–present) is available for download from the FEWS NET data portal (http://earlywarning.usgs.gov/fews/downloads/index.php?regionID=af-e&productID=1&periodID=5).

A time series smoothing technique developed by Swets et al. (1999) was used to smooth NDVI composites for the years 2001–2010. This technique uses a weighted least squares linear regression approach to "correct" noise in the NDVI data as a result of clouds or other atmospheric contamination. This smoothed time series is used to derive a 10-year mean NDVI on a pixel-by-pixel basis for each of 72 composite periods per year. As current-year composites become available, they are added to the time series and smoothed, resulting in a smoothed composite comparable to the historical mean for a given 10-day period. As part of the MODIS standard product, a quality control (QC) band is used to mask out the cloud-affected pixels. Finally, NDVI deviations (anomaly) from the 10-year dekadal average and the previous dekad are produced for each period. Spatial averages for crop districts are generated for temporal traces and compared to average and pervious year performances. Thus, the use of NDVI anomaly offers the simplest and most common vegetation index-based approach for detecting and monitoring droughts. The use of anomaly isolates the variability in the vegetation signal and establishes meaningful historical context for the current NDVI to determine relative drought severity (Anyamba and Tucker, 2012). One of the assumptions in using NDVI and NDVI anomaly is that there is no major land cover change that would result in change in NDVI. Generally, the land cover change in the agricultural (heterogeneous) landscapes is small and limited to a small region that would correspond to a subpixel to pixel change in satellite data. On the other hand, the objective of using NDVI anomaly is to identify negative anomaly over expanse of large areas covering several tens of square kilometers. Hence, small-scale changes in land cover will not affect the use of NDVI anomaly for drought monitoring.

Figure 9.5 show eMODIS NDVI (left) and NDVI anomaly (right) in East Africa for February 16–25, 2014. Although the NDVI distribution for the period shows lower vegetation condition in the northern part of the region than in the southern part, the NDVI anomaly is more meaningful to detecting drought conditions since the large region includes different hydroclimatic conditions with varying rainfall patterns and vegetation cover. Therefore, according to the NDVI anomaly, assuming the absence of a major land cover change in the region, northeastern Tanzania is at a greater drought risk than the normally dry part of northern Sudan, despite the low NDVI. Currently, NDVI

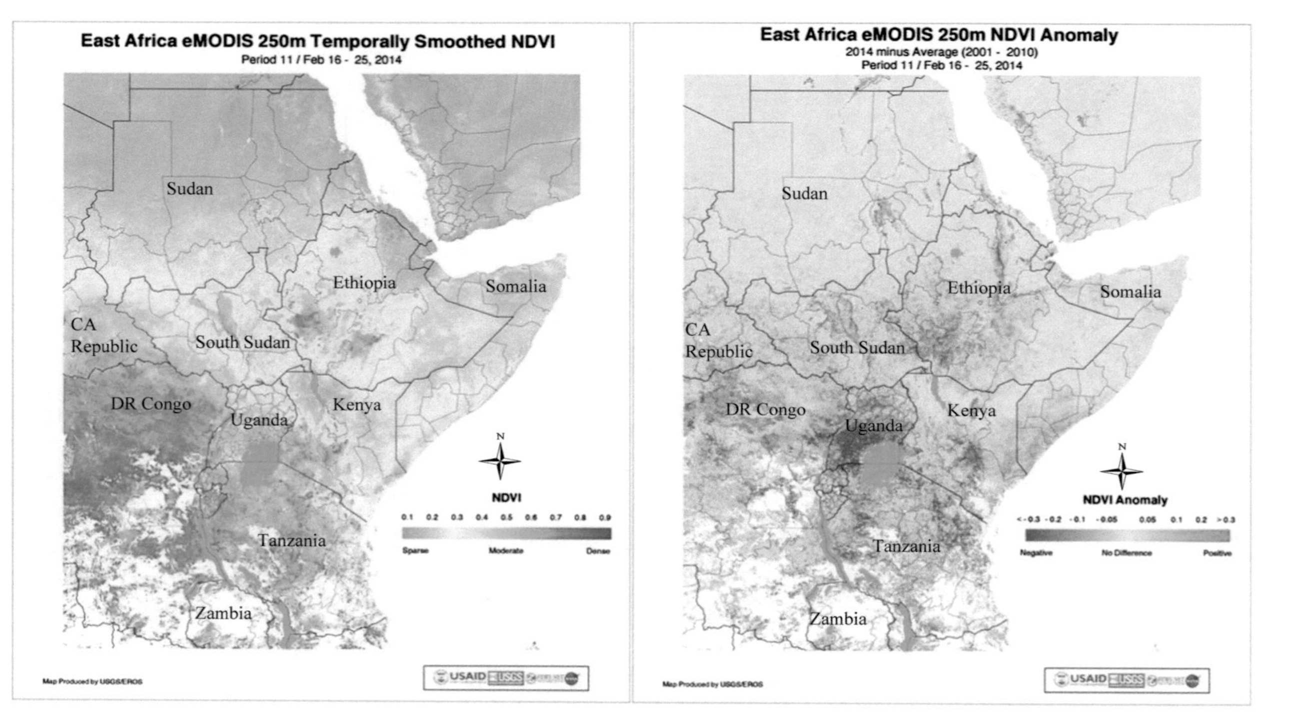

FIGURE 9.5 Drought monitoring using eMODIS NDVI (vegetation condition) and its anomaly (negative anomaly showing drought conditions) for the period of February 16–25, 2014 in East Africa.

and NDVI anomaly information is being disseminated over Africa, Central America, the Caribbean, the Middle East, and Central Asia.

9.4 MODEL-DRIVEN DROUGHT INDICATORS

The FEWS NET decision support system produces the following model-derived agro-hydrologic products for drought monitoring purposes: (1) crop water satisfaction index or Water requirement satisfaction index (*WRSI*); (2) soil moisture index; (3) ET anomalies; and (4) pond water levels. These models are based on the physical principles of energy and/or mass (water) conservation equations (Senay and Verdin, 2003; Senay et al., 2013a,b).

9.4.1 Water Requirement Satisfaction Index (Seasonal Drought Indicator)

The Water Requirement Satisfaction index (*WRSI*) is an index developed by the Food and Agriculture Organization of the United Nations (FAO) in the 1980s (FAO, 1986) to monitor seasonal crop performance. Verdin and Klaver (2002) demonstrated a geospatial implementation of the FAO *WRSI* approach in southern Africa. Senay and Verdin (2003) implemented a modified version of the FAO *WRSI* model in an operational setup for routine application in FEWS NET. The basic idea is to provide an index that shows the percentage of the idealized crop water requirement that is met by rainfall during a crop growing season. *WRSI* for a season is based on the water supply and demand a crop experiences during a growing season. *WRSI* is calculated as the ratio of seasonal actual ET (*AET*) to the seasonal crop water requirement (*WR*):

$$WRSI = \frac{AET}{WR} \times 100 \tag{9.2}$$

where *WR* is calculated from the Penman–Monteith potential ET (*PET*) using the crop coefficient (K_c) to adjust for the growth stage of the crop as

$$WR = PET \times K_c \tag{9.3}$$

AET represents the actual (as opposed to the potential) amount of water withdrawn from the soil water reservoir ("bucket"). Whenever the soil water content is above the maximum allowable depletion (MAD) level (based on crop type), the *AET* will remain the same as *WR*, i.e., no water stress. But when the soil water level is below the MAD level, the *AET* will be lower than WR in proportion to the remaining soil water content (Senay and Verdin, 2003).

The soil water content is obtained through a simple mass balance equation where the level of soil water is monitored in a bucket defined by the water holding capacity (WHC) of the soil and the crop root depth:

$$SW_i = SW_{i-1} + PPT_i - AET_i \tag{9.4}$$

where *SW* is soil water content, *PPT* is precipitation, and *i* is the time step index.

The most important inputs to this model are *PPT* and *PET*. Apart from the RFE estimates for precipitation estimates, FEWS NET at the USGS calculates daily *PET* values for Africa at 1.0-degree resolution from 6 hourly numerical meteorological model output using the Penman–Monteith equation (Shuttleworth, 1992; Senay et al., 2008). In addition, the *WRSI* model uses relevant soil information from the FAO (1986) digital soils map and topographical parameters derived from HYDRO-1K digital elevation data (Gesch et al., 1999).

The WRSI calculation requires an SOS and end of season time (EOS) for each modeling grid-cell. Maps of these two variables are useful in defining the spatial variation of the timing of the growing season and, consequently, the crop-coefficient function, which defines the crop water use pattern of crops. The model determines the SOS using onset-of-rains based on simple precipitation accounting. As explained in Section 9.2.2, the onset-of-rains or start of season is determined using a threshold (Agrhymet, 1996) amount and distribution of rainfall received in three consecutive dekads. The length of growing period (LGP) for each pixel is determined by the persistence, on average, above a threshold value of 0.5 (Agrhymet, 1996) of a climatological ratio between *PPT* and *PET*. Thus, EOS was obtained by adding LGP to the SOS dekad for each grid cell. The *WRSI* model can simulate different crop types (maize (corn), sorghum, millet, and wheat) whose seasonal water use pattern has been published in the form of a crop coefficient.

At the end of the crop growth cycle, or up to a certain dekad in the cycle, the sum of total AET and total *WR* are used to calculate *WRSI* in a geographic information system environment at 0.1-degree (about 10-km) spatial resolution. A case of "no deficit" will result in a *WRSI* value of 100, which corresponds to the absence of yield reduction related to water deficit (FAO, 1986) and a seasonal *WRSI* value less than 50 is regarded as a crop failure condition (Smith, 1992).

Yield reduction estimates based on *WRSI* contribute to food security preparedness and planning. As a monitoring tool, the crop performance indicator can be assessed at the end of every 10-day period during the growing season. As an early warning tool, EOS crop performance can be estimated using long-term average meteorological data.

Due to the difference in the growing season, *WRSI* maps are generated and distributed on a region by region (e.g., the Sahel, southern Africa, and Greater Horn of Africa regions), using dominant crops as indictors for a given region. At the end of every dekad, two image products associated with the *WRSI* are produced and disseminated for the FEWS NET activity. Figure 9.6 shows the *WRSI* (left) and *WRSI* anomaly (right) over the southern Africa region for the 2012–2013 growing season, which spans from September to May in the different parts of the region. Currently, *WRSI* information is being disseminated over Africa, Afghanistan, Central America, and the Caribbean.

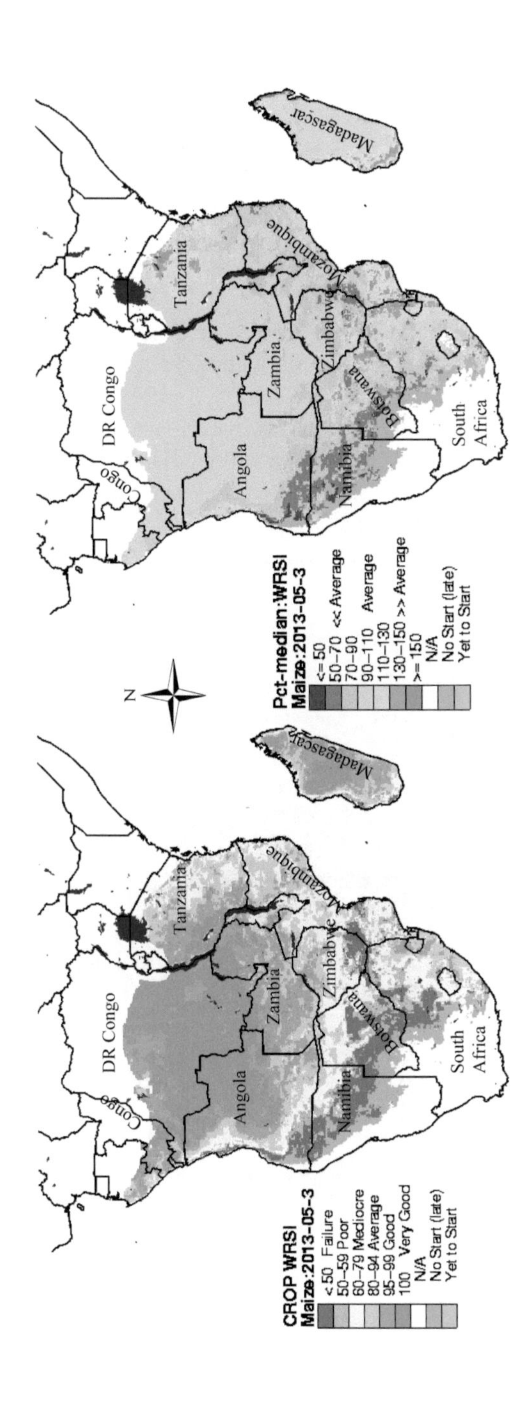

FIGURE 9.6 Modeled water requirement satisfaction index (*WRSI*) data (left) and *WRSI* anomaly data (right) for maize crop over southern Africa for 2012–2013 season.

9.4.2 Soil Water Index (Short-term Drought Indicator)

Soil Water Index (*SWI*) is developed from daily accounting of the input rainfall and ET losses in a control-volume defined by the soil *WHC* in a given modeling pixel (Senay and Verdin, 2003). *SWI* is computed as:

$$SWI = \frac{SW_i}{WHC} \times 100 \tag{9.5}$$

where *SW* is the soil water for the particular dekad *i,* and is calculated using Eqn (9.4)

The available soil water level is expressed as a percentage of the *WHC*. *SWI* values are qualitatively classified as wilting (when $SWI \leq 10\%$), stress (10–50), satisfactory (50–90), and sufficient (90–100). For drought monitoring purposes pixels within the wilting class receive attention when the 7-day rainfall forecast is dry. Figure 9.7 shows that although much of southern Africa is under satisfactory or sufficient condition during the first dekad of February; the eastern part of South Africa and western Angola are under wilting conditions, indicating the need for increased rainfall for these areas. Currently, *SWI* information is being disseminated over Africa, Central America, and the Caribbean regions.

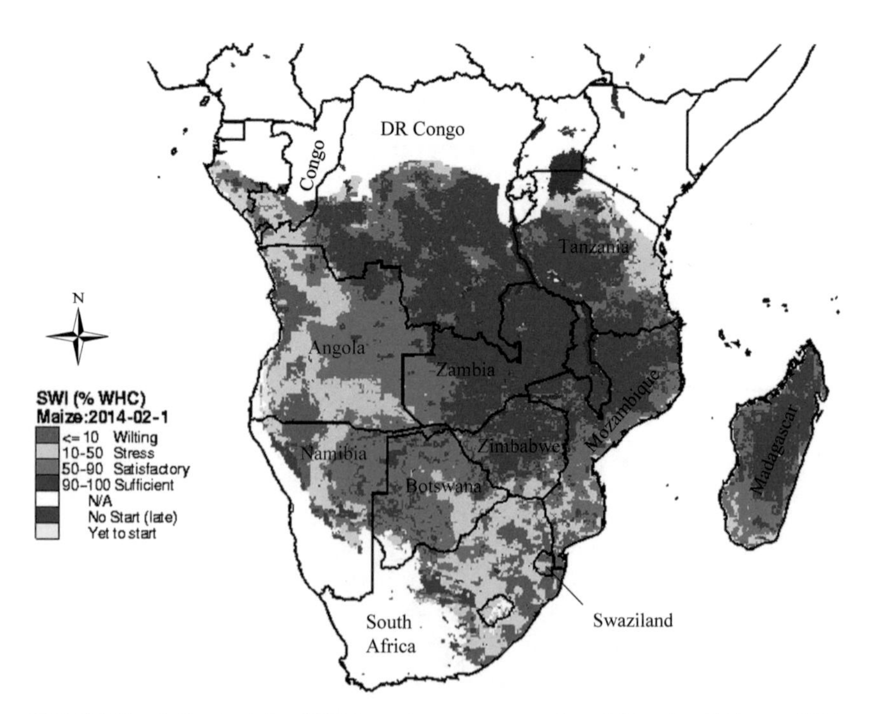

FIGURE 9.7 Soil water index (*SWI*) for maize crop over southern Africa. As of February 2014, western Angola, northwestern Namibia, northeastern South Africa, parts of Swaziland, and the southern tip of Mozambique show drought (crop wilting) conditions.

9.4.3 Evapotranspiration (Multiscale Drought Indicator)

Evapotranspiration (ET) is an important component of the water cycle and is composed of two-subprocesses: evaporation from soil and vegetation surfaces and transpiration that consists of the exchange of moisture between the plant and atmosphere through plant stomata. Because ET can be difficult to measure accurately, especially at large spatial scales, several hydrological modeling techniques have been developed to estimate actual ET using satellite remote sensing. In general, the ET modeling techniques can be grouped into two broad classes that include models based on surface energy balance (e.g., Bastiaanssen et al., 1998; Bastiaanssen et al., 2005; Su et al., 2005; Allen et al., 2007; Anderson et al., 2007; Senay et al., 2007) and water balance (e.g., Allen et al., 1998; Senay, 2008) principles.

Although water balance models focus on tracking the pathways and magnitude of rainfall in the soil-vegetation system, most remote sensing energy balance models use land surface temperature (LST) as a primary constraint in partitioning radiant energy available at the surface between heat and water fluxes. Senay et al. (2013a) formulated an Operational Simplified Surface Energy Balance (SSEBop) approach that is now widely applied by FEWS NET to monitor and assess drought conditions and relative crop performance across varied hydroclimatic regions. One of the main drivers of the SSEBop ET algorithm is the LST, which is obtained from sensors such as MODIS or Landsat. Broadly, SSEBop ET algorithm can be explained using following equations:

$$ETa = ETf \times kETo$$

where ETo is the grass reference ET for the location estimated using Penman–Monteith equation (Allen et al., 1998; Senay et al., 2008) with weather parameters derived from weather stations or model assimilated global fields (Kanamitsu, 1989); k is a coefficient that scales the grass reference ET into the level of a maximum ET experienced by an aerodynamically rougher crop ($k = 1.2$). ETf is the ET fraction that is derived as

$$ETf = \frac{Th - Ts}{dT}$$

where ETf is the ET fraction (0–1); dT is a temperature difference between the hot/dry and cold/wet reference boundary conditions that are predefined (from clear-sky radiation balance calculation); dT is unique for each day and pixel, ranging generally between 5 and 25 k depending on location and season; Th is the hot/dry reference boundary condition, representing the temperature of a dry-bare (hot) surface ($Th = Tc + dT$); Tc is the cold boundary condition, representing the cold/wet-vegetated surface, which in equilibrium with the air temperature i.e. all net radiation is used for latent heat flux. Tc is derived as a fraction of the air temperature (obtained from station or gridded weather fields); Ts is the LST, derived from satellite thermal data such as Landsat or MODIS.

The model has been used to produce 1 km ET using a 14-year MODIS data set over the CONUS, Africa, and Southeast Asia. For drought monitoring purposes, ET and their anomalies are calculated for a given accumulation period that is appropriate for the growing season. Figure 9.8 shows the spatial distribution of annual ET anomaly as a fraction of the average (2003–2013) for the 2012 season showing the severity and extent of drought in various parts of the world, with a prominent drought condition in the central United States. Figure 9.8 shows a new global ET anomaly product that will be operationally posted for global monitoring. Currently, ET data are being disseminated over Africa, and Central Asia. A dekadal global ET and ET anomaly product is also being produced. This dekadal product will be made available on the FEWS NET data portal.

9.4.4 Water Levels (Short-term Drought Indicator)

Precipitation and vegetation cover influence runoff generation and the behavior of the water body (Hass et al., 2011). Information on the availability of surface water resources is not only a direct indicator of drought condition but also a measure of how livelihoods (including agricultural activities and ecosystem) are affected by climate variability and how resilient they are for droughts and floods. Monitoring water level in small to medium surface water bodies across arid to semiarid regions provides indications of the drought situation such as water availability and vegetation condition. Although other drought indicators discussed earlier in the chapter provide information on ecosystem response to drought condition, water levels will provide information on the direct impact of drought to human and pastoral livelihoods.

In order to strengthen existing and already established drought monitoring and assessment activities in Africa, FEWS NET has initiated a large-scale project to monitor small to medium surface water bodies in Africa (Velpuri et al., 2014). Under this project, high spatial resolution satellite data (30-m Advanced Spaceborne Thermal Emission and Reflection Radiometer (ASTER) and 90-m Shuttle Radar Topography Mission Digital Elevation Model) are integrated with coarser intermediate resolution, globally available, satellite-driven data sets (25-km Tropical Rainfall Measuring Mission and 100-km Global Data Assimilation System reference ET) for operational modeling of the condition of water points (water levels) using a water balance approach (Velpuri et al., 2012; Velpuri and Senay, 2012; Senay et al., 2013b).

This project is further planned to cover entire sub-Saharan Africa (Figure 9.9). Operational monitoring of water levels is currently available for Ethiopia and Kenya. Up to date information and historical data (since 1998) on daily rainfall, evaporation, scaled depth and condition of the each water point in East Africa (Kenya and Ethiopia) are currently being disseminated online in near-real time (http://watermon.tamu.edu). Work is currently in progress to expand operational monitoring to Somalia, South Sudan, Sudan, Chad, Niger, and Mali.

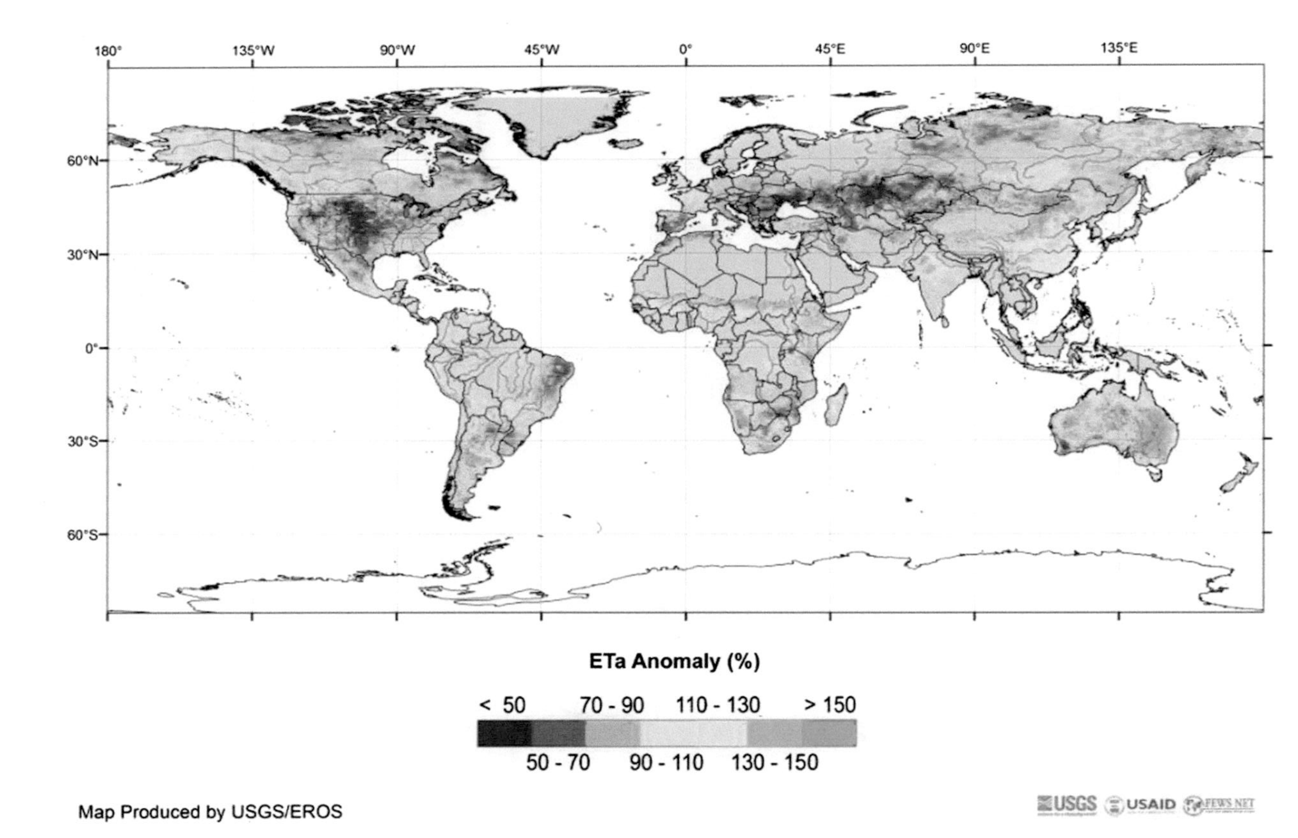

FIGURE 9.8 Global ET anomaly (%) product produced by SSEBop approach for 2012 at 1 km resolution.

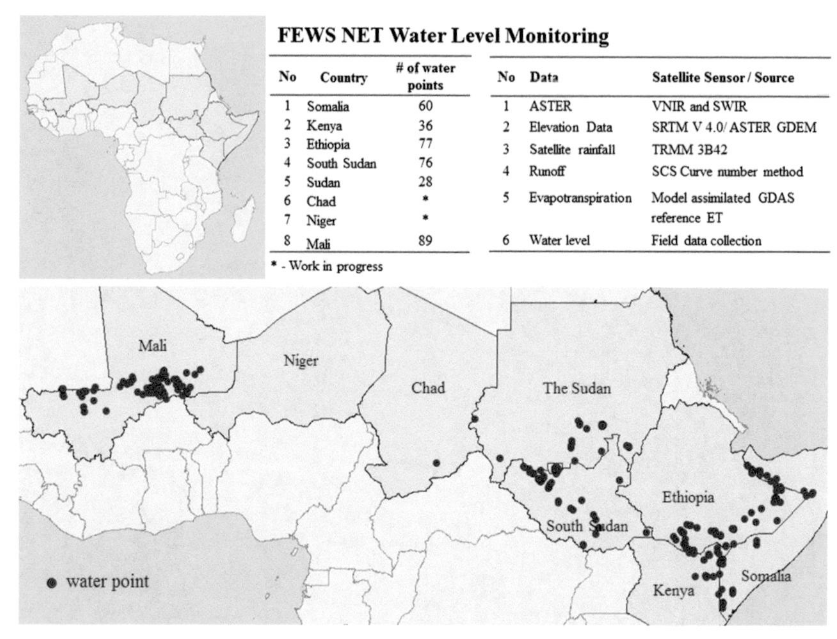

FEWS NET Water Level Monitoring

No	Country	# of water points
1	Somalia	60
2	Kenya	36
3	Ethiopia	77
4	South Sudan	76
5	Sudan	28
6	Chad	*
7	Niger	*
8	Mali	89

* - Work in progress

No	Data	Satellite Sensor / Source
1	ASTER	VNIR and SWIR
2	Elevation Data	SRTM V 4.0/ ASTER GDEM
3	Satellite rainfall	TRMM 3B42
4	Runoff	SCS Curve number method
5	Evapotranspiration	Model assimilated GDAS reference ET
6	Water level	Field data collection

FIGURE 9.9 FEWS NET water level monitoring initiative in sub-Saharan Africa. Detailed information on the water monitoring approach is provided in Senay et al. (2013b) and Velpuri et al. (2014).

9.5 HAZARD OUTLOOK (SHORT-TERM DROUGHT BULLETIN)

FEWS NET food security analysts rely on a convergence of evidence to make drought and food security assessments (Rowland et al., 2005). Based on the information generated from multiple indicators, FEWS food security analysts at the NOAA Climate Prediction Center (CPC) produce hazard outlooks on a weekly basis. These outlooks serve as short-term drought bulletins distributed to the regional and national partners for efficient hazard management. The hazard outlook maps are generally based on current weather and climate conditions synthesized from multiple indicators. For example, Figure 9.10 shows the hazard outlook for Africa identifying areas of severe drought, drought and short-term dryness in regions of southern Africa. These areas are identified based on the information obtained and supported by evidence from other indicators produced by FEWS NET and its partners. The severe drought and drought regions illustrated in Figure 9.10 are supported by the evidence provided by NDVI anomaly, crop *WRSI*, *SWI*, and ET anomaly. These hazard outlooks assesses potential impact of the hazard on crop and pasture conditions. These hazard outlooks are created by consulting various products through a consensus reached by a convergence of evidence from products and

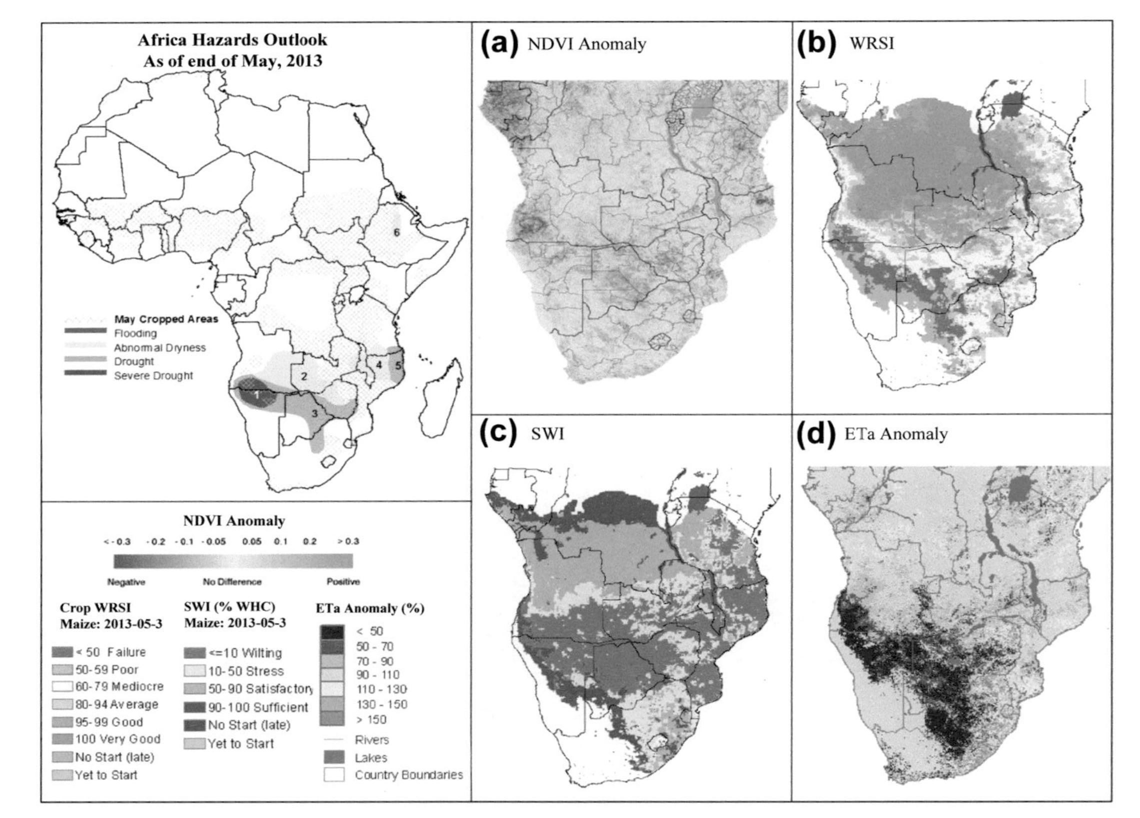

FIGURE 9.10 The Hazard Outlook for Africa (May 2013) generated by the convergence of evidence using FEWS NET drought indicators shown on the right. (a) NDVI – Normalized Difference Vegetation Index; (b)WRSI – Water requirement satisfaction index; (c) SWI – soil water index; (d) ETa – actual evapotranspiration.

field reports and discussions among FEWS NET stakeholders from USGS, NOAA, NASA, and the United States Department of Agriculture as well as other partners. The hazard outlooks are currently available online for Africa, Afghanistan, and Central America countries. Hazard Outlooks are made available on the NOAA CPC Web site (http://www.cpc.ncep.noaa.gov/products/fews/briefing.html).

9.6 VALIDATION OF DROUGHT INDICATORS

All of the drought indicators and modeling approaches presented in this chapter are validated through independent data sets. For example, RFE rainfall data used to produce drought indicators has been thoroughly validated by several researches in Africa (Laws et al., 2003; Dinku et al., 2008; Thiemig et al., 2012); the African rainfall model (CHARMS) used in the production of SPI has been independently validated by Funk et al. (2003). The validity of NDVI and NDVI anomaly products for monitoring the impact of drought on crop growth has been previously demonstrated by several researchers (Kogan, 1997; Peters et al., 2002; Ji and Peters, 2003; Rhee et al., 2010; Anyamba and Tucker, 2012). Validation of *WRSI* and *SWI* products against the crop yields collected from the water-limited and water-unlimited districts in Ethiopia was carried out by Senay and Verdin (2003). Velpuri et al. (2013) performed a comprehensive validation of the ET estimates produced by the SSEBop modeling approach (Senay et al., 2013a) using data from 60 point flux tower eddy covariance stations (data from 221 station months) in the US. The ET estimates were also compared with other global ET data sets and water balance ET (Velpuri et al., 2013) such as MODIS ET data (Mu et al., 2011) and Max Planks Institute's ET data (Jung et al., 2011). The validation results indicated that remote sensing based ET data sets (such as SSEBop ET) can be reliably used for hydrologic applications such as drought monitoring. The water-level monitoring approach presented by Senay et al. (2013b) has already been tested using in situ gauge measurements in East Africa (Senay et al., 2013b). The validation of modeled water levels with field-installed gauge data demonstrated the ability of the model to capture both the spatial patterns and seasonal variations.

9.7 SUMMARY AND CONCLUSIONS

This chapter presents examples of how remote sensing and hydrologic modeling techniques are being used by FEWS NET to assist decision support processes for drought-related hazard assessment, monitoring and management. Regions monitored by the FEWS NET are illustrated in Figure 9.11. All of the products produced under the FEWS NET are available for download from the FEWS NET data portal (http://earlywarning.usgs.gov/fews/index.php#DATA%20PORTALS). The list of drought monitoring indicators and their availability in various geographical regions is provided in Table 9.3.

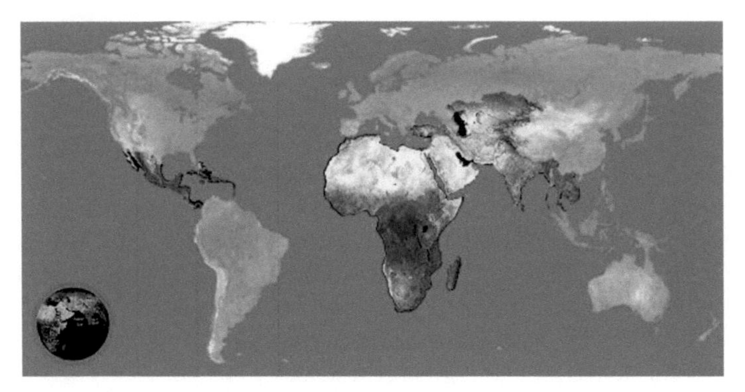

FIGURE 9.11 Regions where FEWS NET drought monitoring products are available.

TABLE 9.3 Drought Monitoring Activities/Products Generated by FEWS NET

Drought Monitoring Indicators	Africa	Central America	Caribbean	Mexico	Middle East	Central Asia	Global
1. Rainfall based							
SPI	Y	–	–	–	–	–	–
SOS and EOS	Y	Y	Y	–	–	–	–
Dekadal RFE and anomaly	Y	–	–	–	Y	Y	–
Short-term dryness indicators	Y	Y	Y	Y	Y	Y	–
2. Vegetation-index based							
NDVI	Y	Y	Y	–	Y	Y	–
3. Model based							
SWI	Y	Y	Y	–	–	–	–
WRSI	Y	Y	Y	–	–	–	–
ET	Y	Y*	Y*	Y*	Y*	Y	Y*
Water levels	Y	–	–	–	–	–	–

Y*- will be available in near future as part of global product.

The FEWS NET activities have used remote sensing satellite data to monitor drought and famine since the late 1980s. NDVI was first used to monitor the green-up in sub-Saharan Africa and to detect anomalous vegetation conditions. Currently, NDVI and NDVI anomaly information is being disseminated over Africa, Central America, the Caribbean, the Middle East, and Central Asia. Since 1995, satellite-derived rainfall estimates have been used to identify rainfall-based drought indicators. The SPI index is produced currently for Africa. However, other rainfall-based drought indicators such as SOS, and monthly and dekadal indicators are available for other regions (Table 9.3). With the availability of information on crop phenology from NDVI and precipitation, agro-hydrologic models were developed to monitor the extent and severity of drought hazard. The *WRSI* crop model is applied to Africa, Central America and Caribbean regions for a variety of crops during their growing seasons. Other model-based indicators such as *SWI* and ET are also being produced and disseminated for other regions. FEWS NET water level monitoring is being focused mainly on sub-Saharan Africa. The choice of producing an index for a country or region is often based on the requirement by the donor agency and availability of products.

Early detection and early warning of drought must be persuasive enough to overcome the risk- avoidance behaviors of decision makers in responsible organizations such as national governments, donor agencies, international organizations, and nongovernmental organizations (Cutler, 1993). It is for this reason that FEWS NET food security analysts rely on a convergence of evidence to make drought and food security assessments (Rowland et al., 2005). No single source of information is sufficiently authoritative and comprehensive to identify potential famine (and drought) area alone (Mason et al., 1987; Kelly, 1993). Therefore, FEWS NET monitors drought situations using a range of drought indicator products and methods. Conclusions on the drought situations are made most confidently when many or all factors indicate a similar situation of dryness in a region. Any reduction of ambiguity associated with data and information used by FEWS NET contributes to confidence in food security assessments and an improved linkage between early warning and early response.

The main objective of this chapter was to present a summary of an operational practice that uses a combination of remotely sensed data and hydrologic modeling for monitoring and assessing drought hazards. With the availability of global data sets such as satellite-based rainfall, NDVI and land-surface temperature, the FEWS NET decision-support system demonstrated the effective application of satellite-derived data and products for monitoring drought conditions worldwide. With the principle of convergence of evidence, the onset, extent and severity of drought can be detected early in the season and monitored throughout the year for mitigating the potential impact of drought hazard on food security and agricultural markets worldwide. FEWS NET drought monitoring system has the capability to assess and summarize drought risks at an administrative level, which can be approximated by an area of around 400 km^2 for most FEWS NET countries. The drought-hazard monitoring approach perfected by USGS FEWS NET through the integration of satellite data and hydrologic modeling can form the basis for similar decision support systems at a national level for producing reliable and useful information that is regional in scope and relevant for local, district-level, decision making.

ACKNOWLEDGMENTS

This work was made possible by the funding of the USGS FEWS NET (GR12D00BRHA100). We thank FEWS NET partners such as NOAA, NASA, USDA, Chemonics, and other international and national partners for their support in providing satellite and field information and their participation in weekly "weather hazard" meetings. Our thanks go to Ronald Lietzow and Ronald Smith of the USGS Earth Resources Observation and Science (EROS) Center for ensuring the timely processing and posting of the various input and derived geospatial products required for the FEWS NET agro-hydrologic modeling. We are grateful to Greg Husak for developing the SPI code. We also thank Thomas Adamson, Lei Ji, Ramesh Singh, Sandra Coopers, and editors of this book for their review and useful comments. Any use of trade, firm or product names is for descriptive purposes only and does not imply endorsement by the U.S. Government.

REFERENCES

AGRHYMET, 1996. Methodologie de suivi des zones a risque. AGRHYMET FLASH, Bulletin de Suivi de la Campagne Agricole au Sahel 2 (0/96), 2. Centre Regional AGRHYMET, B.P. 11011, Niamey, Niger.

Anyamba, A., Tucker, C.J., 2012. Historical perspective of AVHRR NDVI and vegetation drought monitoring. Remote Sens. Drought: Innovative Monit. Approaches 23.

Allen, R.G., Pereira, L., Raes, D., Smith, M., 1998. Crop Evapotranspiration. Food and Agriculture Organization (FAO) publication, 56. FAO of the United Nations, Rome, Italy, 290 pp.

Allen, R.G., Tasumi, M., Morse, T.A., Trezza, R., Kramber, W., Lorite, I., Robison, C.W., 2007. Satellite-based energy balance for mapping evapotranspiration with internalized calibration (METRIC)—Applications. J. Irrig. Drain. Eng. 133, 395—406.

Anderson, M.C., Norman, J.M., Mecikalski, J.R., Otkin, J.A., Kustas, W.P., 2007. A climatological study of evapotranspiration and moisture stress across the continental United States based on thermal remote sensing: 1. Model formulation. J. Geophys. Res. 112, D10117. http://dx.doi.org/10.1029/2006JD007506.

Bastiaanssen, W.G.M., Menenti, M., Feddes, R.A., Holtslag, A.A.M., 1998. A remote sensing surface energy balance algorithm for land (SEBAL): 1) formulation. J. Hydrol. 212 (213), 213—229.

Bastiaanssen, W.G.M., Noordman, E.J.M., Pelgrum, H., Davids, G., Thoreson, B.P., Allen, R.G., 2005. SEBAL model with remotely sensed data to improve water-resources management under actual field conditions. J. Irrig. Drain. Eng. 131, 85—93.

Crafts, A.S., 1968. Water deficits and physiological processes. In: Kozlowski, T.T. (Ed.), Water Deficits and Plant Growth. Plant Water Consumption and Response, vol. II. Academic Press, New York, pp. 85—133.

Cutler, P., 1993. Responses to famine: why they are allowed to happen. In: Field, J.O. (Ed.), The Challenge of Famine: Recent Experience, Lessons Learned. Kumarian Press, West Hartford, CT, pp. 72–87.

Dinku, T., Chidzambwa, S., Ceccato, P., Connor, S.J., Ropelewski, C.F., 2008. Validation of high resolution satellite rainfall products over complex terrain. Int. J. Remote Sens. 29, 4097–4110.

FAO, 1986. Early Agrometeorological Crop Yield Forecasting. FAO Plant Production and Protection paper No. 73, by M. Frère and G.F. Popov. FAO, Rome, Italy.

Funk, C., Michaelsen, J., Verdin, J., Artan, G., Husak, G., Senay, G., Gadian, H., Magadazire, T., 2003. The collaborative historical African rainfall model: description and evaluation. International Journal of Climatology 23 (1), 47–66.

Gesch, D.B., Verdin, K.L., Greenlee, S.K., 1999. New land surface digital elevation model covers the earth: *EOS Transactions*. Am. Geophys. Union 80 (6), 69–70.

Guttman, N.B., 1999. Accepting the standardized precipitation index: a calculation algorithm. JAWRA J. Am. Water Resour. Assoc. 35 (2), 311–322.

Haas, E.M., Bartholomé, E., Lambin, E.F., Vanacker, V., 2011. Remotely sensed surface water extent as an indicator of short-term changes in ecohydrological processes in sub-Saharan Western Africa. Remote Sens. Environ. 115, 3436–3445.

Heim, R.R., 2002. A review of twentieth-century drought indices used in the United States. Bull. Am. Meteorol. Soc. 83 (8).

Huffman, G.J., Adler, R.F., Arkin, P., Chang, A., Ferraro, R., Gruber, A., Janowiak, J., McNab, A., Rudolf, B., Schneider, U., 1997. The Global Precipitation Climatology Project (GPCP) Combined Precipitation Dataset. Bull. Amer. Meteor. Soc. 78, 5–20.

Jenkerson, C.B., Maiersperger, T., Schmidt, G., 2010. eMODIS: A User-friendly Data Source. U.S. Geological Survey. Open-File Report 2010–1055, 10 pp.

Ji, L., Peters, A.J., 2003. Assessing vegetation response to drought in the northern Great Plains using vegetation and drought indices. Remote Sens. Environ. 87 (1), 85–98.

Jung, M., Reichstein, M., Ciais, P., Seneviratne, S.I., Sheffield, J., Goulden, M.L., Bonan, G., Cescatti, A., Chen, J., de Jeu, R., Dolman, A.J., Eugster, W., Gerten, D., Gianelle, D., Gobron, N., Heinke, J., Kimball, J., Law, B.E., Montagnani, L., Mu, Q., Mueller, B., Oleson, K., Papale, D., Richardson, A.D., Rousard, O., Running, S., Tomelleri, E., Viovy, N., Weber, U., Williams, C., Wood, E., Zaehle, S., Zhang, K., 2010. Recent decline in the global land evapotranspiration trend due to limited moisture supply. Nature 467, 951–954.

Kanamitsu, M., 1989. Description of the NMC global data assimilation and forecast system. Weather and Forecasting, Vol. 4, pp. 335–342.

Kogan, F.N., 1997. Global drought watch from space. Bull. Am. Meteorol. Soc. 78, 727–636.

Kelly, M., 1993. Operational value of anthropometric surveillance in famine early warning and relief: Wollo Region, Ethiopia, 1987-88. Disasters 17, 48–55.

Laws, K.B., Janowiak, J.E., Huffman, G.J., 2003. Verification of rainfall estimates over Africa using RFE, NASA MPA-RT, and CMORPH. In: AGU Fall Meeting Abstracts, vol. 1, p. 0731.

Maselli, F., Papale, D., Puletti, N., Chirici, G., Corona, P., 2009. Combining remote sensing and ancillary data to monitor the gross productivity of water-limited forest ecosystems. Remote Sens. Environ. 113 (3), 657–667.

Mason, J.B., Haaga, J.G., Maribe, T.O., Marks, G., Quinn, V.J., Test, K.E., 1987. Using agricultural data for timely warning to prevent the effects of drought on child nutrition in Botswana. Ecol. Food Nutr. 19, 169–184.

McKee, T.B., Doesken, N.J., Kleist, J., January 17–23, 1993. The relationship of drought frequency and duration to time scales. In: Eighth Conference on Applied Climatology. American Meteorological Society, Anaheim CA, pp. 179–186.

McVicar, T.R., Jupp, D.L., 1998. The current and potential operational uses of remote sensing to aid decisions on drought exceptional circumstances in Australia: a review. Agric. Syst. 57 (3), 399–468.

Morid, S., Smakhtin, V., Moghaddasi, M., 2006. Comparison of seven meteorological indices for drought monitoring in Iran. Int. J. Climatol. 26 (7), 971–985.

Mu, Q., Zhao, M., Running, S.W., 2011. Improvements to a MODIS global terrestrial evapotranspiration algorithm. Remote Sensing of Environment. 115, 1781–1800.

NDMC, 2007. National Drought Mitigation Center University of Nebraska, Lincoln, USA. Online http://www.drought.unl.edu/ (accessed Feb 2014).

Peters, A.J., Walter-Shea, E.A., Ji, L., Vina, A., Hayes, M., Svoboda, M.D., 2002. Drought monitoring with NDVI-based standardized vegetation index. Photogramm. Eng. Remote Sens. 68 (1), 71–75.

Rhee, J., Im, J., Carbone, G.J., 2010. Monitoring agricultural drought for arid and humid regions using multi-sensor remote sensing data. Remote Sens. Environ. 114 (12), 2875–2887.

Rowland, J., Verdin, J., Adoum, A., Senay, G., 2005. Drought monitoring techniques for famine early warning systems in Africa. In: Boken, V.K., Cracknell, A.P., Heathcote, R.L. (Eds.), Monitoring and Predicting Agricultural Drought: A Global Study. Oxford University Press, New York, pp. 252–265.

Running, S.W., Nemani, R.R., 1988. Relating seasonal patterns of the AVHRR vegetation index to simulated photosynthesis and transpiration of forests in different climates. Remote Sens. Environ. 24, 347–367.

Senay, G.B., Verdin, J., 2003. Characterization of yield reduction in Ethiopia using a GIS-based crop water balance model. Can. J. Remote Sens. 29 (6), 687–692.

Senay, G.B., Budde, M., Verdin, J.P., Melesse, A.M., 2007. A coupled remote sensing and simplified surface energy balance approach to estimate actual evapotranspiration from irrigated fields. Sensors 7, 979–1000.

Senay, G.B., 2008. Modeling landscape evapotranspiration by integrating land surface phenology and a water balance algorithm. Algorithms 1, 52–68.

Senay, G.B., Verdin, J.P., Lietzow, R., Melesse, A.M., 2008. Global daily reference evapotranspiration modeling and evaluation. J. Am. Water Resour. Assoc. 44, 969–979.

Senay, G.B., Budde, M.E., Verdin, J.P., 2011a. Enhancing the simplified surface Energy balance (SSEB) approach for estimating landscape ET: Validation with the METRIC model. Agric. Water Manage. 98, 606–618.

Senay, G.B., Leake, S., Nagler, P.L., Artan, G., Dickinson, J., Cordova, J.T., Glenn, E.P., 2011b. Estimating Basin scale evapotranspiration (ET) by water balance and remote sensing methods. Hydrol. Processes 25 (26), 4037–4049.

Senay, G.B., Bohms, S., Verdin, J.P., 2012. Remote sensing of evapotranspiration for operational drought monitoring using principles of water and Energy balance. In: Wardlow, B.D., Anderson, M.C., Verdin, J.P. (Eds.), Remote Sensing of Drought: Innovative Monitoring Approaches. CRC Press, pp. 123–144.

Senay, G.B., Bohms, S., Singh, R.K., Gowda, P.H., Velpuri, N.M., Alemu, H., Verdin, J.P., 2013a. Operational evapotranspiration mapping using remote sensing and weather datasets: a new parameterization for the SSEB approach. JAWRA J. Am. Water Resour. Assoc. 49 (3), 577–591.

Senay, G.B., Velpuri, N.M., Alemu, H., Pervez, S.M., Asante, K.O., Kariuki, G., Taa, A., Angerer, J., 2013b. Establishing an operational waterhole monitoring system using satellite data and hydrologic modelling: application in the pastoral regions of East Africa. Pastoralism 3 (1), 1–16.

Shuttleworth, J., 1992. In: Maidment, D. (Ed.), Evaporation. Handbook of Hydrology, vol. 4. McGraw-Hill, New York, pp. 1–4, 53.

Smith, M., 1992. Expert consultation on revision of FAO methodologies for crop water requirements. Food and Agricultural Organization of the United Nations, Rome, Italy, FAO Publication 73.

Su, H., McCabe, M.F., Wood, E.F., Su, Z., Prueger, J., 2005. Modeling evapotranspiration during SMACEX: comparing two approaches for local and regional scale prediction. J. Hydrometeorol. 6 (6), 910–922.

Swets, D.L., Reed, B.C., Rowland, J.D., Marko, S.E., 1999. A weighted least-squares approach to temporal NDVI smoothing. In: In: 1999 ASPRS Annual Conference: From Image to Information, Portland, Oregon, May 17-21. Proceedings: Bethesda, Maryland. American Society for Photogrammetry and Remote Sensing.

Thenkabail, P.S., Gamage, M.S.D.N., Smakhtin, V.U., 2004. The Use of Remote Sensing Data for Drought Assessment and Monitoring in Southwest Asia, 85. International Water Management Institute. Research Report.

Thiemig, V., Rojas, R., Zambrano-Bigiarini, M., Levizzani, V., De Roo, A., 2012. Validation of satellite-based precipitation products over sparsely gauged African river basins. J. Hydrometeorol. 13 (6).

Tucker, C.J., 1979. Red and photographic infrared linear combinations for monitoring vegetation. Remote Sens. Environ. 8, 127–150.

Tucker, C.J., Choudhury, B.J., 1987. Satellite remote sensing of drought conditions. Remote Sens. Environ. 23, 243–251.

Velpuri, N.M., Senay, G.B., Asante, K.O., Thompson, S., 2012. A multi-source satellite data approach for modelling Lake Turkana water level: calibration and validation using satellite altimetry data. Hydrol. Earth Syst. Sci. 16 (1), 1–18.

Velpuri, N.M., Senay, G.B., 2012. Assessing the potential hydrological impact of the Gibe III Dam on Lake Turkana water level using multi-source satellite data. Hydrol. Earth Syst. Sci. 16 (10), 3561–3578.

Velpuri, N.M., Senay, G.B., Singh, R.K., Bohms, S., Verdin, J.P., 2013. A comprehensive evaluation of two MODIS evapotranspiration products over the conterminous United States: using point and gridded FLUXNET and water balance ET. Remote Sens. Environ. 139, 35–49.

Velpuri, N.M., Senay, G.B., Rowland, J., Verdin, J.P., Alemu, H., 2014. Africa-wide monitoring of small surface water bodies using multisource satellite data: a monitoring system for fews net. In: Nile River Basin. Springer International Publishing, pp. 69–95.

Verdin, J., Klaver, R., 2002. Grid cell based crop water accounting for the famine early warning system. Hydrol. Processes 16, 1617–1630.

White, D.H., O'Meagher, B., 1995. Coping with exceptional droughts in Australia. In: Wilhite, D.A. (Ed.), Drought Network News, vol. 7. University of Nebraska, pp. 13–17.

Wilhite, D.A., Glantz, M.H., 1985. Understanding the drought phenomenon: the role of definitions. Water Int. 10, 111–120.

WMO (World Meteorological Organization), 1975. Droughts and Agriculture. WMO Technical Note 138.

Xie, P., Arkin, P.A., 1997. A 17-year monthly analysis based on gauge observations, satellite estimates, and numerical model outputs. Bull. Am. Meteorol. Soc. 78 (11), 2539–2558.

Chapter 10

Hydrological Modeling for Drought Assessment

Shreedhar Maskey and Patricia Trambauer
UNESCO-IHE Institute for Water Education, Delft, The Netherlands

ABSTRACT
Droughts are widespread natural hazards and are reported to have killed more lives than all the other natural hazards combined over the past century. Repeated droughts can advance an area into desertification, severely impact on ecosystems, and cause many other tangible and intangible damages. Droughts can be characterized quantitatively for their spatial extent, intensity, and duration using various drought indices. In this chapter we focus on the assessment of various aspects of hydrology-related droughts and show that a process-based hydrological model has a lot to offer for drought assessment. To illustrate this we present a case study on the Limpopo river basin in Southern Africa. By comparing different types of droughts, we also highlight that various types of droughts need to be analyzed to understand the comprehensive nature of drought impacts.

10.1 INTRODUCTION

Drought is a prolonged period of abnormally dry weather condition leading to a severe shortage of water. The shortage of water can be felt at various levels, such as insufficient soil moisture for plant transpiration, drying out river discharge unable to meet various user demands, substantially lowered water levels in reservoirs and ponds, and lowered groundwater table that may take many years to restore. Four types of droughts are usually referred to in the literature: meteorological, agricultural, hydrological, and socioeconomic (Wilhite and Glantz, 1985). Although the first three indicate droughts as physical phenomena, the last one associates the physical phenomena of droughts to their impacts on people and environment.

A meteorological drought occurs when accumulated precipitation for a defined period (normally on the scale of months) is lower than that in a normal year, which is usually defined by a long-term average. Agricultural droughts are generally developed as a result of a meteorological drought. A few days or

Hydro-Meteorological Hazards, Risks, and Disasters. http://dx.doi.org/10.1016/B978-0-12-394846-5.00010-2

weeks of lack of moisture in the root zone, especially during the growing season, may already create stress on crops resulting in reduced crop yields. Agricultural droughts can also be triggered or aggravated by other meteorological conditions such as high temperatures, winds, and low relative humidity (Heim, 2002). Hydrological droughts are the result of reduced streamflows, lowered water levels on lakes and reservoirs, and lowered groundwater tables. These conditions are normally observed as the meteorological drought continues. Thus there is usually a lag between the onsets of a meteorological drought and a hydrological drought. Similarly, the hydrological droughts generally take longer time than meteorological droughts to terminate as the next storm may be sufficient to recover from the meteorological drought but insufficient to replenish the streams and lakes, for example, and bring them to normal levels. Socioeconomic droughts refer to the impacts of the other types of droughts (meteorological, agricultural, and hydrological) on social and economical aspects of the population affected. Therefore, socioeconomic drought is a result of the physical drought conditions (as a hazard) and vulnerability of the society to drought hazards.

The connection between the different types of droughts is obvious, but their relationships are rather complex as to how one affects the other and, more specifically, if and when one type of drought (e.g., meteorological) may lead to another type of drought (e.g., hydrological). In this chapter we focus on the use of a hydrological model for assessing various aspects of hydrology-related droughts and show that a process-based hydrological model has a lot to offer for drought assessment. By comparing different types of droughts, we also illustrate that various types of droughts need to be analyzed to understand the comprehensive nature of drought impacts.

10.2 DROUGHTS AS A NATURAL HAZARD WORLDWIDE

Droughts are a widespread natural hazard worldwide with a high societal impact (Alston and Kent, 2004; Glantz, 1987). According to the "Emergency Events Database (EM-DAT) the International Disaster Database" of the Centre for Research on the Epidemiology of Disasters, at least 642 drought events were reported across the world from 1900 to 2013 (EM-DAT, 2014). The estimated death toll of these events is about 12 million, while the number of affected people is more than 2 billion, and economic damages were over US $135 billion (Masih et al., 2014). Among all natural hazards, droughts have resulted in the highest number of deaths worldwide causing more losses of lives than all the other hazards combined (Table 10.1).

Recent studies show that the frequency and severity of droughts seem to be increasing in some areas as a result of climate variability and change (IPCC, 2007; Patz et al., 2005; Sheffield and Wood, 2008; Lehner et al., 2006). Moreover, the rapid increase of world population continues to aggravate water shortages on local and regional scales (Trambauer et al., 2014).

TABLE 10.1 Natural Hazards and Their Contribution to Deaths, People Affected, and Economic Damages from 1900 to 2013 (EM-DAT, 2014)

Hazard	Percentage of Deaths (%)	Percentage of Affected (%)	Percentage of Economic Damage (%)
Droughts	51.4	31.6	5.4
Floods	30.5	51.7	24.3
Earthquakes and tsunami	11.3	2.6	30.4
Storms	6.1	13.7	37.4
Volcanic eruptions	0.4	0.1	0.1
Mass movement (landslides/avalanches)	0.3	0.2	0.3
Wildfires	Negligible	0.1	2.1

The impacts of drought go beyond the loss of life and property. Intangible damages such as impacts on environment and ecosystems are mostly unaccounted for. Repeated droughts can advance an area into desertification—a process of land degradation with severe and irreversible loss of productivity. Similarly, severe loss of groundwater tables can have far-reaching consequences (e.g., a long-term reduction in streamflows particularly during dry seasons) and may require many years to decades to replenish them. Other environmental impacts include disappearance of small lakes, springs, and vegetation lands; soil erosion; increased vulnerability to forest fires; etc.

10.3 CHARACTERIZATION OF DROUGHTS: DROUGHT INDICES

The definition of drought presented at the beginning of this chapter is very qualitative: "…prolonged period of abnormally dry weather … leading to severe shortage of water." Obvious questions one would ask are: How long does the condition need to continue to call it a "prolonged"? When do we call a weather condition an "abnormal"? And when do we call the water shortage a "severe"? To answer these questions, drought indices have been proposed by the literature (Table 10.2). A drought index is a standardized numerical value usually based on anomalies of a certain parameter representing the availability of moisture or water (e.g., precipitation, soil moisture, and streamflow) from its long-term mean. Thus a drought index is used to characterize a drought quantitatively—its onset, termination, and intensity. A drought index often measures the intensity of a drought event, which together with its duration (i.e., the date of

TABLE 10.2 Drought Indices

Indices	Input Variable/ Model	Drought Type	References
PDSI (including PHDI and Z-index)	P, T, WBM	M, H	Palmer (1965)
CMI	P, T, WBM	A	Palmer (1968)
NDVI	Remote sensing	A	Tucker (1979)
SWSI	P, Q, snowpack, reservoir storage	H	Shafer and Dezman (1982)
VCI	Remote sensing	A	Kogan (1990)
SPI	P	M	McKee et al. (1993)
SMDI	Hydrological model	A	Hollinger et al. (1993)
ETDI	Hydrological model	A	Narasimhan and Srinivasan (2005)
SRI	Q	H	Shukla and Wood (2008)
GRI	Groundwater storage	H	Mendicino et al. (2008)
SPEI	P, T	M	Vicente-Serrano et al. (2010a,b)
RSAI	RS	A	Trambauer et al. (2014)

M, meteorological; H, hydrological; A, agricultural; P, precipitation; T, temperature; Q, river discharge; WBM, water balance model; PHDI, Palmer Hydrological Drought Index; CMI, Crop Moisture Index; SPEI, Standardized Precipitation Evapotranspiration Index; SMDI, Soil Moisture Drought Index.

termination minus the date of onset) define the severity of the event. Various indices have been developed in the past decades to characterize different types of droughts. In Table 10.2 a number of drought indices are listed that are either commonly used or are relatively recently reported in the literature.

10.3.1 Drought Indices Based on Precipitation

The Standardized Precipitation Index (SPI), proposed by McKee et al. (1993), is one of the most commonly used drought indices that require only precipitation for its computation. The SPI is a meteorological drought index and represents the precipitation deficit (dry condition) or surplus (wet condition) based on its departure from the long-term mean. The computation of the SPI involves fitting the precipitation time series to a probability distribution (usually the gamma distribution) (see Figure 10.1, left panel), which is then transformed to a standardized normal distribution with zero mean and one variance

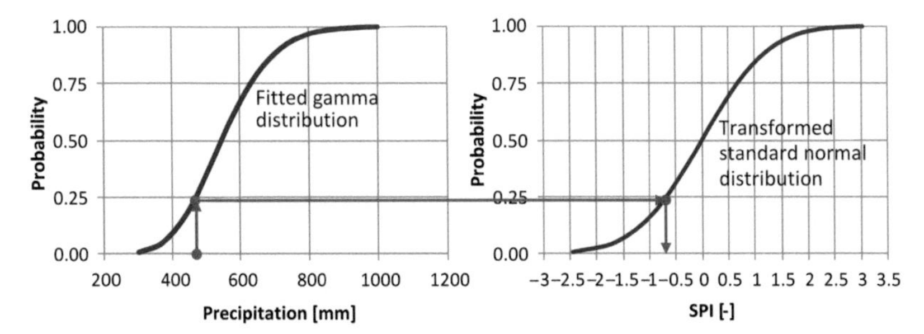

FIGURE 10.1 SPI computation: Gamma distribution fitted with the precipitation time series (left panel) and transformed to standardized normal distribution (right panel). This example is based on an annual precipitation (i.e., SPI-12). The distribution transformation procedure applied here is the same as in Edwards and McKee (1997).

(see Figure 10.1, right panel). The SPI can be computed for different timescales by accumulating the precipitation time series (before fitting to the distribution) over the time period of interest: typically 3 months (SPI-3), 6 months (SPI-6), and 12 months (SPI-12). Based on the SPI values, droughts may be classified into mild ($0 > SPI > -1$), moderate ($-1 \geq SPI > -1.5$), severe ($-1.5 \geq SPI > -2$), and extreme ($SPI \leq -2$) (Lloyd-Hughes and Saunders, 2002).

10.3.2 Drought Indices Based on Precipitation and Temperature

The Standardized Precipitation Evapotranspiration Index (SPEI) proposed by Vicente-Serrano et al. (2010a) is similar to the SPI, but requires precipitation and temperature for its computation. The temperature is used to estimate potential evaporation (PET) using the Thornthwaite method. Then the computation follows the same procedure like SPI, which is fitting a distribution and then transform to a standardized normal distribution, but using the time series of the difference between precipitation and PET instead of precipitation only. Similar to SPI, SPEI can be computed for different timescales (typically 3 months for SPEI-3, 6 months for SPEI-6, and 12 months for SPEI-12). SPEI is primarily a meteorological drought index but because of the addition of PET in its computation, it is more likely to show higher correlation with agricultural and hydrological drought indices that SPI.

10.3.3 Drought Indices Requiring Hydrological Models or Observations

A number of drought indices are based on hydrological fluxes (e.g., evaporation and runoff) or hydrological storages (e.g., soil moisture, groundwater, reservoirs) or a combination of both. Unless a long-term observation data of these hydrological flux or storage variables are available, which is usually not the case except for the streamflow, the computation of such indices relies mostly on hydrological models. Note that if hydrological modeling is involved, precipitation and temperature are required as input to the model. Here we briefly describe some of these indices.

10.3.3.1 Palmer's Drought Indices

Palmer (1965) pioneered the first comprehensive drought index called Palmer Drought Severity Index (PDSI). It integrates precipitation, soil moisture, and temperature in its computation using Thornthwaite evaporation equation and a hydrologic accounting model. The PDSI has other derivatives such as the Palmer hydrological drought index for hydrological long-term droughts, Palmer Z-Index for short-term monthly agricultural droughts, and Crop Moisture Index (Palmer, 1968) for short-term weekly agricultural droughts. PDSI was developed in the United States and is still widely used in there, but is gradually being replaced by other indices internationally (Keyantash and Dracup, 2002). The reason for its decreasing popularity—at least outside the United States—is partly due to the introduction of relatively simple indices in recent years (in particular SPI and SPEI) and partly due to some of its limitations. The limitations discussed in the literature include the timescale of drought it addresses is often not clear (Keyantash and Dracup, 2002), its computation is relatively complex and difficult to interpret (Guttman, 1998), and it required a number of arbitrary assumptions in its development (Alley, 1984). In the Palmer's classification, droughts are categorized into incipient ($-0.5 \geq \text{PDSI} > -1$), mild ($-1 \geq \text{PDSI} > -2$), moderate ($-2 \geq \text{PDSI} > -3$), severe ($-3 \geq \text{PDSI} > -4$), and extreme ($\text{PDSI} \leq -4$).

10.3.3.2 Evapotranspiration Deficit Index

The Evapotranspiration Deficit Index (ETDI), proposed by (Narasimhan and Srinivasan, 2005), is based on potential and actual evaporation. First it defines the monthly water stress, WS, from the potential and actual evaporation (Eqn (10.1)) and computes the anomalies of the water stress, WSA, from its long-term average (Eqns (10.2) and (10.3)) as follows:

$$\text{WS} = \frac{\text{PET} - \text{AET}}{\text{PET}} \tag{10.1}$$

$$\text{WSA}_{y,m} = \frac{\overline{\text{WS}_m} - \text{WS}_{y,m}}{\overline{\text{WS}_m} - \text{minWS}_m} 100, \quad \text{if } \text{WS}_{y,m} \leq \overline{\text{WS}_m} \tag{10.2}$$

$$\text{WSA}_{y,m} = \frac{\overline{\text{WS}_m} - \text{WS}_{y,m}}{\text{maxWS}_m - \overline{\text{WS}_m}} 100, \quad \text{if } \text{WS}_{y,m} > \overline{\text{WS}_m} \tag{10.3}$$

where PET and AET are the monthly reference PET and monthly actual evaporation, respectively; $\overline{\text{WS}_m}$ is the long-term median value of water stress for month m; maxMWS_m and minWS_m are the long-term maximum (minimum) water stress for month m; and $WS_{y,m}$ is the monthly water stress ratio. The WS (−) varies from 0 to 1 and WSA varies from 0 percent to 100 percent. Narasimhan and Srinivasan (2005) defined the ETDI from WSA but scaled between −4 and 4 to be comparable with the PDSI. Trambauer et al. (2014) presented ETDI by scaling between −2 and 2 (Eqn (10.4)) to make it better comparable with the SPI and SPEI.

$$\mathrm{ETDI}_{y,m} = 0.5\mathrm{ETDI}_{y,m-1} + \frac{\mathrm{WSA}_{y,m}}{100} \tag{10.4}$$

10.3.3.3 Root Stress Anomaly Index

The Root Stress Anomaly Index (RSAI), proposed by Trambauer et al. (2014), is based on the "root stress" (RS), and is computed similar to the ETDI described above. The RS is a spatial indicator of the available soil moisture, or the lack of it, in the root zone. The RS varies from 0 to 1. Its value of 0 means that the soil water availability in the root zone is at field capacity (hence no water stress), whereas the value of 1 means that the soil water availability in the root zone is nil and the plant is under maximum water stress. First, the anomaly of the monthly RS, RSA, is computed from its long-term average (Eqns (10.5) and (10.6)) as follows:

$$\mathrm{RSA}_{y,m} = \frac{\overline{\mathrm{RS}_m} - \mathrm{RS}_{y,m}}{\overline{\mathrm{RS}_m} - \mathrm{minRS}_m} 100, \quad \text{if } \mathrm{RS}_{y,m} \le \overline{\mathrm{RS}_m} \tag{10.5}$$

$$\mathrm{RSA}_{y,m} = \frac{\overline{\mathrm{RS}_m} - \mathrm{RS}_{y,m}}{\mathrm{maxRS}_m - \overline{\mathrm{RS}_m}} 100, \quad \text{if } \mathrm{RS}_{y,m} > \overline{\mathrm{RS}_m} \tag{10.6}$$

where $\overline{\mathrm{RS}_m}$ is the long-term median RS for month m and maxRS_m and minRS_m are the long-term maximum (minimum) RS for month m. The RSAI is scaled between −2 and 2 (using the same procedure as in Narasimhan and Srinivasan, 2005) as follows:

$$\mathrm{RSAI}_{y,m} = 0.5\mathrm{RSAI}_{y,m-1} + \frac{\mathrm{RSA}_{y,m}}{100} \tag{10.7}$$

10.3.3.4 Catchment Runoff or Streamflow-Based Indices

Using the concept of SPI, three indices based on either the catchment runoff or the streamflow are proposed in the literature. These are the Standardized Runoff Index (SRI) by Shukla and Wood (2008), Streamflow Drought Index (SDI) by Nalbantis and Tsakiris (2009), and Standardized Flow Index (SFI) by Wen et al. (2011). The SRI uses the catchment runoff (generally simulated by a hydrological model) and SDI and SFI use streamflows (usually from observed data, but can also be model simulated). Although Shukla and Wood (2008) and Wen et al. (2011) used the same procedure as in SPI by McKee et al. (1993), Nalbantis and Tsakiris (2009) applied a slightly different procedure. Wen et al. (2011) argued that compared to SRI, SFI has the advantage of representing the impacts of river regulation and water extraction. On the other hand, SRI is more suitable for representing the impacts of changes in land use and climate. All these indices can be computed for different timescales: typically 3, 6, and 12 months.

10.3.3.5 Groundwater Resource Index

The Groundwater Resource Index (GRI) is a groundwater drought index that was proposed by Mendicino et al. (2008) as a standarization of the monthly values of groundwater storage (detention) without any transformation:

$$\mathrm{GRI}_{y,m} = \frac{S_{y,m} - \mu_{S,m}}{\sigma_{S,m}} \tag{10.8}$$

where $S_{y,m}$ is the value of the groundwater storage for the year y and the month m, and $\mu_{S,m}$ and $\sigma_{S,m}$ are the mean and the standard deviation, respectively, of the groundwater storage S simulated for the month m in a defined number of years. The same classification that is used for the SPI may be applied to GRI (Wanders et al., 2010).

10.3.3.6 Surface Water Supply Index

The Surface Water Supply Index (SWSI) was first developed for major basins in Colorado, USA (Shafer and Dezman, 1982). It takes an integrated approach to indicate the availability of water in the sense that it uses precipitation, reservoir water storage and streamflow (for summer months), and snowpack (for winter months). Like the PDSI, the SWSI is centered on 0 and has a range between −4.2 and +4.2. In its computation, the relative importance of each of the components (precipitation, reservoir storage, and snowpack or streamflow) is assigned through a weighting coefficient. The determination of the coefficient is not fully objective, which is an issue for its application. The fact that the SWSI is unique to each basin makes it difficult to compare values between basins or regions (NDMC, 2014).

10.3.4 Drought Indices Based on Remote Sensing Data

Over the recent years, the increasing advancements made in remote sensing technology have proven the great potential of remote sensing for drought monitoring and assessment. The Normalized Difference Vegetation Index (NDVI), which is derived from remote sensing (satellite) data, is closely linked to drought conditions. Based on NDVI values, Kogan (1995) derived the Vegetation Condition Index (VCI) (Eqn (10.9)) to represent agricultural drought conditions:

$$\mathrm{VCI} = 100\frac{(\mathrm{NDVI} - \mathrm{NDVI}_{\min})}{(\mathrm{NDVI}_{\max} - \mathrm{NDVI}_{\min})} \tag{10.9}$$

where NDVI, $NDVI_{max}$, and $NDVI_{min}$ are the smoothed weekly NVDI, its multiyear maximum, and its multiyear minimum, respectively.

The VCI varies from 0 to 100, corresponding to the changes in vegetation conditions from extremely unfavorable to optimal (Kogan, 1995). Given that VCI is based on vegetation, it is primarily useful during the growing season and has limited utility during the dormant season. During the growing season, the VCI allows detection of drought and measurement of the time of its onset, intensity, duration, and impact of vegetation (Heim, 2002). Moreover, the evaporation derived from remote sensing data may also be used for deriving EDTI.

10.4 SIGNIFICANCE OF HYDROLOGICAL MODELS FOR DROUGHT ASSESSMENT

Hydrological models simulate water flux and storage through various media within a hydrological cycle. Precipitation and temperature are the major driving inputs for all hydrological models. Depending on the details of the physical process representations, a hydrological model may require other inputs such as solar radiation, relativity humidity, and wind speed. The outputs from a process-based or conceptual hydrological model ranges from evaporation, soil moisture, and groundwater recharge to reservoir inflow and river runoff (Figure 10.2). Moreover, the runoff can be obtained in three different components: the surface runoff, interflow and baseflow, or the flow from the groundwater contribution. As we see in the description of drought indices in Section 10.3.3, a lot of the drought indices are based on the outputs that can be obtained from a hydrological model, e.g., evaporation, soil moisture, runoff. Although some observation data may be available for hydrological drought monitoring and analysis of past droughts, such data are generally limited to few locations usually insufficient to cover the spatial variability of droughts. This is clearly where the use of a hydrological model, distributed or semidistributed, is extremely useful. Moreover, a hydrological model is also predicting tool, which is necessary for operational drought forecasting. Thus there is a lot a hydrological model can offer for drought assessment, monitoring, and forecasting.

However, not all hydrological models sufficiently represent hydrological processes that are important for characterizing droughts in a given climatic

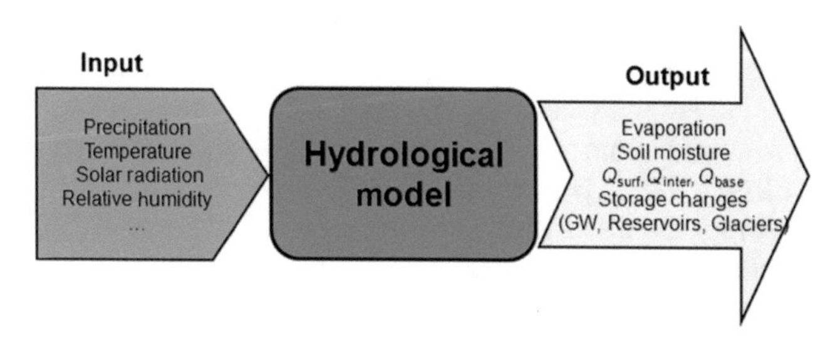

FIGURE 10.2 Input and output to (from) a hydrological model.

condition. In general, processes like evaporation and surface water–groundwater interaction are very important for drought assessment. Generally a detailed process-based, distributed or semidistributed model is preferred with a continuous simulation capability. Note that some models that are aimed at application for floods may not have a good evaporation component or may only be used for an event simulation (contrary to a continuous simulation). Trambauer et al. (2013) presented a decision tree for selecting the right type of hydrological model for drought forecasting in Africa, which is adapted here for selecting an appropriate hydrological model for drought assessment and forecasting in general (Figure 10.3).

10.5 CASE STUDY: HYDROLOGICAL DROUGHT ASSESSMENT FOR THE LIMPOPO BASIN

A comprehensive analysis of the hydrological droughts in the Limpopo river basin, Africa, was presented by Trambauer et al. (2014) using a process-based, distributed hydrological model. Some of the results and discussion presented in this section are largely based on their study but with additional perspective on the added value of hydrological-model-driven drought indices.

10.5.1 Hydrological Model of the Limpopo Basin

The (semi) arid Limpopo basin, located in southern Africa (Figure 10.4), is one of the most water-stressed basins in Africa and is expected to face even more serious water scarcity issues in the future, limiting economic development in the basin (Zhu and Ringler, 2012). Hydrological droughts and their space–time variability are schematized using a process-based distributed hydrological model. The model is based on the global hydrological model PCRaster GLobal Water Balance (PCR-GLOBWB) (van Beek and Bierkens, 2009), but applied with a refined spatial resolution of $0.05° \times 0.05°$. The process-based model was applied making use of the best available input data, e.g.,:

- Meteorological forcing (precipitation, daily minimum and maximum temperature at 2 m) from the ERA-Interim reanalysis data set from the European Centre for Medium-Range Weather Forecasts.
- The Digital Elevation Model from the Hydro1k Africa (USGS EROS, 2006).
- Soil data from the Digital Soil Map of the World (Food and Agriculture Organization (FAO), 2003).
- Vegetation types from Global Land Cover Characterization (GLCC) (USGS EROS, 2002; Hagemann, 2002).
- Lithology from the lithological map of the world (Dürr et al., 2005).
- Irrigated areas from Siebert et al. (2007) and FAO (1997).

The model is tested by identifying historical droughts in the period 1979–2010 with simulated hydrological and agricultural drought indices. The

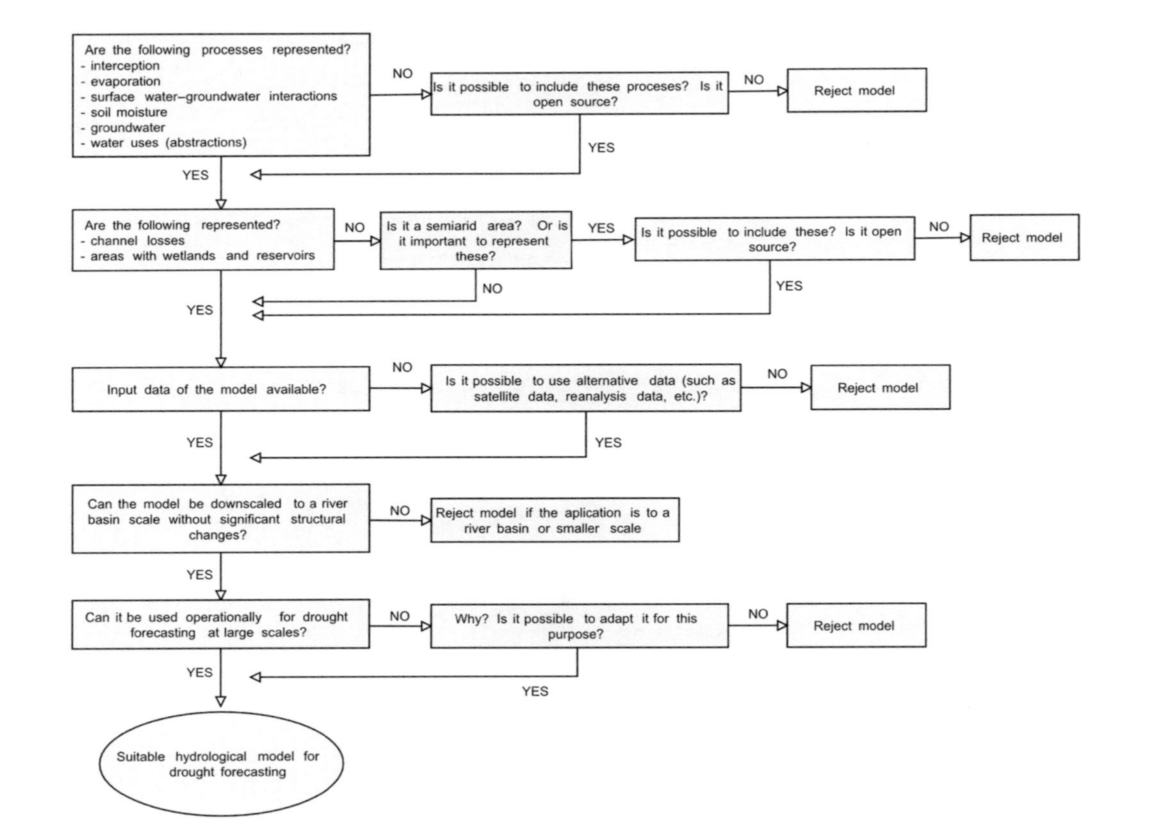

FIGURE 10.3 A decision tree for choosing a hydrological model suitable for using in drought assessment. *Adapted from Trambauer et al. (2013).*

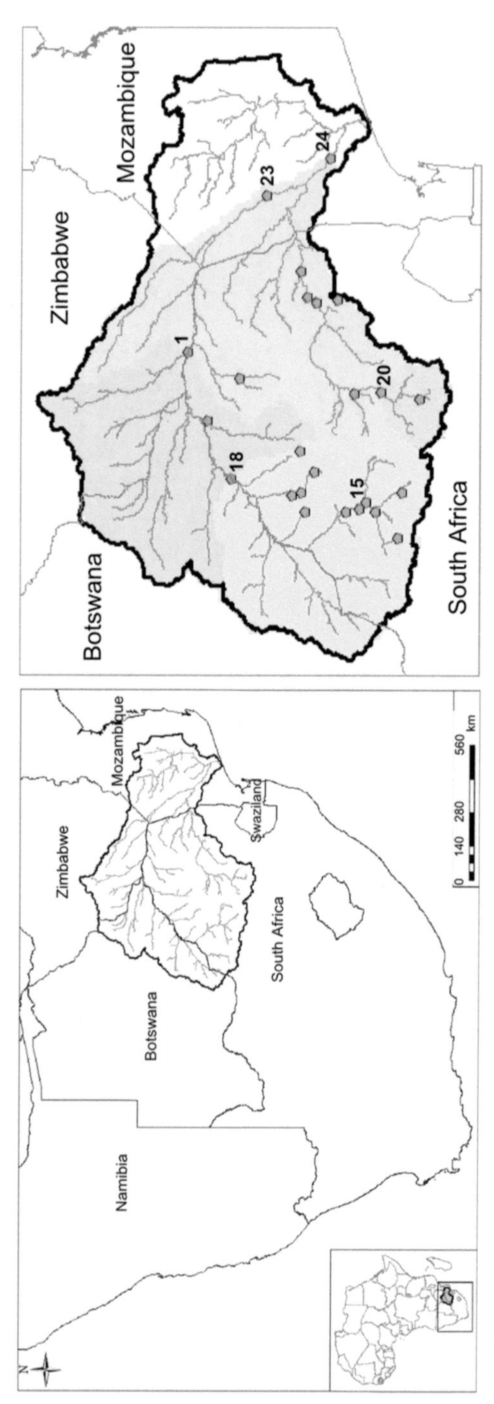

FIGURE 10.4 Limpopo river basin: the location of the basin (left) and the locations of hydrometric stations (right). Selected stations are highlighted. The subbasins draining to each hydrometric station are named after the station number. *From Trambauer et al., 2014*

very low runoff coefficients and high rainfall variability pose major challenges in modeling hydrological droughts in (semi) arid basins. Small errors in the meteorological forcing and estimation of evaporation may result in significant errors in the runoff estimation. The hydrological model was verified for runoff in a number of gauging stations in the basin and its performance was found to be satisfactory based on the evaluation measures and ranges proposed by Moriasi et al. (2007). These results were presented by Trambauer et al. (2014).

DEWFORA (2012) reported that in the period 1980–2000, the Southern African region was struck by four major droughts, notably in the seasons 1982/83, 1986/87, 1991/92 and 1994/95. The drought of 1991/1992 was the most severe in the region in recent history. After the year 2000, important droughts include the years 2002/2003/2004 and 2005/2006. Droughts in the Limpopo river basin also show significant spatial variability. A study covering only the Botswana part of the basin documents a severe drought that occurred in 1984 (Dube and Sekhwela, 2007). However, in that year no documentation of drought in the other parts of the basin was found.

10.5.2 Spatial Variation of Droughts in the Limpopo Basin

The model is able to simulate hydrological drought-related indices reasonably well. We have derived a number of different drought indices from the model results, such as ETDI, RSAI, SRI, and GRI. While the SRI is based on river runoff accumulated for a particular river section, the ETDI, RSAI, and GRI are spatial indices that can be estimated at any location in the basin. Figure 10.5 presents the RSAI and ETDI maps for a very dry season (1982/1983), for a very wet season (1999/2000), and for a season where both dry and wet conditions occurred in the basin (1984/1985).

ETDI and RSAI are directly related to water availability for vegetation with or without irrigation, and GRI is related to the groundwater storage. Moreover, we computed the widely known drought indices SPI and SPEI at different aggregation periods to verify the correlation of the different aggregation periods for these indices and the different types of droughts. We assess the ability of different drought indices to reconstruct the history of droughts in a highly water-stressed, semiarid basin. Moreover, we investigate whether widely used climate indices for drought identification can be complemented with indices that incorporate hydrological processes. We found that all the indices considered (with the exception of GRI) are able to represent the most severe droughts in the basin and to identify the spatial variability of the droughts.

10.5.3 Comparison of Different Drought Indices

Figure 10.6 compares the agricultural, hydrological, and groundwater drought indices for the largest selected subbasin (# 24). The agricultural indices ETDI and RSAI are compared with the meteorological drought indices SPI and SPEI

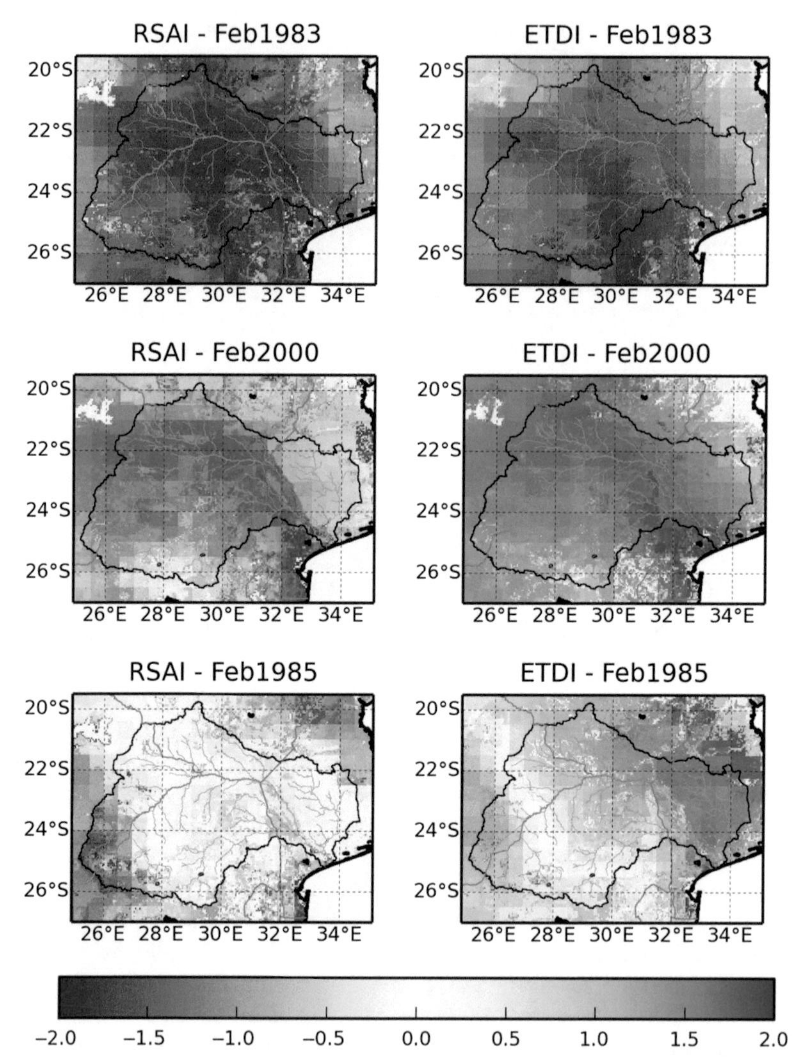

FIGURE 10.5 Root stress anomaly index (RSAI) and Evapotranspiration Deficit Index (ETDI) in the Limpopo basin for selected years. *Adapted from Trambauer et al., 2014*

with a short aggregation period (3 months) that are commonly used as indices of agricultural droughts. The upper plot shows that the indices are mostly in phase, correctly representing the occurrence of dry and wet years, and the intensities of the events are in general quite similar. The hydrological drought index SRI-6 is compared with the meteorological drought indices SPI-6 and SPEI-6 (upper middle plot). All three indices roughly follow the same pattern, but the fluctuation of the SRI seems to be slightly lower than that of the meteorological indices (SPI and SPEI). Moreover, it is clearly visible from

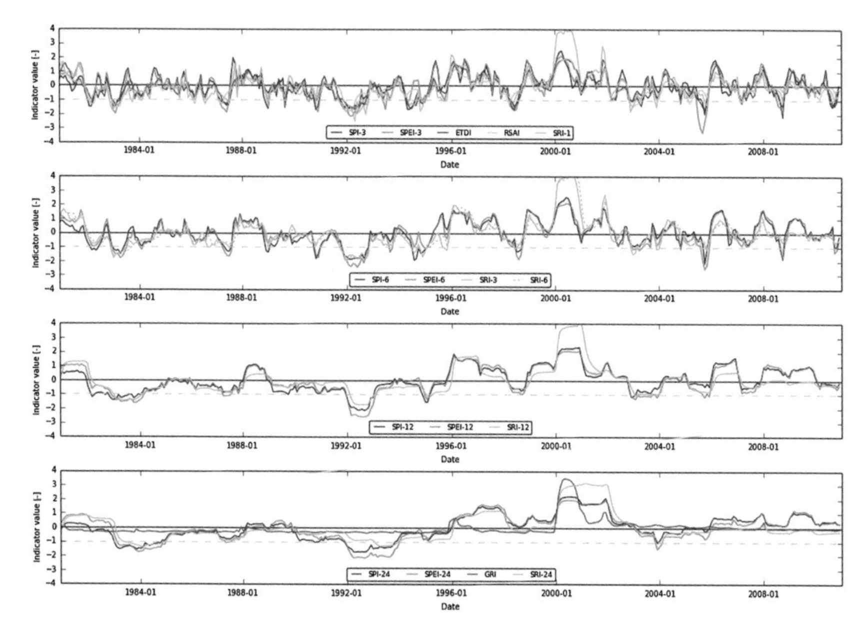

FIGURE 10.6 Time series of aggregated drought indices for subbasin #24. Upper graph, indices used to characterize agricultural droughts (SPI-3, SPEI-3, ETDI, and RSAI); upper middle graph, indices used to characterize hydrological drought (SPI-6, SPEI-6, and SRI-6); lower middle graph, indices used to characterize groundwater droughts (SPI-12, SPEI-12, and SRI-12); and lower graph, indices used to characterize extended groundwater droughts (SPI-24, SPEI-24, GRI, and SRI-24). *Adapted from Trambauer et al., 2014*

Figure 10.6 that the temporal variability or fluctuation of the indices reduces when moving from drought indices associated to agricultural to those associated to hydrological drought. This means that several mild agricultural droughts do not progress further to hydrological droughts. For example, a short but intense agricultural drought is noticeable at the beginning of the 2005/2006 season, but this did not progress to be a hydrological drought. This is coherent with the literature, which indicates that this season was delayed, and after a dry start of the season, good rainfall occurred from the second half of December (Department of Agriculture of South Africa, 2006). Moreover, to identify groundwater droughts, or major drought events, the time series of GRI is compared to the time series of meteorological drought indices with long aggregation periods (SPI-12, SPEI-12, SRI-12, SPI-24, SPEI-24, SRI-24) (see Figure 10.6, lower middle and lower plots). The plots show that as the variability of the index reduces further the number of multiyear prolonged droughts increases. However, for groundwater droughts only two events (1982/1983 and 1991/1992) are identified as moderate to severe droughts ($Iv < -1$). The GRI index shows much less temporal variability than the other indices and does not identify any extreme events with the exception of the flood of 1999/2000. Similar results using GRI were found by Wanders et al. (2010), who indicated that GRI has a very low number of droughts with a high average duration. Moreover, a study of Peters and Van Lanen (2003) investigated groundwater droughts for two climatically contrasting regimes. For the semi-arid regime they found multiannual droughts to occur frequently. They indicated that the effect of the groundwater system is to pool erratically occurring dry months into prolonged groundwater droughts for the semiarid climate.

A similar analysis for other subbasins allowed making some characterization on the spatial variability of droughts. For example, a clear hydrological drought period occurred at subbasin #18 in the years 1985/1986, which is not apparent for the subbasins #24 and 1 (Figure 10.7). These localized drought events that affected the upper part of the basin were not apparent for the lower part of the basin. Moreover, the severe floods that occurred in the lower part of the basin in 1999/2000 are much less severe in the upstream parts of the basin. For subbasin #20 (the smallest subbasin considered) the results show that the flood of 1996/1997 was more severe than that of 1999/2000. Similarly, while the drought of 2003/2004 is quite mild when averaged over the whole basin (subbasin #24), it is quite severe for subbasin #20 (similar to the droughts of 1983/1984 and 1991/1992).

10.5.4 Do Hydrological Drought Indices Bring Additional Information?

The results from the previous section show that even though meteorological indices with different aggregation periods serve to characterize droughts reasonably well, added value occurs in computing indices based on the hydrological model for the identification of droughts/floods and their severity. The hydrological drought indices also help identify the spatial and temporal

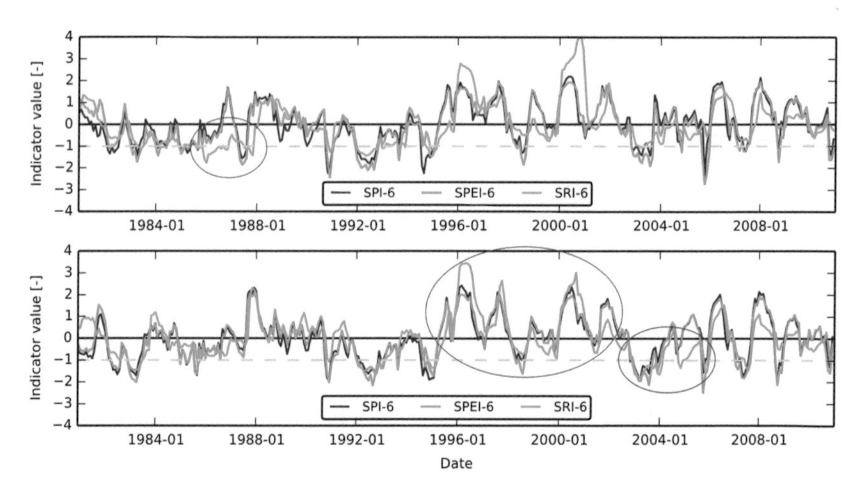

FIGURE 10.7 Time series of aggregated indices used to characterize hydrological drought (SPI-6, SPEI-6, and SRI-6) for subbasin #18 (top panel) and #20 (bottom panel).

evolution of drought and flood events that would otherwise not have been apparent when considering only meteorological indices.

We also completed a contingency-table-type analysis on drought/no drought identification by some of the different indices to evaluate their correlation and have a better comparison on the timing and intensity of the droughts identified by the different indices. For example, Figure 10.8 presents a contingency table between meteorological droughts identified by SPI < -0.5 and hydrological droughts identified by SRI < -0.5 for the subbasin with larger contributing area with available data in the Limpopo basin (subbasin #24). Four cases are presented: the upper plots correspond to the analysis of the unshifted time series of the indices which is completed for aggregation periods of (a) 3 months and (b) 6 months. The lower plots represent the same analysis, but the time series of the hydrological drought index (SRI) is shifted 1 month to represent the lag that normally occurs between the onset of a

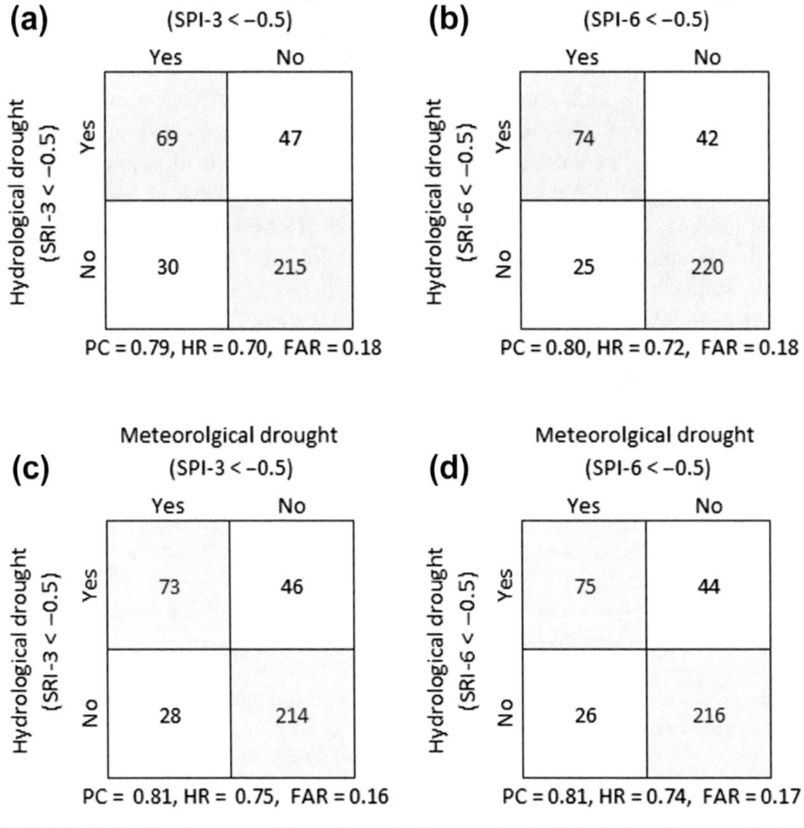

FIGURE 10.8 Contingency tables and standard scores for hydrological and meteorological drought identification. Upper plots: no shift between SPI and SRI time series for (a) 3-months and (b) 6-months aggregation periods. Lower plots: SRI time series shifted 1 month with respect to SPI for (c) 3-months and (d) 6-months aggregation periods.

meteorological and a hydrological drought. Similar to the upper plots, the analysis is completed for aggregation periods of (c) 3 months and (d) 6 months. From the contingency table we computed the standard scores: proportion correct (PC), hit rate (HR), and false alarm rate (FAR) as a measure of accuracy and discrimination of the signal of the indices. All the four cases presented in Figure 10.8 show high PC (around 80 percent), high HR (higher than 70 percent), and low FAR (lower than 18 percent). In general, the scores are slightly better when the time series of the hydrological drought index (SRI) is shifted for 1 month. This result is expected given the known lag time between the onset of a meteorological and a hydrological drought.

The cases that are identified as meteorological droughts but not as a hydrological droughts, or vice versa, are not misses but most probably point to meteorological droughts that did not progress further to hydrological droughts, or hydrological droughts triggered by antecedent conditions.

10.6 CONCLUSIONS

This chapter discussed the challenge of assessing hydrology-related droughts and showed the potentialities (and limitations) of a process-based hydrological model. To this end, a case study on the Limpopo river basin in Southern Africa was presented. It was demonstrated that various types of droughts need to be analyzed to understand the comprehensive nature of drought impacts. Numerous drought indices can be derived from a process-based hydrological model outputs, such as SRI, SFI or SDI, ETDI, RSAI, GRI, and SWSI. Even though SPI and SPEI require only precipitation and temperature for their computations and are relatively simple to derive, computing hydrology-related drought indices provides more comprehensive drought assessments. Moreover, hydrological models are essential to forecast droughts and to analyze the impacts of global changes (climate change, land use change, etc.) on droughts.

REFERENCES

Alley, W.M., 1984. The Palmer drought severity Index: limitations and assumptions. J. Clim. Appl. Meteorol. 23, 1100–1109. http://dx.doi.org/10.1175/1520-0450(1984)023<1100:tpdsil>2.0.co;2.

Alston, M., Kent, J., 2004. Social Impacts of Drought. Centre for Rural Social Research, Charles Sturt University, Wagga Wagga, NSW.

Department of Agriculture of South Africa, 2006. Crops and Markets – First Quarter, vol. 87, No. 927. Directorate Agricultural Statistics – Department of Agriculture. http://www.daff.gov.za/docs/statsinfo/Crops_0106.pdf (last accessed December 2013).

DEWFORA, 2012. WP6-D6.1-Implementation of Improved Methodologies in Comparative Case Studies – Inception report for each case study, DEWFORA Project – EU FP7. www.dewfora.net (last accessed December 2013).

Dube, O.P., Sekhwela, M.B.M., 2007. Community Coping Strategies in Semiarid Limpopo Basin Part of Botswana: Enhancing Adaptation Capacity to Climate Change. http://www.aiaccproject.org/working_papers/Working%20Papers/AIACC_WP47_Dube.pdf (last accessed 1–40.12.13.).

Edwards, D., McKee, T., 1997. Characteristics of 20th century drought in the United States at multiple time scales. Colorado State University Dept. of Atmospheric Science Fort Collins Colorado, Climatology Rep 97-2, 155.

EM-DAT, 2014. The OFDA/CRED International Disaster Database. Universite Catholique de Louvain, Brussels, Belgium. Created on: March 27, 2014 – Data version: v12.07, (accessed 27.03.14.).

Glantz, M. H. e., 1987. Drought and Hunger in Africa: Denying Famine a Future. Cambridge University Press, Cambridge.

Guttman, N.B., 1998. Comparing the Palmer drought index and the standardized precipitation index. J. Am. Water Resour. Assoc. 34, 113–121. http://dx.doi.org/10.1111/j.1752-1688.1998.tb05964.x.

Heim, R.R., 2002. A review of twentieth-century drought indices used in the United States. Bull. Am. Meteorol. Soc. 83, 1149–1165. http://dx.doi.org/10.1175/1520-0477(2002)083<1149:AROTDI>2.3.CO;2.

Hollinger, S., Isard, S., Welford, M., 1993. A new soil moisture drought index for predicting crop yields, Preprints, Eighth Conference on Applied Climatology, American Meteorological Society. Anaheim (CA) 187–190.

IPCC, 2007. Climate change 2007: the physical science basis. In: Solomon, S., Qin, D., Manning, M., Chen, Z., Marquis, M., Averyt, K.B., Tignor, M., Miller, H.L. (Eds.), Contribution of Working Group I to the Fourth Assessment, Report of the Intergovernmental Panel on Climate Change. Cambridge University Press, Cambridge, United Kingdom and New York, NY, USA, p. 996.

Keyantash, J., Dracup, J.A., 2002. The quantification of drought: an evaluation of drought indices. Bull. Am. Meteorol. Soc. 83, 1167–1180.

Kogan, F., 1990. Remote sensing of weather impacts on vegetation in non-homogeneous areas. Anglais 11, 1405–1419.

Kogan, F.N., 1995. Droughts of the late 1980s in the United States as derived from NOAA polar-orbiting satellite data. Bull. Am. Meteorol. Soc. 76, 655–668.

Lehner, B., Döll, P., Alcamo, J., Henrichs, T., Kaspar, F., 2006. Estimating the impact of global change on flood and drought risks in Europe: a continental, integrated analysis. Clim. Change 75, 273–299.

Lloyd-Hughes, B., Saunders, M.A., 2002. A drought climatology for Europe. International Journal of Climatology 22, 1571–1592. http://dx.doi.org/10.1002/joc.846.

Masih, I., Maskey, S., Mussá, F.E.F., Trambauer, P., 2014. A review of droughts on the African continent: a geospatial and long-term perspective, Hydrol. Earth Syst. Sci. 18, 3635–3649. http://dx.doi.org/10.5194/hess-18-3635-2014.

McKee, T.B., Doesken, N.J., Kleist, J., 1993. The relationship of drought frequency and duration to time scales. In: Proceedings of the 8th Conference on Applied Climatology, vol. 17, No. 22. American Meteorological Society, Boston, MA, pp. 179–183.

Mendicino, G., Senatore, A., Versace, P., 2008. A groundwater resource index (GRI) for drought monitoring and forecasting in a Mediterranean climate. J. Hydrol. 357, 282–302. http://dx.doi.org/10.1016/j.jhydrol.2008.05.005.

Moriasi, D., Arnold, J., Van Liew, M., Bingner, R., Harmel, R., Veith, T., 2007. Model evaluation guidelines for systematic quantification of accuracy in watershed simulations. Trans. ASABE 50, 885–900.

Nalbantis, I., Tsakiris, G., 2009. Assessment of hydrological drought revisited. Water Resour. Manage. 23, 881–897.

Narasimhan, B., Srinivasan, R., 2005. Development and evaluation of soil moisture deficit index (SMDI) and evapotranspiration deficit index (ETDI) for agricultural drought monitoring. Agric. For. Meteorol. 133, 69–88. http://dx.doi.org/10.1016/j.agrformet.2005.07.012.

NDMC, 2014. Comparison of Major Drought Indices: Surface Water Supply Index. National Drought Mitigation Center, University of Nebraska, Lincoln (accessed 24.03.14.).

Palmer, W.C., 1965. Meteorological Drought. US Department of Commerce, Weather Bureau, Washington, DC, USA.

Palmer, W.C., 1968. Keeping track of crop moisture conditions, nationwide: The new crop moisture index,. Weatherwise 21, 156–161. http://dx.doi.org/10.1080/00431672.1968.9932814.

Patz, J.A., Campbell-Lendrum, D., Holloway, T., Foley, J.A., 2005. Impact of regional climate change on human health. Nature 438, 310–317.

Peters, E., Van Lanen, H.A.J., 2003. Propagation of drought in groundwater in semiarid and subhumid climatic regimes. In: Hydrology in Mediterranean and Semiarid Regions: International Conference, Montpellier, France. IAHS Press, Wallingford, UK, pp. 312–317.

Shafer, B., Dezman, L., 1982. Development of a surface water supply index (SWSI) to assess the severity of drought conditions in snowpack runoff areas. In: Proceedings of the Western Snow Conference, pp. 164–175.

Sheffield, J., Wood, E.F., 2008. Projected changes in drought occurrence under future global warming from multi-model, multi-scenario, IPCC AR4 simulations. Clim. Dynam. 31, 79–105.

Shukla, S., Wood, A.W., 2008. Use of a standardized runoff index for characterizing hydrologic drought. Geophys. Res. Lett. 35, L02405. http://dx.doi.org/10.1029/2007gl032487.

Trambauer, P., Maskey, S., Winsemius, H., Werner, M., Uhlenbrook, S., 2013. A review of continental scale hydrological models and their suitability for drought forecasting in (sub-Saharan) Africa. Phys. Chem. Earth 66, 16–26. http://dx.doi.org/10.1016/j.pce.2013.07.003.

Trambauer, P., Maskey, S., Werner, M., Pappenberger, F., van Beek, L.P.H., Uhlenbrook, S., 2014. Identification and simulation of space-time variability of past hydrological drought events in the Limpopo river basin, southern Africa. Hydrol. Earth Syst. Sci. 18, 2925–2942. http://dx.doi.org/10.5194/hess-18-2925-2014.

Tucker, C.J., 1979. Red and photographic infrared linear combinations for monitoring vegetation. Remote Sensing of Environment 8, 127–150.

van Beek, L.P.H., Bierkens, M.F.P., 2009. The Global Hydrological Model PCR-GLOBWB: Conceptualization, Parameterization and Verification, Utrecht University. Faculty of Earth Sciences, Department of Physical Geography, Utrecht, The Netherlands.

Vicente-Serrano, S.M., Beguería, S., López-Moreno, J.I., 2010. A Multiscalar Drought Index Sensitive to Global Warming: The Standardized Precipitation Evapotranspiration Index,. Journal of Climate 23, 1696–1718. http://dx.doi.org/10.1175/2009jcli2909.1.

Vicente-Serrano, S.M., Beguería, S., López-Moreno, J.I., Angulo, M., El Kenawy, A., 2010. A New Global 0.5° Gridded Dataset (1901–2006) of a Multiscalar Drought Index: Comparison with Current Drought Index Datasets Based on the Palmer Drought Severity Index,. Journal of Hydrometeorology 11, 1033–1043.

Wanders, N., Lanen, H.A.J., van Loon, A.F., 2010. Indicators for Drought Characterization on a Global Scale. WATCH Water and Global change EU FP6. Technical report No. 24. http://www.eu-watch.org/publications/technical-reports/3 (last accessed December 2013).

Wen, L., Rogers, K., Ling, J., Saintilan, N., 2011. The impacts of river regulation and water diversion on the hydrological drought characteristics in the Lower Murrumbidgee River, Australia. J. Hydrol. 405, 382–391.

Wilhite, D.A., Glantz, M.H., 1985. Understanding: the drought phenomenon: the role of definitions. Water Int. 10, 111–120. http://dx.doi.org/10.1080/02508068508686328.

Zhu, T., Ringler, C., 2012. Climate change impacts on water availability and use in the Limpopo River Basin. Water 4, 63–84.

Index

Note: Page numbers followed by "f" and "t" indicate figures and tables respectively.

H

I

K

L

T

U

V

W

X

Printed in the United States
By Bookmasters